Designed cover image: Yamini Jhanji Dhir

First edition published 2027
by CRC Press
2385 NW Executive Center Drive, Suite 320, Boca Raton FL 33431

and by CRC Press
4 Park Square, Milton Park, Abingdon, Oxon, OX14 4RN

*CRC Press is an imprint of Taylor & Francis Group, LLC*



ISBN: 9781032450421 (hbk)
ISBN: 9781032558837 (pbk)
ISBN: 9781003432739 (ebk)

DOI: 10.1201/9781003432739

Typeset in Times
by codeMantra

# Contents

# About the Editor

**Yamini Jhanji Dhir** belongs to Ludhiana, Punjab; however, she has been residing in Bhiwani, Haryana, for the past decade or so, fulfilling her academic and work commitments. She had a meritorious academic record with B. Tech & M. Tech in Textile Technology with honors. She did her PhD from IIT, Delhi, in 2015, and since then has been ardently involved in research and development activities. Dhir has a rich academic experience of nearly two decades, serving as an assistant professor till October 2023 and thereafter as an Associate Professor. Currently, she is serving as an Associate Professor in the Department of Textile Technology at the Technological Institute of Textile & Sciences, Bhiwani. She has been instrumental in developing and taking undergraduate and postgraduate courses related to textiles, fashion, garments and accessories. She is conferred with several institutional administrative responsibilities related to the admission process, innovation council, etc. She obtained a grant from AICTE for conducting an AICTE-sponsored Faculty Development Programme and International Conference in 2020 and has successfully organized and coordinated both events in the institute.

She has several conference publications, book chapters and research publications in journals of national and international repute to her credit. Her areas of interest include smart textiles, fashion accessories, sportswear and comfort characteristics of textiles.

She has authored a book entitled *Fashion Accessories: A Complete Guide to Raw Materials, Construction Methods & Styles*. She is the recipient of the best paper award in some national and international conferences and has also served as key note speaker, session chair and resource person in several academic events. She chaired a session on Sustainable & Circular Fashion at the FTC Conference at IIT Delhi. Also, she has organized several workshops, faculty development, orientation and refresher programs along with social awareness campaigns.

# Contributors

**Ashish Bhardwaj**
Department of Fashion & Apparel Engineering
The Technological Institute of Textile & Sciences
Bhiwani, India

**Sonjib Bora**
Department of Knitwear
National Institute of Fashion Technology
Bengaluru, India

**Yamini Jhanji Dhir**
Department of Textile Technology
The Technological Institute of Textile & Sciences
Bhiwani, India

**Samridhi Garg**
Department of Textile Technology
Dr B R Ambedkar National Institute of Technology
Jalandhar, India

**Drishti Goel**
Department of Fashion Design
National Institute of Fashion Technology
Bengaluru, India

**Sujit Shrikrushnarao Gulhane**
Centre for Textile Functions
NMIM'S
Shirpur Campus, Dhule, India

**Anu H. Gupta**
University Institute of Fashion Technology & Vocational Development
Punjab University
Chandigarh, India

**Harpreet Kaur**
Department of Fashion Design
Kanya Mahavidyalaya
Jalandhar, India

**Kulbir Kaur**
University Institute of Fashion Technology & Vocational Development
Punjab University
Chandigarh, India

**Amit Madhu**
Department of Textile Chemistry
The Technological Institute of Textile & Sciences
Bhiwani, India

**RajkishoreNayak**
School of Communication & Design
RMIT University
Vietnam

**Rathinamoorthy Ramasamy**
VIT Fashion Institute of Technology
Vellore Institute of Technology
Chennai, India

**Paramita Sarkar**
Department of Fashion Design
National Institute of Fashion Technology
Bengaluru, India

**Anu Sharma**
Department of Textile Design
National Institute of Fashion Technology
New Delhi, India

**Shipra Sharma**
Department of Fashion Design
National Institute of Fashion Technology
Bengaluru, India

**Monica Sikka**
Department of Textile Technology
Dr B R Ambedkar National Institute of Technology
Jalandhar, India

**Nagender Singh**
Department of Fashion & Apparel Engineering
The Technological Institute of Textile & Sciences
Bhiwani, India

**Ruby Kashyap Sood**
Department of Textile Design
National Institute of Fashion Technology
New Delhi, India

**Sudha Subramanian**
Department of Textile Technology
A.C.Tech Campus, Anna University
Chennai, India

**Surjit Ramamoorthy**
Department of Fashion Technology
PSG College of Technology
Coimbatore, India

**Krutika Dnyaneshwar Thakare**
Department of Knitwear
National Institute of Fashion Technology
Bengaluru, India

# Preface

The content of this book is designed to address traditional and innovative concepts of garment manufacturing, spanning from designing, raw material selection, to manufacturing and related domains tailored for engineering students, particularly those in the fields of designing, garment technology, fashion technology, fashion and apparel engineering as well as textile engineering domains. The inclusion of both traditional and the latest technologies implemented in designing, cutting, pre-sewing and post sewing room operations in garment industries for operational efficiency will acquaint readers with the advantages of modern technologies over traditional practices. Furthermore, the concept of sustainability and designing of capsule, modular collection along with the integration of artificial intelligence into the garment and fashion pipeline will provide readers with useful insights into current and upcoming trends. The content pertaining to smart textiles such as wearable electronics and smart textiles for health monitoring and sportswear will keep the readers well informed about the latest developments in the promising and ever-evolving field of technical textiles.

The salient features of this book include case studies, market surveys, field trials and design inputs, which provide an effective and practical approach that is liekly to engage and captivate the readers.

# 1 Introduction to Garment Manufacturing

*Monica Sikka and Samridhi Garg*

## 1.1 INTRODUCTION

The labour-intensive sector of garment manufacturing is characterised by low fixed capital investment, a diversity of product designs and consequently input materials, unpredictable production volumes, fierce rivalry, and often high standards for product quality (Scott, 2006; Hassler, 2003; Forza and Vinelli, 2000). The design, cutting, and stitching of garments from fabric are the main responsibilities in the clothing manufacturing sector. Making garments involves a set of stages or processes on an industrial scale. The textile industry has had a significant impact on the Indian economy. In India, clothing worth a trillion rupees is created. About 33% of its knitwear production and 20% of its woven garment manufacture, both in terms of volume, are exported. The final sale of 25% of the world's manufactured clothing is to foreign markets, leaving 75% for domestic use. The industry, which directly and indirectly employs around 6 million people in a distribution that is almost equal, consists of over 100,000 units. The indirect component of the apparel business serves the direct manufacturing sector and includes items like sewing/embroidery thread, buttons, buckles, zippers, metal plates, cardboard sheets, plastic butterflies, and packaging material. A little over 20% of the whole apparel industry is organised, mostly with an eye on exports. The remaining are often restricted companies and are proprietary or partnership companies. Men's clothing manufacture is predominantly centred in western and southern India, while women's clothing manufacturing is mostly concentrated in northern India. The true birthplace of children's apparel is in eastern India. Cotton makes up 80% of all clothing produced, synthetic, mixed fabrics make up 20%, and silk and wool make up the remaining 20%. For men, women, and children, the sector produces more than 100 distinct styles of clothing. In addition to shawls/scarves, handkerchiefs, gloves, and other clothing items, this category also contains overcoats/raincoats, ensembles, jackets, dresses, skirts, and trousers. It also includes shirts, blouses, inner garments, T-shirts, jerseys/pullovers, and newborn clothes. Fabric accounts for 65%–70% of manufacturing costs, followed by labour (15%), overhead (10%–15%), and the manufacturer's profit.

In this chapter, an overview of the garment manufacturing industry and global market share is explained, followed by market outlook and future potential of the industry, highlighting the necessary technological interventions.

DOI: 10.1201/9781003432739-1

## 1.2 GARMENT MANUFACTURING STEPS

### 1.2.1 Design, Pattern Making, Fit Sample Making

The design is provided by the client. The buyer provides the merchandiser with the technical sheet and the order's artwork after placing an order. For performing the operation, either the manual method or the computer method is used. The technical paper and artwork should be used as a reference for creating the patterns for each kind of apparel. To finish it, both a manual and a computerised procedure are employed. The major objective of making a fit sample is to follow the precise guidelines about the garment's design. After being created, it is sent to the customer for adjustment.

### 1.2.2 Production Pattern Making, Grading, Marker Making

A net dimension is added here as an allowance for bulk manufacturing. Both manual and digital methods are used in the pattern-making process. The customer offers the order's size ratio during the order confirmation process. To grade that order in accordance with the buyer's instructions, either a computer or a human can grade assignments. Markers are sheets of paper that are so thin that they can be folded into the exact shape of a certain garment. It must be necessary to facilitate the cutting process. Both manual and computerised methods can be used in the marker-making process.

### 1.2.3 Spreading

The fabric rolls are first unrolled and laid down layer by layer on a cutting table. You can carry out this operation manually or using an automated spreading device. Based on the marker size and marker length, the layer's length is determined. On the table, various numbers of fabric layers are stretched out, and they are all cut at the same time. The number of plies in a lay is determined by the fabric thickness. The height of the lay is typically maintained in accordance with the cutting machine blade height.

### 1.2.4 Cutting

The most significant activity that takes place in the cutting room is cutting. After completing distribution, placement, and marking, workers move on to the crucial cutting stage. When it comes to the cutting room, there is just one complete rule: "Once you cut it, there is no turning back." It is expected that any issues with the cut will have an impact on both the stitching procedure and the finished product. Nothing that can be done will make the situation better. To complete the cutting operation, a sharp knife is utilised. These are the following cutting techniques:

- Manual cutting
- Heat cutting
- Laser cutting

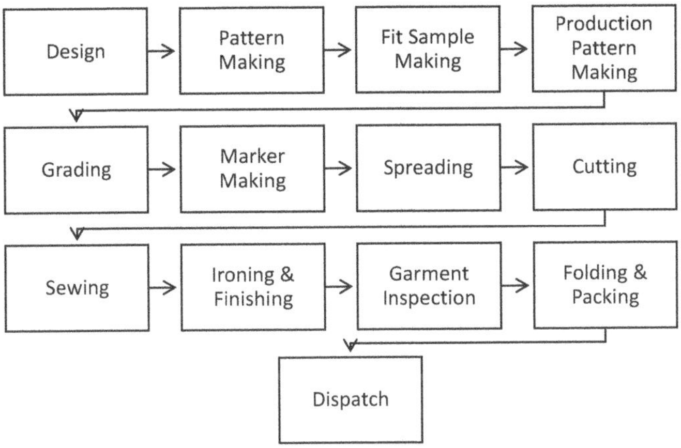


**FIGURE 1.1** Flow chart of garment manufacturing process.

### 1.2.5 Sewing

After the cut components are packaged according to the sizes, colours, and numbers decided upon by the sewing room, stitching or sewing is completed (Figure 1.1). On an assembly line, clothes are finished as they go down the line after being sewn. Before handing the finished piece to the next operator, sewing machine operators repeatedly stitch the same area of the garment. As an example, the first worker may sew the collar to the body of the garment while the second worker sews the sleeve to the body. At the conclusion of the sewing line, quality inspection is carried out to make sure the garment has been made correctly and is free of any manufacturing flaws. The clothing should be patched or repaired at authorised sewing locations. Designer apparel is created out of leftover fabric during this time-consuming process.

The primary stage of clothing manufacturing is component assembly, and stitching is done in line with the requirements of the customer.

High-powered single-needle or automatic sewing machines are used to complete the stitching operation. Buttonholes and buttons are typically sewn using sewing machines, while collar components are fused using fusing equipment.

### 1.2.6 Ironing and Finishing, Garment Inspection, Folding and Packing, Dispatch

Here, clothing is cured using steam, and any finishing that is required should be done as well. This process is manually carried out. The whole wardrobe is now inspected in line with the buyer's specifications. Final inspection is done by hand. The buyer's poly bag is used to package the whole order of clothing in this instance. The packing of garments is done manually. To minimise garment damage, all clothes must be cartooned while following the buyer's instructions. It is a manual for this process. It is eventually supplied to the consumer once all required processes have been taken.

## 1.3 ORGANISATION OF GARMENT MANUFACTURING INDUSTRIES

Depending on the types of market sectors on which the business concentrates, the apparel or garment sector is a labour-intensive, low-pay industry yet a dynamic, inventive one. The high-end fashion industry is seen as a modern technology since it has highly flexible employees and designers who are compensated comparably well. The clothing industry has to deal with a vast range of different garment types, which may be essentially split into two categories: exterior garments and inside garments.

The majority of businesses that serve this market sector are based in developed countries, frequently in specific regions within these countries. The manufacture of uniforms, undergarments, and other standard goods in large quantities is another type of important market sector.

The mainstream of the producers in this sort of standard product market sector is located in emerging nations. The role of the retailer has been more significant in the structuring of the supply chain for low- to medium-priced items on the market. The retail market sector has grown more competitive, giving international merchants more market sway.

The primary source of raw materials for the clothing business is textiles, which have created vertical supply chains with each other that include functions for sales and distribution. In the textile and garment industries, there are obtaining and processing of raw resources, or getting ready and making textile fibres, production of textile fabrics and yarns.

Dyeing and finishing textile materials to give them the desired aesthetic, physical, and visual qualities, such as bleaching, printing, dyeing, and coating, transforming fabrics into clothing that may be either fashionable or practical.

Figure 1.2 shows the various departments of garment manufacturing industries. Planning, designing, executing, and shipping the order (product) to the customer are all part of this crucial process known as **Merchandising**. Helping a customer through the ordering process is part of the merchandising process. Marketing merchandising

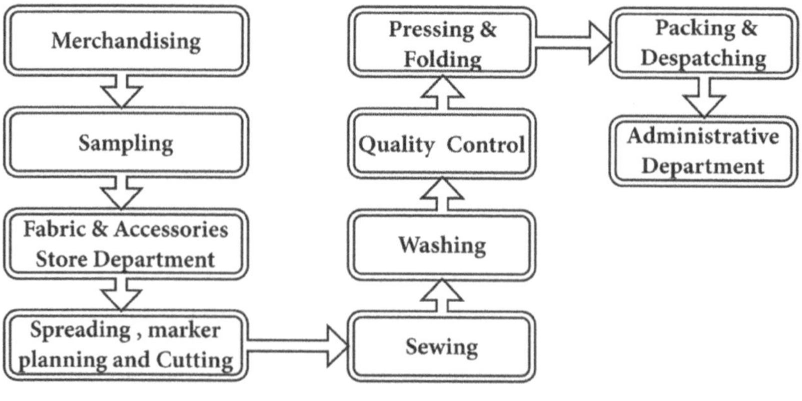


**FIGURE 1.2** Various departments of garment manufacturing industries.

and product merchandising are the two forms of merchandising carried out in a clothing unit (Tyler, 1992; Banumathi and Nasira, 2012). Development of the product, pricing, and ordering are the primary goals of marketing merchandising, which also involves direct interaction with the consumer. The appropriate clothing unit handles product merchandising, which includes all duties from procurement through completion (Chuter, 1995). **Sampling** department works with merchandising and production. It predicts bulk product look and fit and checks for pattern irregularities according to the buyer's request (Banumathi and Nasira, 2012). It helps calculate fabric, thread, and accessory use. **Fabric & accessories store department** involves sourcing of the raw material for fabric production and completing the work related to documentation of the purchased raw materials. **Production planning** schedules orders. They follow all industrial procedures. Manufacturing planning and scheduling are necessary to buy raw materials, finish production, and ship on schedule. The **cutting room** cuts cloth to design. The **production department** receives the garment pattern from the cutting department. The cutting master spreads the pattern elements above the cloth. After that, a cutting machine cuts all the layers after chalking off the design. The **sewing section** sews clothes. Sewing floors employ numerous production processes and line layouts. Factories use assembly lines or groups. **Washing** items after sewing removes dust, tracing marks, and gives them a cleaned appearance. This section washes, cuts, and tests clothing. **Quality control's** key duties are similar across organisations. Quality control does the following:

- Standardising quality
- Quality SOPs(Standard Operating Procedure)
- Quality assurance
- Pre-production quality check
- Checking inbound fabric and trimmings for quality
- Participation in product development and sample quality
- Avoid sending substandard cloth for cutting
- Mark minor fabric issues and notify the cutting department
- Fabric and trim quality audit report
- Pre-production meeting.

Documents linked to shipments are prepared by the **shipping and documentation** department. They interact with customers to arrange shipment delivery and then send the cargo to customers.

## 1.4 CONTRIBUTION TO ECONOMY

India's textile and clothing sector offers advantages along the whole value chain, from fibre, yarn, and fabric to clothes. The traditional handloom, handicrafts, wool, and silk goods, as well as the organised textile industry in India, are just a few of the many components that make up the very varied textile and apparel sector in India. Spinning, weaving, processing, and garment manufacture are all part of the organised textile industry in India, which is distinguished by the employment of capital-intensive equipment for mass-producing textile goods. By 2027, India's exports of readymade clothing are expected to reach $30 billion with a compound annual growth rate (CAGR) of 12%–13%. In the domestic market, $150.5 billion was spent on textiles and garments in 2019–2020.

India's economy has been severely impacted by COVID-19. Numerous transportation options, including trains and aircraft, are stopped as a result of the nationwide shutdown, which is anticipated to harm domestic business in India. The virus has been damaging and disrupting supply chains and the world economy for approximately three to six months. The epidemic has impacted the bulk of India's export markets (the US and European Union [EU] combined account for around 60% of the country's total garment exports in value terms), resulting in order cancellations and order delays that have increased inventory levels and increased working capital needs. The fourth quarter of FY20 saw a dry up in apparel orders, problems with working capital, and a lack of clarity on tariffs and incentives, particularly as exporters from Bangladesh, Sri Lanka, and Vietnam are given privileged access. The lockdown in India also impacted domestic consumption.

In terms of market attractiveness, India is predicted to come as second by 2025, contributing up to $121 billion, while China is predicted to come as first, contributing up to $378 billion. India is one of the economies that is growing the quickest, with a GDP growth rate of 7.2% from 2017 to 2018. As a consequence, the general people have more buying power, which increases demand for goods made by the textile sector. The upshot of this expansion is a wide range of manufacturing capabilities for various goods that may be transported both inside and outside of India. Like other businesses, the textile industry in India confronts a lot of challenges while having a lot of opportunities for investment. The government's continual policy adjustments at the federal, state, and local levels put the textile industry under a lot of stress. Due to the implementation of the GST, clothes and accessories are more expensive.

Another challenge is that the textile sector cannot access the newest and most sophisticated technology and cannot satisfy the requirements of the international export market. Along with these issues, the Indian textile industry also faces challenges including child labour, competition from other countries for cheap apparel, and personal safety rules (Chiripal, 2022).

## 1.5 TECHNOLOGICAL INTERVENTION IN CONVENTIONAL PROCESSES

The progress and innovation of technology have always pushed the industrial sector. The revolutionary benefits of modern technology are assisting manufacturers to decrease human interaction, increase plant productivity, and gain a competitive edge. Examples include big data analytics and sophisticated robots.

The future of manufacturing will be shaped by sophisticated technologies like artificial intelligence (AI), the internet of things (IoT), and 3D printing, among others, which will reduce production costs, increase productivity, and reduce mistakes. A manufacturing facility's success depends on productivity; hence, each company is required to spend a lot on these technologies.

### 1.5.1 Wearable Technology

This technological trend allows for the direct exploration and creation of wearable intelligent devices by integrating multimedia, sensor, and wireless communication technologies into people's clothing and accessories that facilitate gesture and eye movement operations. A few examples of wearable textile technology are Levi's Commuter x Jacquard Jacket, Sensoria Fitness Socks, Neviano Swimsuits, Nadi X Yoga Pants, UA Recover Clothing, Smart baby hat by Neopenda, and others.

Industry 4.0: The manufacturing and industrial sectors are fast embracing the IoT, allowing plant owners the possibility to increase efficiency and decrease complexity. The Industrial Internet of Things (IIoT) incorporates a variety of technologies, such as automated manufacturing, big data, sensor data, machine learning, and cloud connectivity. Utilising technology for proactive and predictive maintenance, real-time monitoring, resource optimisation, supply chain visibility, cross-facility operations research, and safety may help plant managers decrease downtime and improve process efficiency. The regular upkeep and repairs that plants need to function effectively are essential. However, not all equipment and devices need the same level of maintenance. Plant managers can monitor the health of their equipment and carry out proactive maintenance by using the IIoT. They can schedule maintenance around when it's essential due to their real-time performance monitoring, which lowers the possibility of unanticipated outages and productivity loss (Figure 1.3).

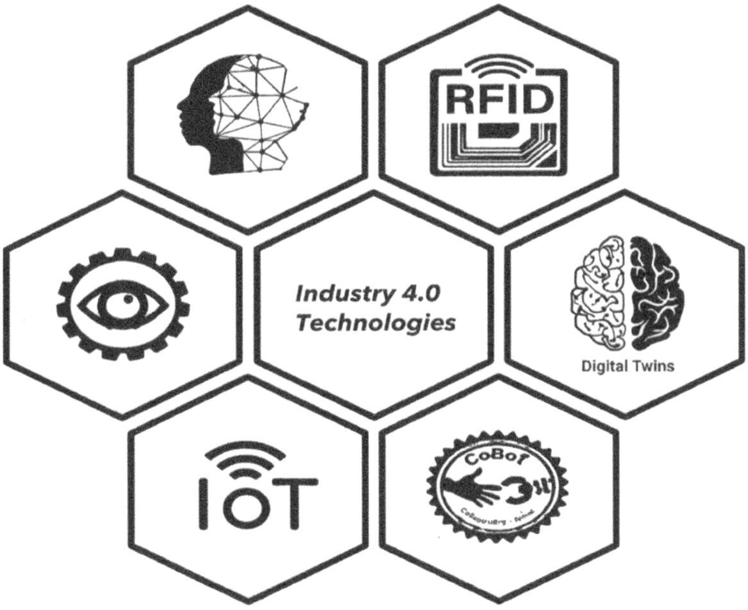


**FIGURE 1.3** Industry 4.0 technologies.

Big Data Analytics: Big data analytics may offer a variety of approaches to enhance the efficiency of industrial operations, enable product customisation, and enhance the performance of assets. According to a recent Honeywell poll, 68% of American firms have already made big data analysis investments. Big data analysis may help these firms make decisions by providing information on productivity and waste efficiency, which will save operating costs and improve overall performance.

Machine Learning and AI: For many years, firms have employed robots and mechanisation to boost productivity and lower manufacturing costs per unit. Production teams are using AI to analyse data, gain insights, replace inventory, save operating costs, and provide seamless quality control across the whole manufacturing process. The day of mindless robots doing repetitive factory jobs has passed. Robots and humans may now work together to create agile manufacturing processes that continuously learn, develop, and make informed production decisions due to AI and machine learning. Therefore, manufacturers may manage regular chores using industrial robots and intelligent automation while concentrating their time and resources on activities that increase revenue, such as research and development, adding new product lines, and enhancing customer service.

3D Printing: High-end sectors including aircraft, mining, machineries, cars, weapons, commercial and service machines, and other industrial equipment are predicted to be significantly impacted by 3D printing or additive layer manufacturing technology. With the aid of this ground-breaking technology, producers can create real objects using intricate digital blueprints saved in 3D computer-aided design (CAD). Real things may also be printed using materials including rubber, nylon, plastic, glass, and metal. For medical research, 3D bioprinting made it possible to produce living tissues and functional organs. The 3D printers provide more than traditional production methods since they can make intricate forms and patterns for no extra expense. Additionally, manufacturing as a service is becoming more and more popular as a result of the expanding uses of 3D printing in the manufacturing industry. Maintain a present infrastructure that can serve additional customers without requiring the acquisition of new hardware.

Virtual Reality (VR): By doing away with the necessity for intricate prototypes, VR streamlines the process of product design. In order to visualise their designs digitally and address any issues before manufacturing begins, designers and engineers use VR to generate realistic product models. These digital drawings, simulations, and integrated devices can all be reviewed and interacted with by customers, which drastically cuts down on design time. For instance, automakers are already adopting VR to test their vehicles early in the process of vehicle development, cutting down on the time and expenses involved in modifying designs, tolerances, and safety features (Niekerk, 2022).

## 1.6 PRESENT SCENARIO OF WORLD TEXTILE AND APPAREL INDUSTRIES IN GLOBAL MARKET

This textile market research report offers a thorough overview of everything you need, along with a thorough analysis of the current and future state of the industry. The global textile market expanded from $573.22 billion in 2022 to $610.91 billion in 2023, with a CAGR of 6.6%. The confrontation between Russia and Ukraine, at least briefly, limited the likelihood of a COVID-19 pandemic-related global economic recovery. The dispute between these two countries has resulted in economic sanctions on a number of countries, a rise in commodity prices, and disruptions in the supply chain, all of which have increased prices for goods and services and had an effect on several global markets. The worldwide textile market is anticipated to reach $755.38 billion in 2027 at a CAGR of 5.5% (Textile Market Size, Trends and Global Forecast To 2032).

The present worldwide garment market is projected to be worth $1.7 trillion, or around 2% of the $73.5 trillion global GDP. About 70% of the world's apparel consumption is concentrated in the top eight economies. With a combined share of almost 23%, the four BRIC countries are among the top markets. Although they only account for 11% of the global population, the US and the EU together consume 40% of all clothes, indicating exceptionally high levels of per capita spending on apparel (PEAP) in comparison to the rest of the world. The consumption of clothes in an emerging or developing market would expand more quickly than its economic growth if the disposable income of developing nations increased (Global textile apparel industry fortunes to depend on five major trends). By 2025, it is anticipated that the global clothing consumption would reach $2.6 trillion due to the predicted GDP growth rate and its relationship to the expansion of the garment sector. Market share is shifting from previously dominant industrialised nations like the US and the EU to locations like China and South Asia, which includes India, Bangladesh, and Vietnam. Due to the ongoing downturn in regional economic growth, these industrialised nations are presently encountering difficulties with their consumption, while China, India, and other South East Asian nations are seeing growth. While only accounting for 41% of the global garment market in 2017–2018, the EU and the US are home to 11% of the world's population. But pressure from sluggish demand is anticipated to further reduce their market share going forward, shifting it to nations like China and India that are anticipated to expand at a considerably quicker CAGR of 10%–11% (2017–2025). A rise in per capita income will increase demand in both nations with faster economic development, which will lead to a combined 23% share of clothing consumption by 2025–2026 from a 16% share in 2017–2018 (Global Textile & Apparel Industry – India's Position – Blogs – Televisory). The Indian textile and apparel market is dominated by the clothing or readymade garments segment, which accounts for over half of the textile and apparel market and approximately three-fourths of the domestic market. The Indian domestic market has recently outperformed the world's two largest consumers of textiles and clothing, the EU and the US, and between 2005 and 2017, it saw a strong CAGR of 10%. India's demand for textiles and clothing has been driven by the expansion of the middle and lower middle classes as well as increased disposable income. The domestic textile and apparel industry in India is predicted to develop at a CAGR of 12% to reach $220 billion by 2025–2026 as a result of this forecasted expansion.

India is rising to become one of the most competitive nations in the world for producing textiles and apparel for export (Figure 1.4). India ranks second globally in terms of textile exports with 6% of the market share and fifth in terms of apparel exports with 4% of the market share. With 5% of all worldwide textile and apparel exports, India is ranked second overall. India's textile and apparel exports climbed to $39 billion in 2018–2019, and it is expected that they will reach $70 billion by

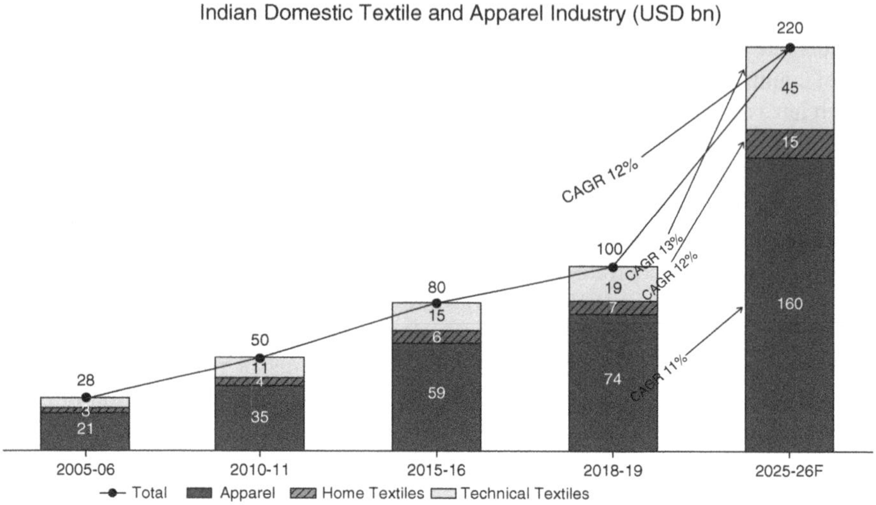


**FIGURE 1.4** Indian domestic textile and apparel industry (USD bn) – Ministry of Textiles, Wazir analysis, televisory's analysis.

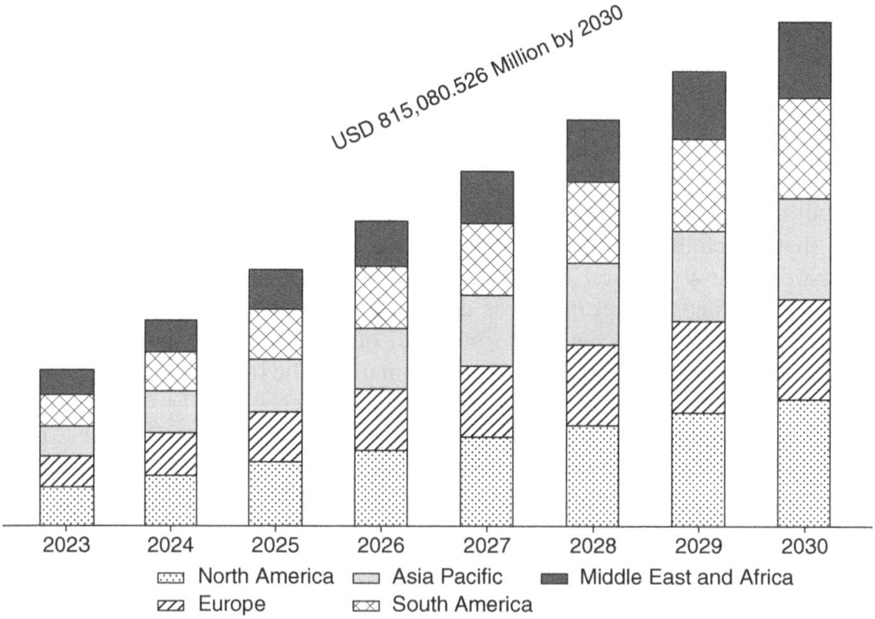


**FIGURE 1.5** Global textile garment market – data bridge marketing research.

2025–2026, growing at a CAGR of 10%. Moreover, half of all textile and apparel exports are cotton, with man-made or synthetic fibres following in second with a 28% share (2018–2019).

The market for textile clothing is anticipated to expand significantly between 2023 and 2030. According to Data Bridge Market Research's analysis, the market is estimated to reach USD 815,080.526 million by 2030 and will increase at a CAGR of 4.2% from 2023 to 2030 (Figure 1.5). The inclination towards the newest fashion trends among the younger generation is the main element anticipated to fuel the market expansion for textile garments.

## 1.7 MARKET OUTLOOK OF THE GARMENT MANUFACTURING AND FUTURE POTENTIAL

### 1.7.1 Market Outlook

The method of producing clothing continues to be impacted by technological advancements. Today, a lot of clothing producers use computers to create patterns, develop things, and display their goods to customers. However, due to the wide range of materials and the intricate cutting and sewing required for most styles, many processes at apparel factories are challenging to automate; therefore, the fashion sector is likely to continue to be labour-intensive. However, according to the Department of Labour, some of the occupations that will see the fastest employment declines between 2018 and 2028 are pressers, textile, garment, and related materials; operators and tenders of textile bleaching and dyeing machines; and setters, operators, and tenders of textile knitting and weaving machines. The best employment opportunities will go to those who are knowledgeable with CAD and computer-aided manufacturing.

Large retailers and manufacturers, including H&M and Nike, suffered significant defeats early in the 2020 coronavirus pandemic as limits on public activity led them to close the doors of their brick-and-mortar stores. Additionally, supply chain disruptions made it harder to get the raw materials needed for clothing production, and trade and travel restrictions made it harder to reach foreign suppliers. Conventional, in-person conferences, fashion displays, and other events switched to virtual platforms, or in some circumstances were postponed or cancelled. While some businesses closed to assist in reducing the spread of the coronavirus, some jobs in this industry were lost, particularly at the manufacturing and retail levels.

### 1.7.2 Future Potential

Market participants must concentrate on the prospects since they generate income. In order to acquire a competitive edge in this fiercely competitive industry, market participants should capitalise on these revenue pockets as well.

#### 1.7.2.1 Increasing Demand in Developing Nations

The Asia Pacific region's capacity for producing clothes has increased as a result of the cheaper cost of producing clothing in emerging nations like China, India, Bangladesh, Vietnam, and many more. The availability of raw resources, government assistance, laxer environmental rules, and cheaper labour costs are a few of the main causes of the reduced manufacturing costs.

In these developing nations, a sizable portion of consumption is made up of clothing. The youth-dominated population, rising disposable income, rising purchasing power, and increased fashion awareness in rural regions are the main drivers of growth in emerging nations. Additionally, the youth-dominated population and rising per capita garment consumption will present a sizable market potential for clothing and fashion companies.

#### 1.7.2.2 Growing Interest in Used Clothes

The expansion of second-hand is being driven by Millennials and Generation Z. With 33% and 31% of all resale customers, respectively, millennials and baby boomers are currently the largest demographics of second-hand buyers.

The second-hand market is presently worth $24 billion, and it is anticipated to increase to $51 billion in the next five years and to $64 billion by 2028.

Retailers are keen to participate in the circular fashion trend since the second-hand market appears to be so vibrant. The distinction between new and second-hand clothing is becoming increasingly hazy for today's fashion customers. And it is this that has significantly changed the retail industry (https://www.fibre2fashion.com/industry-article/8698/global-apparel-market-outlook-an-overview).

## 1.8 APPAREL MARKETING AND MERCHANDISING

The merchandiser serves as a bridge between businesses and consumers. He is responsible for managing every task, including purchasing the raw materials needed to complete the product, creating and finishing the garment, documenting the process, and sending the finished product. He is the one in charge of producing the item. A garment export unit typically has a number of departments, including stores, cutting, production, packaging, checking, etc., with the merchandising department serving as the connecting mechanism between all of them. A merchandiser's responsibility is to coordinate with every office department and customer. A positive working relationship between the exporter and the buyer is established by the merchandiser.

Within the fashion sector, merchandising is a specialist managerial position. The fashion industry is responsible for getting products from designers' showrooms to store sales floors and ultimately into consumers' hands. A retail organisation's internal planning guarantees that there is enough amount of inventory on hand to be sold at prices that customers are ready to pay in order to ensure a profitable operation (Kiron, 2021).

The idea behind the term "merchandising" has roots in early history, at the time of the beginning of international trade. The existence and development of trade between the larger civilisations of ancient times, including Greek, Roman, Indian, Chinese, and Egyptian, are extensively documented. However, in those days, the meaning of the term "merchandising" was limited to the trade of materials or goods that were obtained from nature or generated from nature and were not developed for a particular use or for a particular audience. Materials that satisfy a person's basic needs for survival, such as commodities, live animals, textiles, and opulent objects like gold, diamonds, and valuable stones, were exchanged. In the past, things were manufactured by families and at home, rather than in large quantities. Home production was supplanted by factory systems during the industrial age, and manual labour was replaced by machinery. Large-scale consumption came after mass manufacturing.

## 1.9 RECENT TRENDS AND FORECASTING ANALYSIS OF APPARELS

In the last four to five decades, the global garment sector has undergone significant developments in terms of both technological advancements and volume growth. Today's textile and apparel manufacturing facilities are transitioning from conventional production techniques towards highly targeted design and manufacture of garments with added value. This shift has brought an increase in the popularity and economic success of a vast range of intelligent and smart textiles, which offer more functionality to clothes.

Smart textiles refer to fabrics that have the ability to recognise, react to, and carry out certain tasks that an ordinary cloth is incapable of. These fabrics are made using smart materials or finishes, such as shape-memory polymers, phase-changing materials, or electronic or e-textiles. Applications for smart and intelligent textiles include physiological monitoring (such as measuring heart rate, respiration rate, temperature, activity, and posture), data acquisition for sports training, monitoring potential external hazards or hazards associated with handling hazardous materials, and tracking the location and status of soldiers engaged in combat (Nayak and Padhye , 2015). To evaluate these smart fabrics and apparel, it is necessary to develop new instrumental analysis techniques and test methodologies. Additionally, it is crucial to ensure that fashion and textile designers are educated on the most recent technology used in smart textiles. This will equip them with the necessary knowledge to design and produce smart garments that meet the evolving needs of consumers. In conclusion, the textile and garment industry is evolving, and the advent of smart textiles is set to revolutionise the sector. The industry must embrace these developments by investing in the necessary technology and educating designers and manufacturers on the latest trends and advancements. This will allow them to stay ahead of the curve and remain competitive in a rapidly changing market.

## 1.10 CONCLUSION

Like other businesses, the textile industry in India confronts a lot of challenges despite having a lot of opportunities for investment. The government's continual policy adjustments at the federal, state, and local levels put the textile industry under a lot of stress. Due to the implementation of the GST, clothing and accessories are more expensive. Another challenge is that the textile industry cannot access the newest and most sophisticated technologies and cannot satisfy the requirements of the international export market. Along with these issues, the Indian textile industry also faces challenges like child labour, competition from other countries for cheap apparel, and personal safety rules.

India's garment industry must implement a number of changes and adopt certain new procedures to improve its competitiveness in order to overcome the aforementioned challenges and achieve the anticipated global market aim. A strategy to increase production includes putting more of an emphasis on technological advancements and expanding the weaving industry's capacity. In order to boost the overall business industry, state governments should also give approval for wastewater treatment plants.

If both the federal and state governments adequately support both its small- and large-scale players, the Indian textile industry would reach incredible heights. The Indian textile industry should take into account lowering the taxes imposed on exports that receive government subsidies in addition to educating their personnel to meet the evolving needs of the modern market. The demand for goods in the Indian textile industry has increased as a result of an increase in disposable income, creating a huge demand in both domestic and international markets. Consequently, the rapid growth of the retail sector, government support, and investments bode well for India's textile industry's future.

## REFERENCES

Banumathi, P.N. and G.M. Nasira (2012), "Fabric inspection system using artificial neural networks", *International Journal of Computer Engineering Science*, Vol. 2, pp. 20–27.

Chiripal, R. (2022), "Textile sector seeing a structural turnaround in India: Opportunities & challenges", *The Times of India*. Available at: https://timesofindia.indiatimes.com/blogs/voices/textile-sector-seeing-a-structural-turnaround-in-india-opportunities-challenges/ (Accessed: 7 January 2023).

Chuter, A.J. (1995), *Introduction to Clothing Production Management*, Blackwell Scientific Publications, Oxford.

Forza, C. and Vinelli, A. (2000), "Time compression in production and distribution within the textile-apparel chain", *Integrated Manufacturing Systems*, Vol. 11 No. 2, pp. 138–146. https://doi.org/10.1108/09576060010314134

Hassler, M. (2003), "The global clothing production system: commodity chains and business networks", United Nations Industrial Development Organization.

Kiron, M.I. (2021), "Role of merchandiser in garment industry", Textile Learner. Available at: https://textilelearner.net/role-of-merchandiser-in-garment-industry/ (Accessed: 7 January 2023).

Nayak, R., Padhye, R. (2015), *Garment Manufacturing Technology*, Woodhead Publishing Series in Textiles, ISBN-1782422323

Niekerk, F.V. (2022), "Technology interventions in the manufacturing segment", *The Times of India*. Available at: https://timesofindia.indiatimes.com/blogs/voices/technology-interventions-in-the-manufacturing-segment/ (Accessed: 5 January 2023).

Scott, A.J. (2006), "The changing global geography of low-technology, labor-intensive industry: clothing, footwear, and furniture", *World Development*, Vol. 34, No. 9, pp. 1517–1536.

Tyler, D.J. (1992), *Materials Management in Clothing Production*, Blackwell Scientific Publications, Oxford.

# 2 Design Ideas and Conceptualization

*Paramita Sarkar, Shipra Sharma, and Sonjib Bora*

## 2.1 ELEMENTS OF DESIGN AND PRINCIPLES OF DESIGN CREATING VISUAL INTEREST

Absolutely, everything you see around you in the visual world is composed of the fundamental elements of design, which include points, lines, shapes, forms, and colors. These elements are the building blocks of visual composition, and they work together to create the rich tapestry of the visual world. Let's explore how each of these elements contributes to what you see. The elements of design encompass the fundamental components that designers manipulate to convey their ideas, solve problems, and craft visually compelling and functional solutions. These elements serve as raw materials, much like the colors on an artist's palette or the notes in a musician's composition. When combined and manipulated with skill and intention, they give birth to the rich tapestry of designs that shape our world. We are all surrounded by dots, lines, shapes, forms, and colors as shown in Figure 2.1.

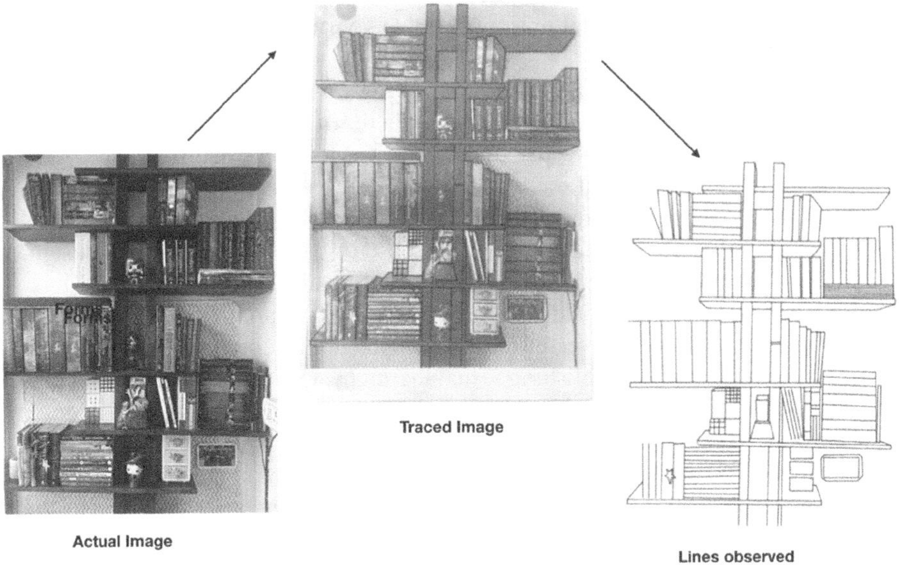


**FIGURE 2.1** Observing lines in the surroundings.

DOI: 10.1201/9781003432739-2

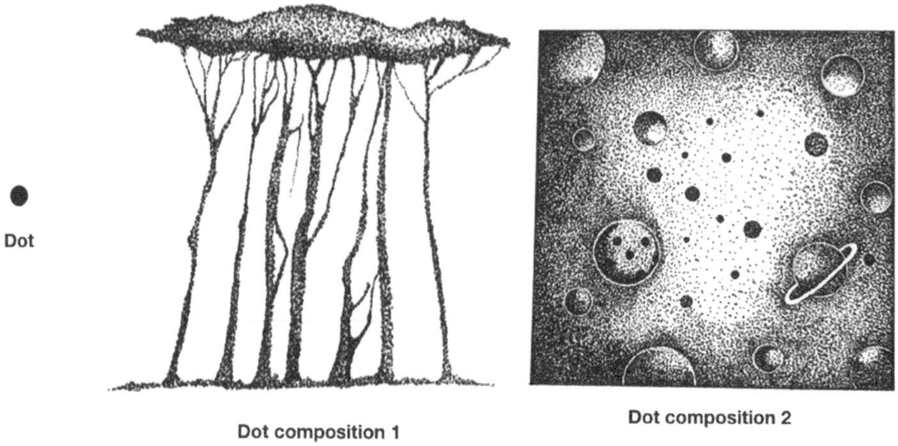


**FIGURE 2.2** Point of dot compositions.

I. **Points:** Points or dots are the smallest and simplest elements in the design. In the physical world, they might correspond to individual pixels on a screen, ink dots on paper, or even the atoms that make up physical objects. Points are often used as markers, focal points, or as part of patterns and textures as shown in Figure 2.2.
II. **Lines:** Lines are created by connecting points. They can be straight or curved, thick or thin, horizontal or vertical. In the world around you, lines can represent physical boundaries, edges, or outlines. They can also be used to create the structure of objects or guide your eye within a composition as shown in Figure 2.3.
III. **Shapes:** Shapes are formed by connecting lines to enclose areas. In your visual environment, shapes are everywhere, from the geometric shapes of buildings and objects to the organic shapes found in nature. Shapes give form to everything you see, and they can convey meaning and aesthetics.
IV. **Form:** Forms are three-dimensional objects with depth and volume as shown in Figure 2.4. They provide a sense of solidity and space in design. The interplay of light and shadow on forms creates depth and realism, making them essential in fields like product design and architecture.
V. **Color:** Color is a powerful tool in design, influencing emotions, conveying meaning, and establishing visual hierarchy. Understanding color theory, including hue, saturation, and value, is key to effective color use. A color wheel is a visual tool used in art, design, and color theory. It consists of a circular arrangement of colors, typically divided into primary, secondary, and tertiary hues. The primary colors—red, blue, and yellow—serve as the foundation, while secondary colors result from mixing two primary colors, such as green, purple, and orange. Tertiary colors further refine the spectrum by blending a primary with a neighboring secondary color. The color wheel aids in understanding color relationships, harmonies, and contrasts,

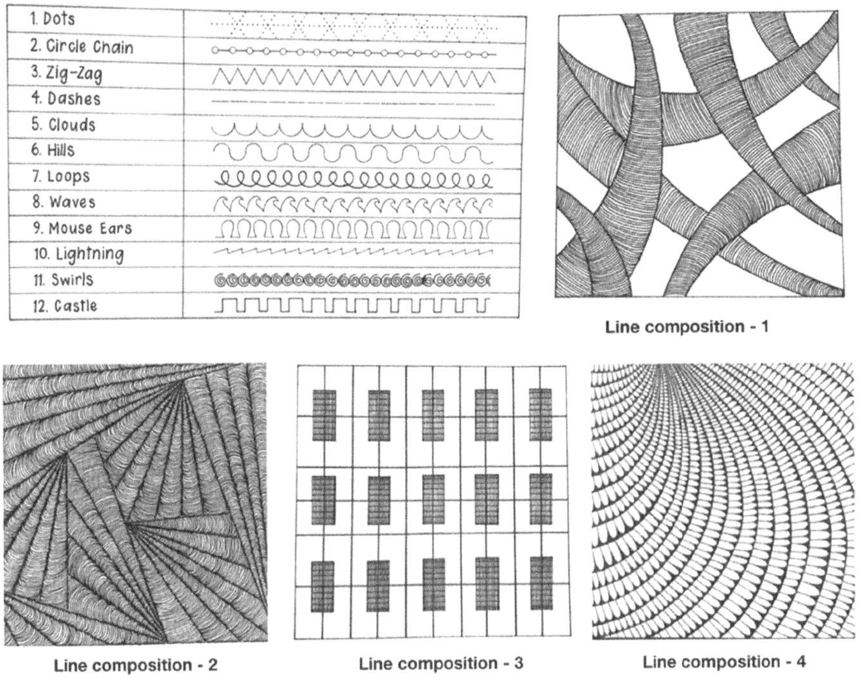


**FIGURE 2.3** Different types of lines and line compositions.

**FIGURE 2.4** Shapes and forms.

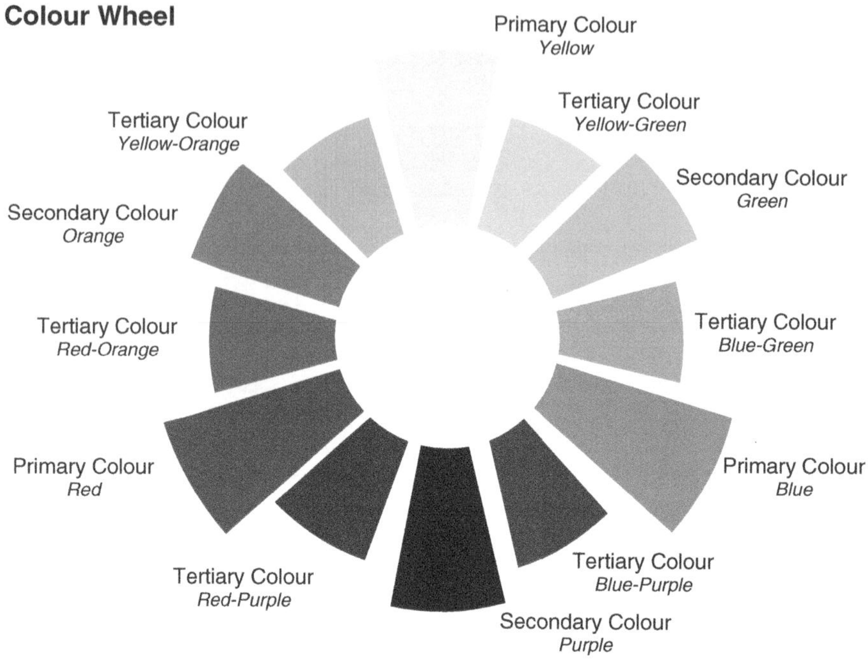


**FIGURE 2.5** Color wheel.

helping artists and designers choose complementary or analogous colors, achieve balance, and create visually appealing compositions in various mediums, from painting to graphic design as shown in Figure 2.5.

VI. **Texture:** Texture adds a tactile quality to the design. It can be visual or tactile and can create a sense of depth, realism, or mood. Texture is particularly significant in fields like interior design, fashion, and digital media as shown in Figure 2.6.

VII. **Space:** Space refers to the area within and around design elements. It includes positive space (occupied by objects) and negative space (the empty areas around objects). Manipulating space is vital for achieving balance, harmony, and a clear visual hierarchy.

### 2.1.1 Principles of Design

The principles of design encompass a range of concepts and techniques that influence the arrangement of elements within a composition. They enable designers to make informed decisions about layout, color, typography, and more, ultimately shaping the user's perception and experience.The principles of design such as balance, rhythm, proportion, emphasis, and unity form the foundation of visual composition (Bevlin, 1994). While the specific principles may vary slightly depending on the design discipline, several core principles are universally recognized and applied. Throughout this chapter, we will delve into these key principles of design, providing a comprehensive overview of each one. We will discuss how these principles interact with

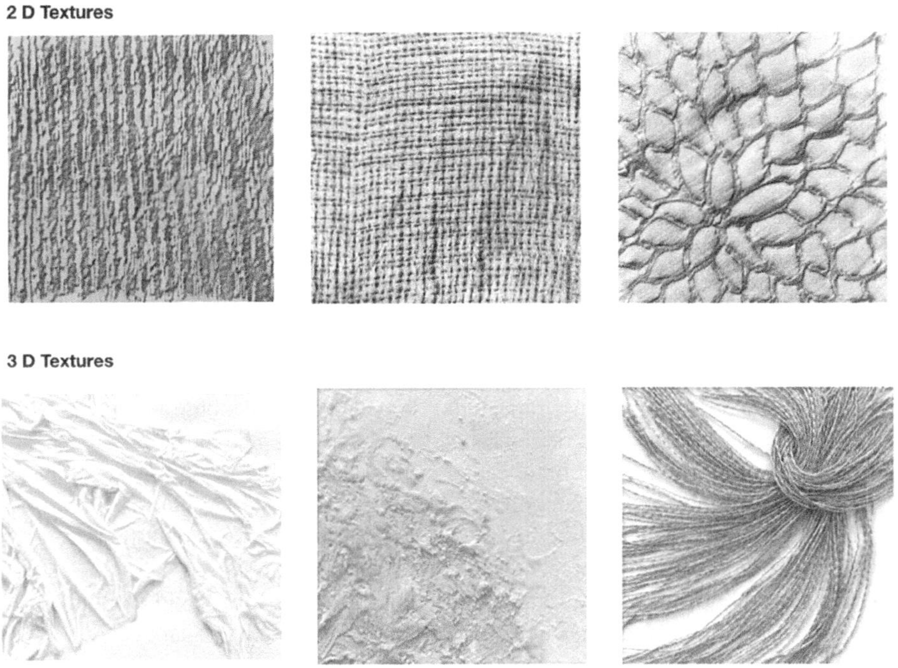


**FIGURE 2.6** 2D traced and 3D structures created by design students.

one another and how they can be harnessed to achieve specific design goals. The principles of design covered in this chapter include:

I. **Balance:** The equilibrium is achieved through the distribution of visual weight in a composition. Balance can be symmetric or asymmetric. A classic example of symmetrical balance is a formal evening gown with a symmetrical design, where the left and right sides mirror each other in terms of design elements, such as embellishments, seams, and patterns. An example of an asymmetrical balance top with a single shoulder strap or an uneven hemline creates a sense of dynamic balance by intentionally avoiding perfect symmetry.

II. **Unity and Variety:** The delicate balance between consistency and diversity in design elements. An example of unity is a collection of clothing items in a cohesive color palette or theme that creates unity, ensuring that the pieces work well together when combined into outfits. Example for variety is within a collection, designers may introduce variety by using different fabrics, textures, or prints to add interest while maintaining an overall sense of unity.

III. **Proportion and Scale:** The relative size and dimensions of elements in relation to each other and the overall composition. An example for proportion is a designer may use proportion to elongate the appearance of a person's legs by pairing high-waisted pants with a cropped top, creating the illusion of longer legs. An example of scale is designers may play with scale by incorporating oversized accessories or prints on clothing items to make a bold fashion statement.

IV. **Contrast:** The juxtaposition of elements to create emphasis and visual interest. An example of color contrast is the use of contrasting colors in clothing, like a black and white outfit, which can create striking visual effects and draw attention to specific areas. Example for Textural Contrast: Combining contrasting textures, such as pairing a smooth satin blouse with rugged denim jeans, adds tactile interest.

V. **Rhythm and Repetition:** The deliberate use of patterns and repetition to create a sense of continuity and rhythm. An example of rhythm is a repeating pattern on a dress, like polka dots or stripes, that creates a rhythmic visual flow and can convey a sense of movement. An example of repetition is **that a** designer may use repeated design elements, like buttons, ruffles, or pleats, to create a sense of consistency and structure in a garment.

VI. **Emphasis:** The focal point or center of interest within a design. An example of a focal point is designers often use embellishments, like sequins, embroidery, or a unique neckline, to create a focal point on a garment that draws attention to a specific area, such as the shoulders or waist.

VII. **Seam Alignment:** Aligning seams along the natural lines and contours of the body ensures that garments fit well and maintain their shape.

These principles are applied by fashion designers to create clothing that not only looks aesthetically pleasing but also functions well, fits comfortably, and reflects the designer's creative vision. Pattern development requires an understanding of body proportions, measurement systems, and garment construction (Aldrich, 2008). Fashion design is a fascinating blend of art and functionality, where the principles of design play a pivotal role in shaping the industry.

## 2.2 DESIGN IDEAS AND SOURCES OF INSPIRATION

Fashion designers draw inspiration and generate design ideas from a variety of sources and experiences. The process of finding inspiration is highly individual and can vary from one designer to another. Here are some common ways fashion designers seek inspiration and develop their design ideas.

Designers often stay up-to-date with current fashion trends by studying fashion magazines, attending fashion shows, and following fashion blogs and social media accounts. They analyze the latest styles and trends to understand consumer preferences.

Cultural elements, including art, music, literature, and history, can serve as rich sources of inspiration. Designers may draw from specific time periods, cultural traditions, or artistic movements to create unique designs. Nature and the environment can provide endless inspiration for color palettes, patterns, and textures. Elements like flora, fauna, landscapes, and weather patterns often find their way into fashion collections. Travel and Exploration: Traveling to different regions and countries exposes designers to diverse cultures, textiles, craftsmanship, and fashion styles. Immersing oneself in new environments can spark creative ideas. Designers often draw from their own life experiences, memories, and emotions. Regular drawing practice helps designers improve observation, visualization, and presentation skills (Stewart, 2011). Sketching is an effective tool for visual thinking, rapid ideation, and communicating design concepts (Eissen & Steur, 2011). Personal stories and journeys can be translated into unique and meaningful fashion designs. Artistic movements, architectural structures, and iconic artworks can inspire clothing silhouettes, patterns, and color schemes. Designers might reinterpret elements from famous artworks into fabric prints, for example.

Street Style and Subcultures: Street fashion and subcultures like punk, hip-hop, or bohemian styles can be influential sources for designers looking to capture a particular mood or aesthetic. Vintage and Retro Styles: Vintage clothing, whether from past decades or historical periods, can serve as inspiration. Designers often reinterpret and modernize classic styles for contemporary fashion collections. Materials and Textiles: Exploring new or unconventional materials and textiles can trigger innovative design ideas. The tactile qualities of fabrics can inspire unique textures and garments. Interacting with other designers, artists, and creative professionals can lead to collaborative projects that spark new design concepts. Sharing ideas and perspectives can be highly inspirational. Social Issues and Movements: Fashion designers may use their platform to address social and political issues, and these themes can become central to their design concepts. Ethical and sustainable fashion, for example, is a growing movement in the industry. For designers working in commercial fashion, understanding the needs and preferences of their target audience is crucial. Consumer research and feedback can influence design decisions. It is important to note that inspiration can strike at any moment, and designers often keep open minds and receptive senses to capture ideas when they arise. The process of turning inspiration into a concrete design involves research, sketching, prototyping, and refining concepts until a collection or garment takes shape. Fashion designers continuously adapt and evolve their creative processes to stay relevant and innovative in a dynamic industry.

## 2.3 MOOD BOARD AND STORYBOARD

Mood boards and storyboards are essential tools in the fashion industry that help designers and creatives conceptualize and communicate their ideas. They serve different purposes and are used at different stages of the design process.

A mood board is a visual collage or presentation that conveys the overall mood, tone, and aesthetic direction of a fashion collection or project. It typically consists of a combination of images, fabric swatches, color samples, text, and other visual elements that capture the desired atmosphere, theme, or inspiration for the collection. Mood boards help designers and teams establish a cohesive visual direction and ensure that everyone involved in the project has a clear understanding of the intended look and feel. They can be physical boards with cutouts or digital presentations created using software like Adobe Photoshop, Illustrator, or even online tools (Figure 2.7).

A storyboard, on the other hand, is a sequence of illustrated or visual panels that outline the narrative or progression of a fashion show, photo shoot, or marketing

**FIGURE 2.7** Image of a mood board.

campaign. It helps fashion professionals plan and visualize how a collection or fashion story will unfold, from start to finish. Storyboards can include sketches of clothing designs, hair and makeup concepts, model poses, and even the set or location where the shoot or show will take place. Storyboards are particularly important for fashion shows and editorial photo shoots, where every detail needs to be carefully choreographed and executed to convey a specific message or story. They are often created to ensure that all team members, including designers, stylists, photographers, and models, are on the same page regarding the creative vision.

In summary, mood boards set the visual and emotional tone for a fashion project, while storyboards provide a structured visual narrative or plan for executing that vision. Both tools are crucial for effective communication, collaboration, and the successful realization of fashion concepts.

## 2.4 FASHION ILLUSTRATION

Fashion illustration in fashion design is a crucial and artistic aspect of the design process that involves creating visual representations of clothing, accessories, and fashion concepts. These illustrations serve as a means of visually communicating design ideas, translating the creative vision of a fashion designer into a tangible form that can be shared with others involved in the design and production process. Fashion illustration serves as a visual language for communicating design ideas, silhouette, texture, and movement (Gill, 2009). In the world of fashion illustration, the depiction of the human figure differs significantly from how it is represented in traditional or "normal" figure drawing. Fashion figures are stylized and elongated, emphasizing certain proportions and characteristics to convey a specific aesthetic and to serve as an idealized template for illustrating clothing.

It is to redefine the beauty of the human body to make it more expressive in illustrations. Here are some key differences between fashion figures and normal figures:

I. **Proportions:** Fashion illustrations often feature elongated proportions, which is ideally 1–10 and 1–12 head-to-body ratio. The legs are longer, the neck is elongated, and the overall figure tends to be taller and slimmer. These exaggerated proportions create a sense of elegance and sophistication and make it easier to showcase clothing. In traditional figure drawing, the proportions are typically more realistic and the body proportion of an average human generally comes between 1 – 6.5 inches and a 1– 8 inches head-to-body ratio. Artists aim to capture accurate body proportions and anatomy.

II. **Body Shape:** Fashion illustrations often feature a more idealized body shape with elongated limbs and a slender waist. This idealized body serves as a canvas to showcase clothing designs and allows for a more dramatic and glamorous presentation. Traditional figure drawing may focus on representing a variety of body shapes and sizes, aiming for a more natural and realistic depiction of the human form.

III. **Poses:** Fashion illustrations frequently depict dynamic and expressive poses to showcase the movement and flow of clothing. Poses can be exaggerated and theatrical to create visual interest and drama. Traditional figure drawing often emphasizes static or natural poses that accurately represent the human body in a relaxed or neutral state.

IV. **Emphasis on Clothing:** The primary purpose of a fashion figure is to showcase clothing designs. Therefore, the figure's pose, proportions, and features are adjusted to highlight the garment's design elements, such as drapery, silhouette, and details. In traditional figure drawing, the emphasis is on capturing the human form itself, with less focus on clothing or accessories.

V. **Simplification:** Fashion illustrations tend to simplify facial features, such as the eyes, nose, and mouth, to draw attention to the clothing. Facial expressions are often minimized or stylized. Traditional figure drawing often pays close attention to capturing the nuances of facial expressions and features.

VI. **Gesture and Fluidity:** Fashion illustrations often emphasis fluid lines and gestures that convey movement and energy. This helps create a sense of dynamism in the illustration. Traditional figure drawing may focus on achieving a sense of stillness and tranquility in the artwork.

In summary, fashion figures are stylized representations of the human form designed specifically to showcase clothing and fashion designs. They prioritize elongated proportions, idealized body shapes, dynamic poses, and simplified facial features to create an aesthetic that complements and highlights the garments being illustrated. These differences between fashion figures and normal figures are essential for fashion illustrators as they convey the essence and allure of the fashion world.

## 2.5 PORTFOLIO

A fashion portfolio, created by fashion designers, is a curated collection of their work that showcases their design skills, creativity, and expertise in the field of fashion. It serves as a visual and often physical representation of a designer's capabilities and style. Fashion portfolios are essential tools for aspiring designers seeking employment, freelance opportunities, or admission to fashion schools. Here's an in-depth explanation of what a fashion portfolio typically includes:

I. **Design Projects:** The core of a fashion portfolio consists of design projects. These projects may encompass a range of categories, such as womenswear, menswear, children's clothing, accessories, or even specialized niches like bridal or sportswear. The fashion design process typically moves from research and concept development to design exploration and final presentation (Jones, 2005). Each project typically includes a series of designs, from initial concept sketches to finalized drawings or digital renderings. Designers may also include photographs of completed garments if available.

II. **Concept Development:** Alongside the final designs, a portfolio often includes documentation of the design process. Research is a critical stage in the design process, providing inspiration and supporting informed creative decisions (Seivewright, 2012). Costume and fashion reflect the cultural, social, and historical contexts in which they develop (Laver, 2012). This may include mood boards, inspiration images, fabric swatches, and concept sketches that demonstrate how the designer arrived at their final creations. This insight into the creative process can be just as valuable as the finished designs.

III. **Technical Drawings:** Technical drawings are essential for communicating design details to pattern-makers and manufacturers. These drawings typically include measurements, stitching lines, seam placements, and other technical specifications. They show that the designer understands how garments are constructed.

IV. **Color and Fabric Selection:** Designers often include color palettes and fabric samples in their portfolios to convey their expertise in color theory and textile knowledge. These elements showcase the designer's ability to make informed choices regarding materials and color schemes.

V. **Fashion Illustrations:** Fashion illustrators may include a selection of their fashion illustrations to demonstrate their ability to visually communicate design concepts and capture the essence of clothing and accessories.

VI. **Garment Photos:** Photographs of completed garments, if available, can provide tangible evidence of the designer's ability to bring their designs to life. These photos are particularly valuable if the designer has produced their own clothing lines or collaborated with manufacturers.

VII. **Look-Books and Editorial Layouts:** Some designers include look-books or editorial layouts that showcase their designs in a styled and visually appealing manner. This helps potential clients or employers envision how the designer's creations can be presented to the public.

VIII. **Resume and Bio:** A fashion portfolio often includes a designer's resume, highlighting their education, relevant work experience, internships, and any awards or accolades they have received. A brief biography or artist statement may also be included to provide context for the work.

IX. **Contact Information:** It's essential to include contact information, such as an email address or website, so that potential clients, employers, or collaborators can reach out for inquiries or opportunities.

In summary, a fashion portfolio is a comprehensive presentation of a fashion designer's skills, creativity, and experience. It's a visual representation of their body of work and serves as a powerful tool for showcasing their talent and securing opportunities within the fashion industry. A well-crafted portfolio not only reflects a designer's aesthetic but also their professionalism and commitment to their craft.

## 2.6 COMPUTER-AIDED DESIGN—THE PORTICO OF DESIGNERS

In the fashion apparel industry, computer-aided design (CAD) software plays a crucial role in streamlining the design and production processes. Fashion design combines creativity with technical knowledge to transform ideas into wearable products (Evans & Thornton, 2010). CAD software allows fashion designers and industry professionals to create, modify, and visualize designs with precision and efficiency. Here are some commonly used CAD software programs in the fashion apparel industry. It is known for its speed and efficiency in pattern development. These CAD software programs provide fashion designers and professionals with the tools needed to bring their creative visions to life, streamline the design process, and ensure accuracy in pattern development and garment production. The choice of software often depends on the specific needs and preferences of the designer or fashion company.

I. **Adobe Illustrator:** Adobe Illustrator is a versatile vector graphics software widely used in fashion design. Designers use it for creating detailed flat sketches, technical drawings, and textile patterns. It offers a range of tools for precise design work and pattern-making.

II. **Adobe Photoshop:** Adobe Photoshop is another Adobe product often used in fashion design for image manipulation, fabric texture design, and creating mood boards. Designers can use it to experiment with color palettes and create digital collage patterns and virtual sampling.

III. **Marvelous Designer:** Marvelous Designer specializes in 3D clothing design and simulation. It is frequently used for creating realistic 3D clothing models and animations.

IV. **CorelDRAW:** CorelDRAW is a vector graphics software similar to Illustrator. It's favored by some fashion designers for its intuitive interface and robust features for creating illustrations, patterns, and technical drawings.

V. **Clo3D and CLO Virtual Fashion:** Clo3D is a specialized 3D garment design software. It allows designers to create 3D virtual clothing samples, which can be useful for prototyping and visualizing how a garment will look when worn as shown in Figure 2.8.

VI. **Optitex:** Optitex is a CAD software widely used in the fashion industry for pattern-making and 3D garment visualization.

VII. **Lectra:** Lectra is a comprehensive fashion software suite that includes solutions for pattern-making, marker-making, and 3D virtual prototyping. It is used by many large fashion companies for end-to-end product development.

VIII. **Browzwear:** Browzwear's software, like VStitcher and Lotta, is popular in the 3D apparel industry at that time simulation. It is frequently used for creating realistic 3D clothing models and animations.

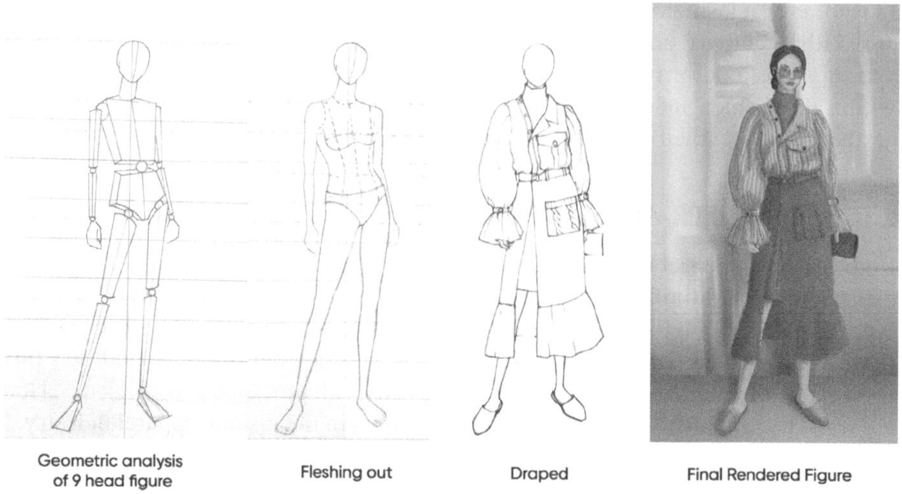


**FIGURE 2.8** Steps followed by designers to make fashion illustration.

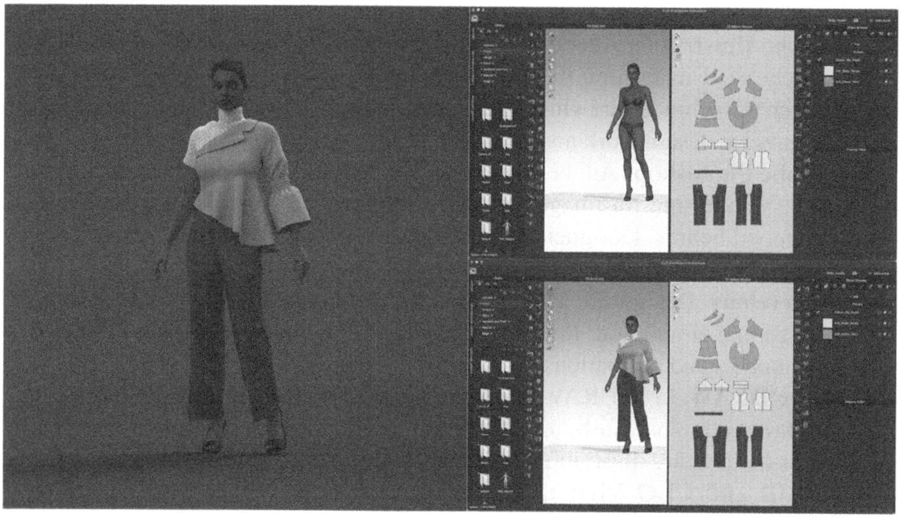

**FIGURE 2.9** 3D illustrations on Clo3D.

IX. **Tukatech:** Tukatech offers a range of CAD and computer-aided manufacturing solutions for fashion design, pattern-making, grading, and marker.

X. **Gerber Technology:** Gerber's software, such as AccuMark, is renowned for pattern design, marker-making, and production planning. It's utilized in various stages of garment production. It allowed designers to create highly realistic 3D garment prototypes, reducing the need for physical sampling and speeding up the design and production process (Figure 2.9).

## 2.7 LATEST SOFTWARE FOR GARMENT DESIGNING

There were several notable software programs for garment design in the fashion and the software landscape in the fashion industry is continually evolving, and new tools and updates to existing software are regularly released. Some of the latest software that are gaining popularity in the apparel Industry are CLO Virtual Fashion, Opitex, and Gerber. In recent years, the apparel industry has witnessed a revolutionary transformation with the integration of artificial intelligence into various aspects of the design, production, and retail processes. The impact of these software programs on the apparel industry was multifaceted. To increase speed and efficiency, reduce physical sampling, improve accuracy, and enhance collaboration, cloud-based collaboration platforms facilitated remote work and collaboration among global teams, allowing for seamless communication and sharing of design assets. By minimizing physical prototypes and optimizing patterns for minimal fabric wastage, these software tools contributed to sustainable practices within the industry. Overall, the use of CAD software resulted in cost savings through improved efficiency and reduced material waste. Please note that the software landscape may have evolved further with new tools and technologies emerging to meet the evolving needs of the fashion and apparel industry.

In conclusion, the process of generating design ideas and conceptualizations is a dynamic and essential aspect of the creative journey in various fields, including art, architecture, fashion, and product design. This chapter has provided insights into the multifaceted nature of design inspiration and the methods by which designers translate abstract concepts into tangible creations. Successful fashion design integrates creativity, technical skills, material knowledge, and market awareness (Sorger & Udale, 2017). Designers draw inspiration from diverse sources, including nature, culture, personal experiences, and social issues. These sources inform their creative thinking and ignite the imaginative spark needed to develop innovative and meaningful designs. This chapter has also highlighted the significance of the research, mood boards, sketching, and prototyping in the idea generation process. Furthermore, the ability to conceptualize design ideas into concrete forms is a hallmark of a skilled designer. This chapter has explored the importance of visualization, iteration, and collaboration in refining and transforming ideas into well-defined concepts. Ultimately, the journey from initial inspiration to final realization is a deeply rewarding one. Designers play a pivotal role in shaping the visual, functional, and emotional aspects of our world, and their ability to generate and conceptualize ideas drives progress, innovation, and creativity in our society. Designers and industry professionals should stay updated with the latest software developments to remain competitive and efficient in the market.

## ACKNOWLEDGMENTS

We acknowledge the efforts of Illustrators, Ms. Drishti Goel and Ms. Krutika Dnyaneshwar Thakare who contributed to all the illustrations in this chapter.

## REFERENCES

Aldrich, W. (2008) *Metric Pattern Cutting for Women's Wear*. 5th edn. Oxford: Blackwell Publishing.

Bevlin, M.E. (1994) *Design through the Ages: An Introduction to the Elements and Principles of Design*. 10th edn. New York: Holt, Rinehart and Winston.

Eissen, K. & Steur, R. (2011) *Sketching: The Basics*. Amsterdam: BIS Publishers.

Evans, C. & Thornton, M. (2010) *Fashion Design*. 2nd edn. London: Laurence King Publishing.

Gill, R. (2009) *Creative Fashion Illustration*. London: Batsford.

Jones, S.J. (2005) *Fashion Design: Manual and Coursebook*. London: Laurence King Publishing.

Laver, J. (2012) *Costume and Fashion: A Concise History*. 5th edn. London: Thames & Hudson.

Seivewright, S. (2012) *Basics Fashion Design 01: Research and Design*. 2nd edn. London: AVA Publishing.

Sorger, R. & Udale, J. (2017) *The Fundamentals of Fashion Design*. 3rd edn. London: Bloomsbury.

Stewart, M. (2011) *Launch Pad: A Creative Guide to Fashion Illustration*. London: Batsford.

# 3 Raw Materials for Apparel Production

*Yamini Jhanji Dhir, Nagender Singh,
and Ashish Bhardwaj*

## 3.1 FABRIC PROPERTIES AND PERFORMANCE

The choice of fabric is a critical decision in garment manufacturing. Fabric properties play a pivotal role in determining the final product's performance, comfort, and overall quality. Whether it's a casual t-shirt, a high-end evening gown, or technical sportswear, understanding fabric properties is essential for designers, manufacturers, and consumers [1–3].

### 3.1.1 Physical and Mechanical Properties

Fabrics have various physical and mechanical properties that determine their suitability for various applications. In garment construction, bending, tensile, compression, and shear characteristics are considered significant. These characteristics affect the sewed cloth's form and its sewability. From a performance perspective, other mechanical characteristics of the fabric, such as drape, tear strength, abrasion resistance, wrinkle and crease recovery, and pilling behaviour, are assessed [4–6]. Depending on the demand, the following factors are established for a complete evaluation of the fabric quality: moisture regain value, crimp in the warp-and-weft yarn, fabric thickness, and fabric density [7, 8]. Measurements of bow and skewness are made, especially for textiles with check and stripe patterns [9, 10]. Here are some fundamental properties:

*Fibre Type:* The fibre used in a fabric greatly influences its properties. Standard fibres include cotton, polyester, nylon, wool, silk, and more, each with unique characteristics [11]. *Strength:* Fabric strength is a crucial mechanical property. It is measured in terms of tensile strength, which is the force required to break a fabric when pulled in opposite directions. Strong fabrics are durable and suitable for heavy-duty applications. *Tear Resistance:* Tear resistance measures a fabric's ability to withstand tearing forces. Fabrics with high tear resistance are less likely to develop holes when subjected to stress. *Abrasion Resistance:* This property determines how well a fabric withstands wear and friction. Fabrics used in applications that experience repeated rubbing or friction, like upholstery or workwear, need good abrasion resistance [12]. *Flexibility and Drape:* Some fabrics are stiff and have low flexibility, while others are soft and drape well. This property is important for clothing, curtains, and other applications where the way a fabric hangs is essential. *Elasticity:* Elastic fabrics can stretch and return to their original shape. This property is important for sportswear, swimwear, and any application requiring a snug fit. *Shrinkage*: Fabrics may shrink

DOI: 10.1201/9781003432739-3

when exposed to heat or moisture. Understanding the potential for shrinkage is crucial when selecting fabrics for specific uses or during garment construction. *Crease Resistance:* Some fabrics are more prone to wrinkling than others. Crease resistance is important in applications where a neat appearance is required. *Colourfastness:* The ability of a fabric to retain its colour over time and with exposure to various environmental factors, such as light, washing, and chemicals. *Weight:* The weight of a fabric is typically measured in grams per square metre or ounces per square yard. It can affect the warmth and comfort of clothing. *Density:* The density of the fibres in a fabric affects its warmth and insulation properties. Higher-density fabrics tend to be warmer [7, 9, 13]. *Thermal Properties:* Fabrics can have varying insulation levels, heat retention, and breathability. These properties are essential for clothing used in different climates. *Moisture Absorption:* Fabrics can absorb and wick moisture away from the body, essential for sportswear and outdoor clothing. *Dimensional Stability:* The ability of a fabric to maintain its size and shape over time, especially after washing and use. *UV Resistance:* Some fabrics are more resistant to UV rays, making them suitable for outdoor applications or sun protection clothing. *Antistatic Properties:* Fabrics can be engineered to minimize the build-up of static electricity, which can be crucial in applications like electronics manufacturing [14, 15].

These properties vary widely depending on the fabric construction and fibre types used. It's important to consider these characteristics when selecting fabrics for different applications, whether it's clothing, upholstery, industrial uses, or technical textiles. Figure 3.1 shows the essential fabric properties required for garments used for various activities [7].

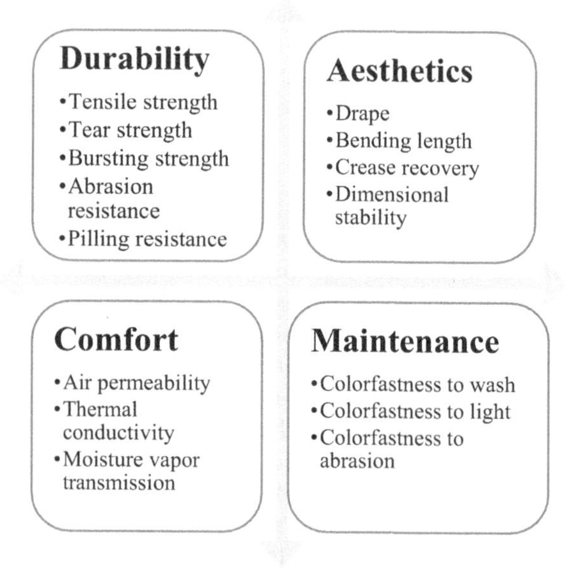


**FIGURE 3.1** Essential fabric properties.

### 3.1.2 Overall Performance

The qualities of the mixed fabric influence the overall functionality of the finished garment. While long-lasting materials enhance the garment's longevity, breathable fabrics enhance comfort. It is essential for designers, pattern makers, and manufacturers to comprehend how fabric characteristics interact with making garments. In addition to being aesthetically beautiful, it helps guarantee that the fabric used is appropriate for the garment's intended design, manufacturing processes, and anticipated performance attributes. Clothing that is well-designed, useful, and aesthetically pleasing is the outcome of the harmonious fusion of fabric characteristics and the garment production method.

The following specifications that a garment requires from the fabric must be met [8–10]:

1. The ability to function in a way that offers the person sufficient physiological protection.
2. Comfort performance ensures that clothing fits the human body and provides a suitable level of comfort.
3. Aesthetic performance enhances the wearer's outward look by emphasizing some anatomical features while hiding others.
4. The performance of the fabric as a material used in apparel engineering.

When used as garment materials, textiles protect the human body from the elements and prevent skin damage from mechanical contact with other bodies. Superior materials should also have a pleasing look and be mechanically and thermally pleasant for the human body.

## 3.2 GARMENT MAKEUP PROCESS AND FABRIC PROPERTIES

A garment is made by several complex stages that are mostly dependent on the characteristics of the selected cloth. The fabric's type, weight, texture, and durability may greatly influence how an item is cut, stitched, and finished. This section will discuss the relationship between fabric characteristics and the process of making garments. Figure 3.2 shows the stages of the garment makeup process [8].

### 3.2.1 Design and Conceptualization

In this process phase, designers develop ideas for the garment's style, silhouette, and general appearance. Sketching, selecting fabrics, and producing design samples or prototypes are all included.

### 3.2.2 Pattern Making

Templates called patterns are used to cut cloth pieces according to a blueprint. To produce patterns for every garment component, pattern designers use design

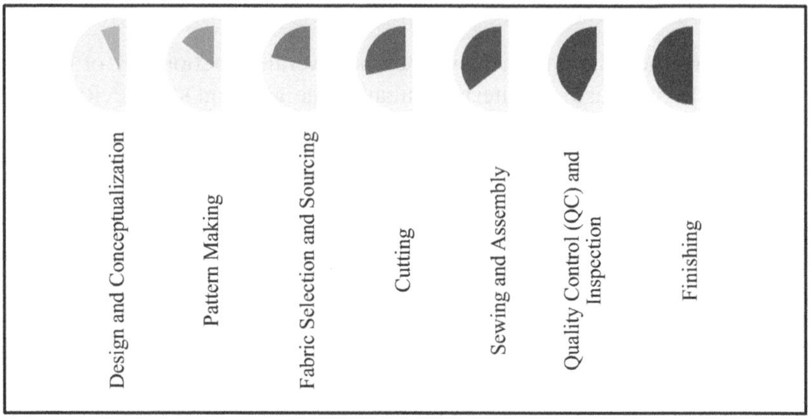


**FIGURE 3.2** Stages of the garment makeup process.

specifications. The basic design of the pattern is influenced by the type of fabric selected. How patterns are designed varies depending on how various fabrics drape and keep their shape. For instance, an inflexible blazer and a flowing evening gown would call for different patterns.

### 3.2.3 Fabric Selection and Sourcing

Once the design and patterns are finalized, the correct fabric is chosen and purchased. Selecting the right textile type, colour, and quality is necessary for this.

### 3.2.4 Cutting

The selected fabric is cut into parts and assembled to make the garment, using the patterns as a guide. Cutting with precision is necessary to reduce waste and ensure the garment fits well. The weight of the fabric is an important factor to consider when cutting. Robust cutting tools are necessary for heavier textiles like denim, whereas sharp blades and careful handling are necessary for lightweight fabrics like chiffon to avoid fraying.

### 3.2.5 Sewing and Assembly

Skilled machinists and sewing workers stitch the cut fabric pieces together. At this step, buttons, zippers, and linings must be attached, along with sewing and serging. Sewing convenience is impacted by fabric texture. While textured materials may require handling to accomplish exact stitching, smooth, slippery textiles could require special needles and thread to prevent snagging. Different sewing techniques are required for flexible fabrics, such as spandex blends, to ensure optimum fit and avoid seam distortion. Stitching heavy or highly abrasion-resistant materials may require heavy-duty sewing machines and stronger thread to endure the stitching stress.

### 3.2.6 QC and Inspection

Garments undergo quality inspections at various stages of production. This entails inspecting the item for flaws and following the stitching, fit, and design requirements. The way a garment drapes on the body depends on the type of fabric. To get the ideal fit and silhouette, this is essential. For instance, a dress made of a rigid, heavyweight fabric will drape differently than one made of a flowing, lightweight fabric.

### 3.2.7 Finishing

To give clothing a finished, polished appearance and to get rid of wrinkles, garments are pressed and ironed. Final touches such as labels, tags, and accessories are added after any loose threads, extra fabric, or defects are cut. Certain blends of synthetic materials, which are inherently wrinkle resistant, could require less ironing and finishing following clothing assembly. Because heat and moisture can affect colour retention, ensuring that the fabric's colour holds true during finishing procedures like steaming and ironing is essential.

## 3.3 TEXTILE MATERIALS FOR APPAREL PRODUCTION

Textile materials are the fundamental building blocks of apparel production, and their properties directly influence the quality, functionality, and aesthetics of garments. These materials are derived from natural or synthetic sources and are processed to form fibres, yarns, fabrics, and finished products. Textile materials serve as the foundation of the textile and fashion industry, influencing the characteristics, appearance, and performance of the final products [11]. This section explores several common textile materials, their key properties, and their applications in apparel production.

**Natural Textile Materials:**

- **Cotton:** One of the most popular natural fibres is cotton. It has a reputation for being soft, absorbent, and breathable, which makes it appropriate for a variety of uses, from bed linens to casual apparel.
- **Silk:** The luxurious and organic protein fibre known as silkworms generate silk. It is highly valued for its sleek, shiny, and refined appearance and is frequently utilized in upscale apparel and accessories.
- **Wool:** The warm and insulating qualities of wool, which comes from sheep, make it extremely valuable. Winter apparel and accessories frequently include it.
- **Linen:** Flax fibres are used to make linen, which is prized for its breathable properties and organic texture. It's frequently utilized in clothing for warm climates.

**Synthetic Textile Materials:**

- **Polyester:** Polyester is a multipurpose synthetic fibre that wicks away moisture and is noted for its resilience and resistance to wrinkling. It is utilized in many apparel categories, including sportswear.

- **Nylon:** Another synthetic material that is very strong and resistant to abrasion is nylon. It is frequently utilized in sportswear, swimwear, and hosiery.
- **Acrylic:** Wool in knitwear is frequently replaced by acrylic fibres, which are valued for their warmth and softness.
- **Spandex (Lycra):** In swimsuits, sportswear, and form-fitting apparel, spandex—a highly elastic synthetic fibre—is a must-have because it offers stretch and recovery.

**Semi-Synthetic Textile Materials:**

- **Rayon (Viscose):** A semi-synthetic fibre derived from regenerated cellulose is called rayon. It is used in many different types of clothing and combines the best features of synthetic and natural fibres.
- **Modal:** Another semi-synthetic fibre made from beech trees is called modal. It is utilized in loungewear, casual apparel, and lingerie because of its softness and breathability.

**Innovative and Sustainable Textile Materials:**

- **Bamboo:** The sustainable, pliable, and moisture-wicking qualities of bamboo fibres, which come from bamboo plants, are highly regarded.
- **Hemp:** Hemp fibres are robust and friendly to the environment. They are utilized in tough, environmentally responsible apparel.
- **Recycled Textiles:** In the textile business, recycled materials—such as recycled polyester and post-consumer fabrics—are becoming more and more viable substitutes.
- **Smart Textiles:** Textile innovations include "smart" materials with integrated technology, including fabrics with inbuilt heating elements or moisture-wicking sensors [16, 17].

## 3.4 SEWING THREAD

Sewing is a crucial step in garment making process that involves joining individual garment components (patterns) to develop a three-dimensional wearable end product. Sewing performance affects the durability and appearance of garment during use and in after care. Sewing performance is largely dependent on the correct selection of seams and stitches, sewing machine, and associated components along with sewing thread. The pragmatic selection of sewing thread in accordance with fabrics to be sewn, desirable strength of seam, and sewing type (manual or machine) is of utmost importance since sewing thread beds into fabric structure and becomes part of the garment [18–20]. The factors demanding due consideration while making sewing thread selection are enlisted below:

- **Type and Weight of Fabric**
  The sewing thread should be chosen according to fabric type and its weight. Finer fabrics require finer sewing threads while coarser sewing threads should be chosen for thicker fabrics.

- **Sewing Technique Deployed**
  Z twisted sewing threads are preferred for lockstitch machines while hand sewing requires S twisted sewing thread. Likewise, air entangled sewing thread is preferred for multineedle sewing machines.
- **Endurance of Stresses, Abrasion, and Needle Heating during Sewing**
  The sewing thread should be able to endure abrasion, stretch, and needle heating during sewing for enhanced sewability. Accordingly, polyester-cotton core spun sewing threads are preferred to avert mechanical damage to sewing thread or textile substrate as a result of needle heating during sewing.
- **Durability**
  The application of garments will greatly influence the sewing thread type since durable synthetic sewing threads are preferred for garments that are subjected to harsh washing and high stress while less-durable ones may be suitable for decorative ensembles.
- **Colour Fastness**
  The sewing thread should be uniformly dyed in order to avoid its fading or bleeding during garment washing.
- **Free from Imperfections**
  The sewing thread used for garment construction should be free from any imperfections like slots, slubs, knots, and thick and thin places.
- **Performance Properties**
  The sewing thread should possess good performance properties such as tensile strength, loop and knot strength, and elongation and abrasion resistance to ensue durable seam and stitch formation. It should also be resistant to heat, light, and chemical exposure. Likewise, stretchability, washing, light, and rubbing fastness of sewing thread are crucial as the garments will be worn, washed, and dried in sunlight.

Thus, the selection of sewing thread should address the type of seam and stitch to be applied on fabric along with application areas, washing and care labelling requirements of garments [21–25].

### 3.4.1 Sewing Thread Classification

Sewing threads can be classified based on several factors, namely (Figure 3.3):

- Fibre type
- Construction and twist direction
- Thread finishing
- Thread sizing
- Thread packages
- Application areas

*Fibre Type*: Both natural and synthetic sewing threads are increasingly being utilized in sewing room operations. Cotton is the most frequently used natural sewing thread;

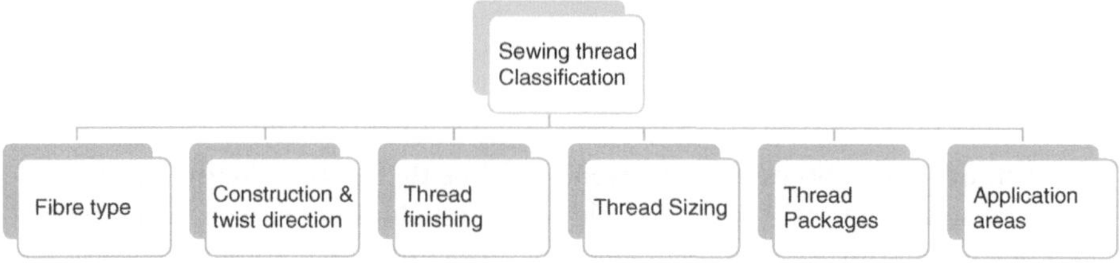


**FIGURE 3.3** Sewing thread classification.

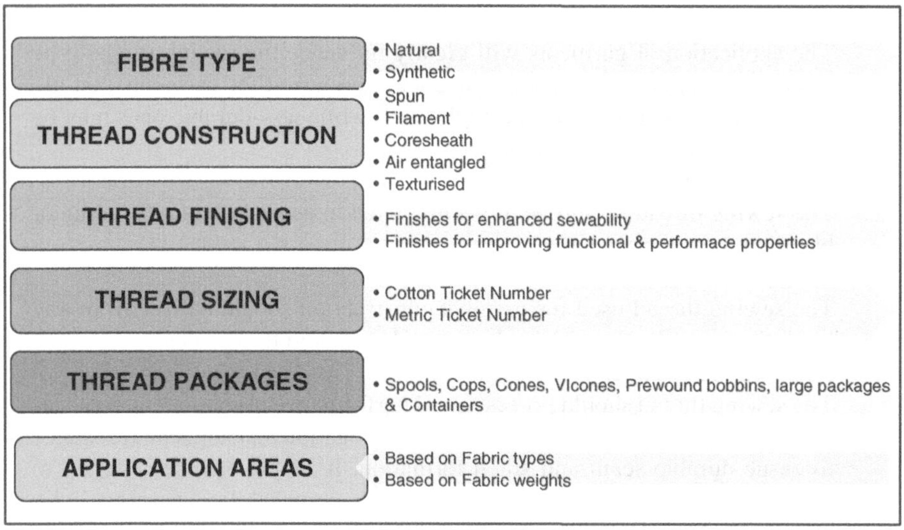


**FIGURE 3.4** Various aspects of sewing threads.

however, linen and silk are also being used based on specific application area. The advantage offered by natural sewing threads is their ability to withstand needle heating due to their high temperature resistance. Synthetic fibres namely polyester and nylon can also be used for making sewing threads. When compared to their natural counterparts, the synthetic sewing threads exhibit higher strength and better resistance to abrasion, rot, mildew, and bacteria. However, their low melting points and hence lower temperature resistance poses problems as the needle gets heated up during sewing. Various aspects of sewing thread are depicted in Figure 3.4.

*Construction & Twist Direction:* The sewing threads can be classified as spun, filament, core sheath, air entangled, and texturized based on their construction. Spun sewing threads can be produced from both natural and synthetic yarns. Spun sewing threads feature a hairy surface thereby rendering them better lubrication properties and exceptional sewing performance. Nevertheless, filament sewing threads outshine in appearance and strength compared to their spun counterparts. Continuous filament threads are produced by filament extrusion from synthetic polymers and twisted for imparting strength. For the same thread size, filament sewing threads exhibit higher strength compared to their spun sewing threads. Filament sewing threads can further be classified as monofilament and multifilament based on number of filaments in yarn cross-section. Monofilament sewing thread produced from a single continuous filament of a specific fineness is stronger, more uniform, and cheaper compared to multifilament sewing threads. However, the former feature rough texture and are less flexible compared to multifilament sewing threads. This limitation restricts the application of monofilament sewing threads to sewing of upholstery, draperies, and garment hemlines. Multifilament sewing thread with multiple filaments in the yarn cross-section, on the other hand, is widely used where high strength is the main requirement such as sewing of leather garments, footwear, and industrial products.

Texturized sewing threads undergo the process of texturization in order to enhance softness, bulkiness, and hence overall texture of the thread. Slight twisting and heat setting of filaments is accomplished to render the aforesaid properties permanently to the sewing thread. The texturized sewing threads offer exceptional seam coverage and extensibility, but these threads are more prone to snagging unlike their untextured counterparts.

Air entangled sewing threads utilize the technique of air entanglement wherein the constituent yarn strands or plies are entangled together instead of twisting as in case of twisted sewing threads. Air entangled sewing threads are particularly suitable for multineedle sewing machines where the traditional sewing threads are prone to untwisting during the sewing process.

Core sheath sewing threads are preferred choice for varied sewing applications owing to their exceptional properties in terms of strength, abrasion, and high temperature resistance compared to their virgin counterparts. Polyester filament yarn in core provides strength while cotton spun yarn in sheath or wrapper arrangement offers high temperature resistance and prevents melting of sewing thread due to needle heating. Thus, the advantages of both natural and synthetic sewing threads can be obtained in core sheath construction of sewing thread.

*Twist Direction:* Sewing thread is a plied structure composed of two or more yarn plies together with the resultant twist in sewing thread opposite to the constituting parent strands. Z twisted and S twisted sewing threads are used on commercial sewing machines. Single-needle lockstitch machine mainly utilizes Z twisted sewing thread for various sewing operations.

*Thread Finishing:* Sewing threads can further be classified based on the chemical finishes applied on them. Thread finishes are used for enhancing the thread sewability and to render a particular functionality to the sewing thread. The finishes not only improve the strength, abrasion resistance, and lubrication properties of sewing threads but also contribute to particular functionality such as flame retardance, water repellancy, soil release, and antimicrobial properties in sewing threads utilized for functional clothing.

*Thread Sizing:* Sewing thread sizes are indicative of coarseness and fineness of sewing thread and thus a determining factor for application-based sewing thread selection. Sewing thread sizes are expressed as ticket number that follows both direct and indirect yarn numbering system. Accordingly, metric ticket number (based on metric yarn count, Nm) and cotton ticket number (based on cotton yarn count, Ne) are commercially used for synthetic, spun, and core (Nm) and cotton sewing threads (Ne).

*Thread Packages:* Sewing threads are supplied in sewing room as thread packages. A variety of thread packages are used for sewing operations based on sewing thread type and construction, thread carrying capacity, type of winding, sewing machine types, and application areas of garments.

Spools are small thread packages with relatively short length of parallel-wound sewing thread and thus suitable for domestic sewing projects.

Cops, cross-wound packages offer a range of colours with approximately 100–200 m thread wound and thus suitable for lockstitch machines.

Cones, the tapered-shape packages with about 1,000–25,000 m of sewing thread wound on paper or plastic cone are economical packages suitable for high-speed industrial sewing machines requiring high thread consumption.

Vicones are parallel tube with flanged bottom containing approximately 1,000–5,000 m thread lengths.

Containers are packages particularly suitable for smooth and lively monofilament sewing threads that pose unwinding problems with traditional sewing thread packages.

Cocoons are self-supporting, centre-less sewing thread package ideal for multi-needle quilting and embroidery machines.

Prewound bobbins are the precision wound packages with consistent and more length of sewing thread compared to traditional metal bobbins that pose problem of frequent change overs due to their limited thread carrying capacities. The productivity and off winding performance are enhanced with prewound bobbins replacing traditional bobbins thereby eliminating the frequent rewinding of bobbins [26].

*Application Areas:* Sewing threads can also be classified based on their preferential areas of application. The application areas of sewing threads are determined by fabric type and its weight. Fine, mercerized cotton, silk, and synthetic sewing threads are used for sewing delicate and light weight fabrics like chiffon, organza, satins, knits, and deep pile fabrics while heavy-duty mercerized cotton and synthetic sewing threads are preferred for medium-weight fabrics such as gabardine, denim, and home furnishing materials. Mercerized cotton and synthetic thread are selected for heavy-weight fabrics like canvas and upholstery [26, 27].

## 3.5 NON-TEXTILE MATERIALS

The various textile raw materials and their properties were discussed in the preceding section. Apart from textile materials, several non-textile materials such as leather, fur, metals, and stones are extensively utilized to create fashion ensembles. This section shall cover the properties and application areas of leather and fur while metals and stones generally utilized as sundries in garment design and development process will be covered under trims and notions in the next section.

The aesthetic, functional, and economic attributes associated with non-textile materials prompt the designers and manufacturers to consider these materials for the assortment of apparels and accessories. Leather enjoys lion's share of market with massive popularity among fashion buffs. However, the extensive utilization of leather has deleterious environmental impact since a lot of unsustainable practices are followed during production process of leather. In the quest to transition towards sustainable goals and principles, eco-friendly and recyclable leather alternative such as mushroom and corn leather are being explored aiming to create a similar visual impact and penchant among the consumers as the virgin leather. Nevertheless, animal-based leather still outshines its sustainable counterparts and mainstream utilization of vegan leather demands more relentless efforts.

Fur, another animal-based non-textile material utilized for garment making is obtained from fur bearing animals having hair, fleece, or fur fibres attached to their hides or skin. The inherent insulation properties of fur makes it ideal candidate for winterwear clothing.

Fur can be classified on the basis of animal families from which it is obtained such as mink, fox, bear, dog, cat, and rodent. The salient fur characteristics like hair length, warmth, durability, and rarity are another basis of fur classification (Figure 3.5). Durability of fur is an indicative of degree of continuous shedding or hair loss of these pelts once fur is transformed into wearable entities. Furs can also be classified as durable and perishable based on their resistance to shedding or hair loss.

Figure 3.5 depicts the fur classification in terms of animal family (from which fur is obtained like mink, fox, bear, dog, cat, and rodent), its characteristics (hair length, warmth, durability, and rarity), and resistance to shedding or hair loss (durable and perishable).

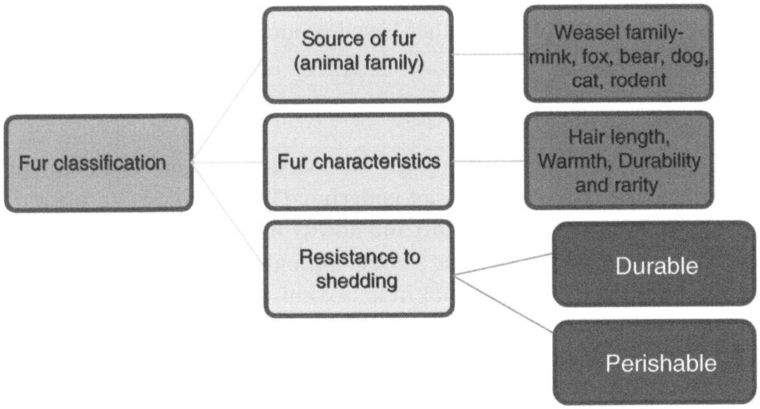


**FIGURE 3.5** Fur classification.

### 3.5.1 Properties of Non-Textile Materials

#### 3.5.1.1 Leather Properties

Leather has several unique and salient properties making it a preferred choice for garment making

- **Tensile and Tear Strength**
  The extremely high tensile strength of leather is crucial for apparels and accessories that demand durability. Leather can withstand significant amount of stress without tearing apart and thus finds application in end products intended for rough and tough usage. Moreover, the high tensile and tearing strength of leather prompts its acceptance for accessories that should offer substantial load carrying capacities such as luggage bags and laptop bags.
- **Elongation and Flexibility**
  Leather exhibits high elongation, ability to stretch to varying lengths without structural breakdown, and flexibility over a wide temperature range.
- **Absorption**
  Leather showcases effective moisture and liquid absorption property making it suitable for footwear design and development where rapid sweat absorption from wearer's feet is key to his comfort and performance.
- **Breathability and Insulation**
  Leather on account of its breathability can adjust to ambient environmental condition. The breathability and insulation properties are imparted to leather during different operations performed to convert animal hide to wearable product.
- **Lasting Mouldability**
  Leather showcases unique ability of conforming and retaining its shape during the service life of fashion products such as footwear and hats. This property is imparted to leather during lasting process when leather shoe upper or crown of hats are pulled over or forced onto last (wooden replica of footwear, hats) for the leather-based products to permanently retain the acquired form [28, 29].

#### 3.5.1.2 Fur Properties

Fur is a bilayered material with top layer of coarse, long, and lustrous outer guard hair while the lower layer comprises of underfur. The insulation and warmth properties inherent to fur are credited to soft and downy underfur. The specific fur properties vary widely on the basis of their source (animal family), habitat of animals, and processing of pelt. Pelt is the animal hide or skin with the fur fibres attached to it and generally obtained from mink and foxes. The pelts obtained from wiesel family (male and female minks) are soft and light weight and feature exceptionally high warmth-to-weight ratio. Mink is a very durable fur with long shelf life; however, its durability depends on product usage and care to prevent its unhairing. Some salient properties of fur that make it a potential candidate for myriad of winterwear apparels and accessories are enlisted below:

- **Insulation and Warmth**
  Fur is delegated as warm and highly insulating material owing to its dense structure, specifically the underneath layer comprising of underfur that can trap a layer of dry air in the skin microclimate thereby preventing heat loss and ensuing thermoregulation for wearer during winters.
- **Texture and Appearance**
  Fur exhibits a luxurious and distinct texture ranging from silky softness to the coarse resilience. The fur from different animal sources is naturally found in a variety of colours and patterns.
- **Durability and Longevity**
  Fur exhibits considerable durability and longevity depending on the care and handling it receives during processing and usage.
- **Breathability**
  Fur is a breathable material and allows thermal and moisture regulation to offer wearer comfort during cold weather conditions.

### 3.5.2 Applications of Non-Textile Materials

The salient properties exhibited by leather and fur make them preferred raw material for apparel and accessory designing. The moisture and liquid absorption property for footwear to prevent foot dampness and ensure wearer comfort is pragmatically tailored in leather during processing thereby making it suitable for a variety of men and women footwear styles. Likewise, the versatility of leather as all-weather outerwear and accessories is credited to leather's breathability and insulation properties imparted during processing. Accordingly, a range of unisex, men and women bomber, biker jackets, and overcoats are designed using leather as the major raw material.

Fur is also significantly contributing to fashion world marking its presence in a spectrum of applications ranging from winter outerwear, coats, and jackets to multitude of fashion accessories such as footwear, scarves, stoles, mittens, gloves kaftans, capes to hair accessories, caps, forehead, and ear coverings [30].

## 3.6 TRIMS AND NOTIONS

The post sewing room operation, namely attachment of trims and notions on the developed garments, is done with utmost care in garment industry because trims and notions are an indispensable component that can enhance or diminish the garment's aesthetic and functional attributes. Trims and notions together referred to as sundries are additional components that are attached on the garment either externally or internally once the sewing room operations are accomplished [31].

Trims are the decorative components that are attached to garment primarily externally by stitching, embroidery, or any other surface embellishment technique. Trims are essential for garments to render:

- Visual appeal to garments
- Aesthetic appreciation
- Value addition

Colourful threads, ribbons, and bows are increasingly being used as edge neatening, finishing, or surface ornamentation on fashion accessories, casual wear, kids wear, and winterwear. A gamut of other decorative components such as artificial flowers, ribbons, laces, appliques (embroidered patches), sequins, and mirrors is explored by designers to create a visually appealing garment. Also, the consumers are generally willing to pay exorbitantly given the distinct and unique style statements rendered to garments by attachment of colourful, vibrant, and quirky decorative components (Figure 3.6).

Figure 3.6 presents the various types of trims (artificial flowers, net, tassels, pompom) and notions (zipper, snap fasteners, eyelets, buckles, chain links, locks, etc.) used for apparel and accessory designing.

Notions, in contrast, are functional attachments on garments that are generally applied internally (labels, elastic) but can also appear on exterior surface of textile substrate (motifs, zippers, buttons) depending on garment design and functionality. The notions are attached to garments by sewing, fusing, welding, drilling, and thread marking process based on types and weight of notion along with its location on the garment. The inclusion of notions on garments primarily results in:

- Improved garment drape and fit
- Ease of garment usage and handling
- Convenient fastening and unfastening
- Provide crucial information regarding care labelling instructions, constituents of garment, country of origin, and sizes
- Identification of brand through unique brand logo

Buttons, toggles, studs, snap fasteners, eyelets, buckles, zippers, velcro/hook and loop, frogging, hooks and eyes, elastic, magnets, and grommets are numerous functional components extensively used in design and development of apparels and accessories. The closures like zippers, buttons, velcros, and snap fasteners are extensively utilized for accessories like cardigans and pullovers, trousers, handbags, footwear, and hair accessories [30, 31].

## 3.7 CONCLUSION

The performance, appearance, and sustainability of the garments can be enhanced or limited depending on the pragmatic selection of textile materials, which is a crucial factor in the design and manufacture of textiles and clothing. Therefore, the correct selection of all raw materials ranging from fibres, yarns, fabrics, non-textile materials to trims and notions is a governing and dominating factor in influencing the quality and performance of the garments so developed. A range of textile materials namely natural and synthetic fibres and yarns and variety of fabrics such as woven and knitted fabrics are explored by designers and manufacturers for garment design and development in accordance with consumer's preferences and economic considerations. Likewise, the role of sewing thread and additional components namely trims and notions cannot be undermined in garment making process. Sewing threads should also be selected giving due consideration to type and weight of textile

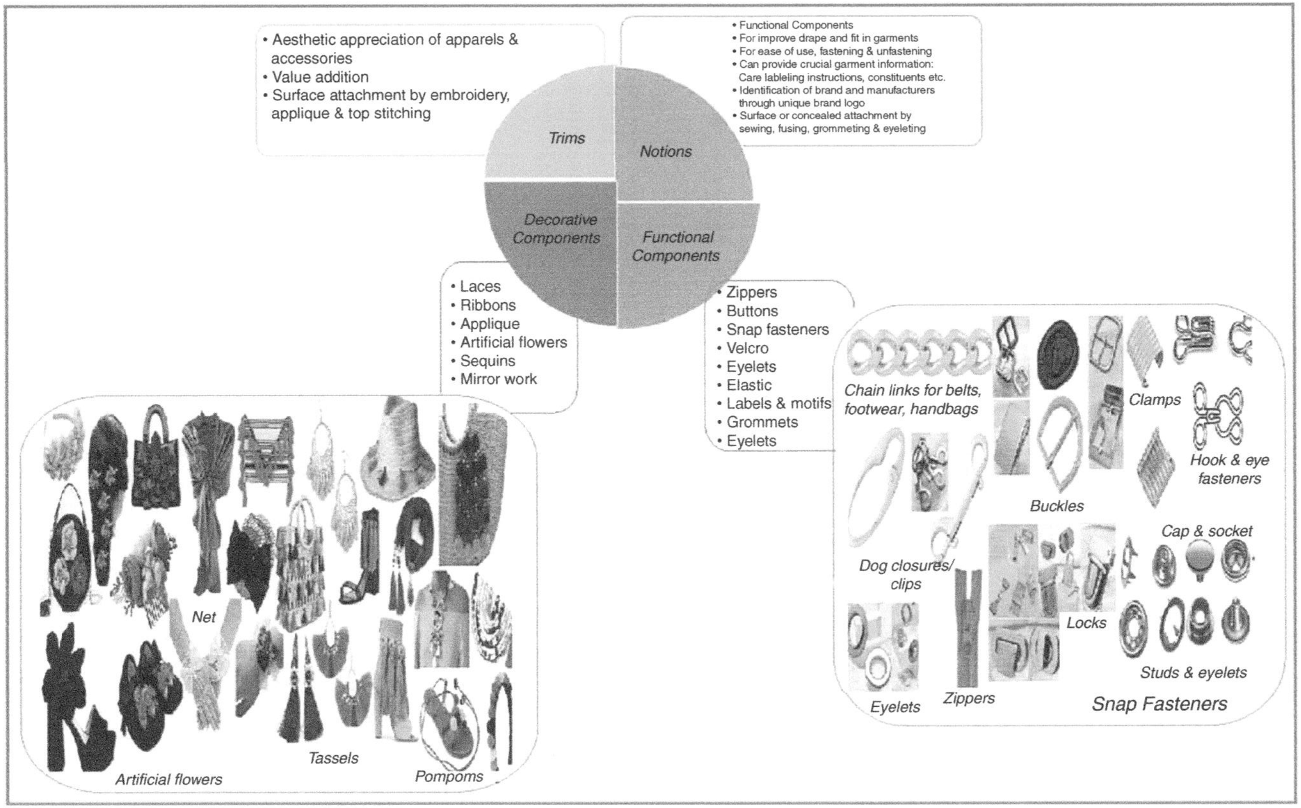


**FIGURE 3.6** Various types of trims and notions used in apparels and accessories.

substrate to be sewn. The trims and notions are essential not just for aesthetic appreciation and value addition of garments but also for enhanced functionality.

The mindful procurement of textile raw materials for garment making process is spearheading garment industry towards a world of endless possibilities of innovations and developments.

## REFERENCES


1. Zhang, B., Zhang, Y. and Zhou, P., 2021. Consumer attitude towards sustainability of fast fashion products in the UK. *Sustainability*, 13(4), p.1646.
2. Papasolomou, I., Melanthiou, Y. and Tsamouridis, A., 2023. The fast fashion vs environment debate: Consumers' level of awareness, feelings, and behaviour towards sustainability within the fast-fashion sector. *Journal of Marketing Communications*, 29(2), pp.191–209.
3. Manickam, P. and Duraisamy, G., 2019. 3Rs and circular economy. In Subramanian Senthilkannan Muthu (Ed.), *Circular economy in textiles and apparel* (pp.77–93). Woodhead Publishing.
4. Esmerij van Loon, Global Fashion Industry Statistics, 2023. Fashion United, https://fashionunited.com/global-fashion-industry-statistics.
5. Jhanji, Y., 2021. Sportswear: Acumen of raw materials, designing, innovative and sustainable concepts. In Bipin Kumar (Ed.), *Textiles for Functional Applications* (pp.261–286). DOI: 10.5772/intechopen.99808.
6. Fletcher, K. and Grose, L., 2012. *Fashion & sustainability: Design for change*. Hachette UK.
7. Venkatraman, P., 2015. Fabric properties and their characteristics. In Steven George Hayes and Praburaj Venkatraman (Eds.), *Materials and technology for sportswear and performance apparel* (pp.53–86). Florida: CRC Press.
8. Behera, B.K., 2015. Role of fabric properties in the clothing-manufacturing process. In Rajkishore Nayak and Rajiv Padhye (Ed.), *Garment manufacturing technology* (pp.59–80). Woodhead Publishing.
9. Kawabata, S. and Niwa, M., 1989. Fabric performance in clothing and clothing manufacture. *Journal of the Textile Institute*, 80(1), pp.19–50.
10. Shishoo, R.L., 1995. Importance of mechanical and physical properties of fabrics in the clothing manufacturing process. *International Journal of Clothing Science and Technology*, 7(2/3), pp.35–42.
11. Grishanov, S., 2011. Structure and properties of textile materials. In M. Clark (Ed.), *Handbook of textile and industrial dyeing* (pp.28–63). Woodhead Publishing.
12. Causa, A. and Netravali, A., 2008. Characterization and measurement of textile fabric properties. In Peter Schwartz (Ed.), *Structure and mechanics of textile fibre assemblies* (pp.4–47). Woodhead Publishing.
13. El Mogahzy, Y.E., 2009. Development of traditional textile fiber products. In Yehia E. El Mogahzy (Ed.), *Engineering textiles: Integrating the design and manufacture of textile products* (pp.329–397). Cambridge: Woodhead Publishing Limited.
14. Choudhury, A.R., Majumdar, P.K. and Datta, C., 2011. Factors affecting comfort: Human physiology and the role of clothing. In Guowen Song (Ed.), *Improving comfort in clothing* (pp.3–60). Woodhead Publishing.
15. Di Domenico, I., Hoffmann, S.M. and Collins, P.K., 2022. The role of sports clothing in thermoregulation, comfort, and performance during exercise in the heat: A narrative review. *Sports Medicine-Open*, 8(1), p.58.

16. Patnaik, S., 2019. 17 Smart textiles and recent developments. In Asis Patnaik and Sweta Patnaik (Eds.), *Fibres to smart textiles: Advances in manufacturing, technologies, and applications* (p.331). Florida: CRC Press.
17. Fletcher, K., 2013. *Sustainable fashion and textiles: Design journeys.* Routledge.
18. Sreenivasa Murthy, H.V., 2015. *Introduction to textile fibres* (Revised Edition). Woodhead Publishers in India.
19. Kiron, M.I., 2021. Textile fibres & their classifications, https://textilelearner.net/classification-of-textile-fibers/.
20. Martin, C., 2022. Textile fibres & their characteristics, https://www.jamesdunloptextiles.com/journal/tips-how-to/textile-fibres
21. Tariq, S., 2024. Twenty types of lace fabrics and trims for making clothes, https://sewguide.com/types-of-lace-fabric-trims/
22. Zugno, L., 2010. Leather in use, https://internationalleathermaker.com/leather-naturally-publishes-guide-to-modern-leather-making/
23. Truth about Fur, 2026. Fur processing techniques, https://www.truthaboutfur.com/how-fur-is-produced/processing/fur-processing-techniques/
24. Truth About Fur. 2026. Facts about fur farming, https://www.truthaboutfur.com/how-fur-is-produced/farming/facts-about-fur-farming/
25. Fur Commission. USA, 2026. Fashion – Natural Fur is the responsible choice, Fur Commission, https://www.furcommission.com/sustainability
26. PETA (People for the Ethical Treatment of Animals). 2026. Wool, fur and leather: Hazardous to the environment, https://www.peta.org/issues/animals-used-for-clothing/animals-used-clothing-factsheets/wool-fur-leather-hazardous-environment/
27. Chidambaram Rameshkumar, 2006. Fancy yarns for fashion, https://www.fibre2fashion.com/industry-article/751/fancy-yarns-for-fashion
28. Amir, 2020. The different materials used in watchmaking, https://www.hsamuel.co.uk/webstore/watches/materialsAndComponents.cdo
29. Deepwear. (2024, June 21). Raw materials & Sourcing in the Indian Textile Industry. https://deepwear.info/blog/raw-materials-and-sourcing-in-the-indian-textile-industry/
30. What are the main raw materials for making clothes, https://www.baldingbeards.com/best-shoe-brands/
31. Frank Jackson, 2026. Men's accessories guide: Ties, Cufflinks, Belts, and Hats to elevate your style, men's accessories guide: Ties, Belts, Hats & Cufflinks. https://www.frankyfashion.com/blogs/news/the-ultimate-men-s-prom-and-graduation-clothing-guide

# 4 Sustainable Principles in Garment Supply Chain

*Rajkishore Nayak*

## 4.1 INTRODUCTION

Garment manufacturing was a major activity in some of the developed countries such as the USA, UK, China and other European countries during the 1960–2000 period (Brun et al., 2008; Ha-Brookshire, 2017). Due to the rise in labour cost, energy cost and change in the policies, especially in the USA, it was hard for the country to survive in the apparel trade. However, China emerged as a leading exporter of apparels during the period 2000–2010, with significant contributions from the European Union (EU) and India (Zhang, Kong, & Ramu, 2016). In recent years, garment manufacturing has been a major activity in several developing countries such as Vietnam, India, Cambodia, Laos, Indonesia and Bangladesh in addition to other countries such as China and the EU. The major factors that are contributing to the growth of garment manufacturing in these countries can be attributed to cheap labour cost and favourable government policies.

Despite of technological developments, the manufacturing of apparels is still a labour-intensive process in many developing countries (Nayak & Padhye, 2015a, 2015b; Thee, 2009). The availability of cheap labour, financial challenges and technical complexity in the automation has prevented the apparel sector to adopt new technologies. Some of the technological advancements include laser applications, digital printing, virtual prototyping, automatic fabric inspection and use of radio frequency identification (RFID) technology. A number of large industries involved in manufacturing of apparels are switching to semi-automatic and automatic processes to cope with the increasing demand. In the future, there will be large demand for various new technologies involved in garment manufacturing that facilitate higher productivity, better efficiency, improved quality, reduced waste and reduced carbon footprint (Nayak & Padhye, 2017).

The garment manufacturing industries, especially in developing countries, are struggling with varying degrees of environmental problems. The inherent nature of the manufacturing processes required for garments largely impacts the environment due to a large amount of energy usage and water consumption, greenhouse gas (GHG) emission, hazardous waste generation and discharge of toxic effluent containing dyes, finishes and auxiliaries into the ecosystem (Nayak & Patnaik, 2021). It has been shown that about 20% of all water pollutions are caused by the textile treatments such as dyeing and printing, which greatly impacts the environment. Although the carbon emission intensity in garment manufacturing (between 1990 and 2015) has been decreased for grey cloth, it has not achieved the desired results in many


DOI: 10.1201/9781003432739-4

other areas to save the planet earth. For example, cotton yarn production has shown the highest increase in the carbon emission intensity by 7.37% during this period. This indicates that cotton yarn has continued to be produced without caring for the environment, due to the consumption of harmful pesticides and synthetic fertilizers, which release hazardous wastes into the nearby land and water systems. Hence, there is a need for green practices in all the processes used for garment manufacturing (Goyal & Nayak, 2020).

There has also been a strategic change in the demand for sustainable products in the global fashion and textile industry. Increasing pressure to become sustainable in the global marketplace due to increased customer awareness and peer pressure has led fashion brands to emphasize sustainable materials, cleaner processes and green supply chain (Gregoratti & Miller, 2011). The use of new sustainable materials, which are renewable and biodegradable, the use of advanced technologies and the use of ethical practices are gaining increased attention in the manufacturing sector of fashion. However, the complexity in the supply chain process in several tiers and lack of stricter regulations in some of the developing countries make it difficult to achieve complete sustainability in the fashion and textile sectors.

This chapter will highlight the principles of sustainable supply chain, and the terminologies relating to fashion sustainability. The concept of fast fashion and associated problems will be discussed in relation to the traditional fashion. Approaches to becoming sustainable from raw material stage to apparel manufacturing stage will be discussed briefly. Further, various approaches in sustainable logistics and warehouse management will be discussed. Waste management and the concept of upcycling and recycling will be discussed in brief in addition to the life cycle assessment (LCA) of fashion. Some of the approaches taken by fashion brands to become sustainable will be highlighted. The findings from the case study conducted in Vietnamese fashion industries to become sustainable will be discussed in this chapter.

### 4.1.1 Definitions of Sustainable Supply Chain

Environmental pollution is a major concern in most of the manufacturing sectors including fashion and textiles (Nayak, 2020). The fashion businesses are facing increased pressure from the stakeholders and government to become environmentally friendly and practice corporate social responsibility (CSR) (De Brito, Carbone, & Blanquart, 2008; Fletcher, 2012; Marcuccio & Steccolini, 2005). In several instances, unethical practices in supply chains of large brands come into focus as they are popular in the global marketplace. These practices can be related to the excessive use of resources, not taking care of society and generation of unnecessary waste. Because of this, sustainability has become a focal point of attraction in many business operations. Several companies have initiated to evaluate the total environmental and social impacts of their products from the cradle to the grave stage or simply during the entire life cycle.

Sustainability in the supply chain relates to the green and ethical practices while manufacturing products or offering services so that the impact on the environment and society is reduced. The impacts can be realized throughout the supply chain process, starting from the design development through the production, storage, delivery and transport till it reaches the consumers or even after the product is reaching its end of life (EOL) (Khurana & Ricchetti, 2016; Li, Zhao, Shi, & Li, 2014). The major objective of sustainability is to reduce the environmental damage caused to our planet while having a positive impact on people around it. These concerns are in addition to the traditional corporate supply chain concerns about performance and profits. Further, sustainability also includes taking care of society and saving resources by using renewable resources.

The term sustainability was coined in 1987 in Brundtland report, which means "satisfying the current needs without compromising the future generation's needs" (Keeble, Topiol, & Berkeley, 2003). The sustainable practices are viewed from three perspectives such as environmental, economic and social, which are known as the "Triple Bottom Line (TBL)" of sustainability (Hacking & Guthrie, 2008). Almost all three TBL of sustainability are neglected during the processes involved for garment manufacturing. However, the environmental aspect of sustainability is the major concern among the TBL of sustainability followed by the social aspects. The environmental aspect is more critical due to severe climate change and the unprecedented changes brought by the fashion and textile manufacturing and supply chain activities (Di Benedetto, 2017). The social aspects are not very far behind as society is not taken care of during the activities related to fashion and textiles.

As the term "green manufacturing" is becoming important around the globe, fashion manufacturers are paying increased attention to adopting various ecofriendly technologies to reduce environmental pollution. Vachon and Klassen (2008) mentioned that the "green manufacturing" practices can help the garment industries to achieve the economic benefits of having a higher number of target customers and to achieve "the edge" over their competitors. Several approaches can be taken to achieve environmental sustainability in garment manufacturing starting from the selection of raw materials to the sewing of a garment. In fact, the first stage of the garment manufacturing, the "conceptualization stage," lays the foundation for the degree of environmental impact of the garment manufacturing process (Nayak & Padhye, 2018). This stage can alter the selection of raw materials, which are sustainable and need ecofriendly process routes for conversion into final garments. Sustainable garment manufacturing involves non-depleting and ecofriendly raw material selection, ecofriendly manufacturing technologies and green production methods (Choudhury, 2014; Shen, Zheng, Chow, & Chow, 2014).

## 4.2 BASIC TERMINOLOGIES PERTAINING TO SUSTAINABILITY

Fashion sustainability can also be referred to as green fashion (Cervellon & Wernerfelt, 2012), biodegradable fashion (Moorhouse & Moorhouse, 2017), ethically made (Haug & Busch, 2016) or ecofriendly or environmentally safe fashion (Jang, Ko, Chun, & Lee, 2012). The term green fashion indicates the practices used in fashion so that the total environmental impact is minimized during manufacturing, transportation, warehousing, distribution and other logistic activities. Biodegradable fashion refers to the fashion that can be decomposed by bactericidal or other micro-organism activities. Fashion that is ethically made has the least environmental impact and takes care of the working conditions, work safety and wages of the employees in the entire supply chain. Ecofriendly or environmentally safe fashion indicates the fashion that is being manufactured considering the total environmental impact of selecting raw materials till the products reach consumers.

The numerous tactics and procedures used to lessen the negative impacts on the environment and enhance the social and ethical standards in the fashion industry are referred to as sustainable principles in the garment supply chain (Nayak, Panwar, & Nguyen, 2020; Shen, 2014). These principles ensure that the fashion industry functions in a way that is not only profitable but also sustainable over the long term. Various practices for sustainable fashion include the use of ecofriendly materials, responsible production techniques, waste management, reduced environmental impacts and ethical labour practices (Nayak, Akbari, & Far, 2019). The use of environmentally friendly materials is a fundamental tenet of sustainable fashion. This covers the use of recycled or biodegradable materials such as Lyocell, recycled polyester and nylon, as well as natural fibres like organic cotton, linen and organic wool in addition to the new fibres derived from renewable resources (e.g., Casein, lotus, coffee and PLA [polylactic acid]) (Nayak, 2022; Nayak, Jajpura, & Khandual, 2023). As they need less resources to create and have a smaller carbon footprint, using these materials helps to lessen the environmental impacts of the fashion industry.

Responsible production practices are another crucial tenet in the fashion sustainability journey. This covers the application of waste reduction and recycling initiatives, the use of energy-efficient production techniques and the application of water conservation measures (Moon, Youn, Chang, & Yeung, 2013). These procedures make sure that the fashion sector functions more sustainably and lessen its negative effects on the environment. The idea of circularity, which refers to the process of designing and creating clothing in a way that minimizes waste and maximizes the use of resources, is one important component of sustainable fashion. This includes employing closed-loop production techniques, which recycle materials and lessen the demand for fresh material supplies, as well as repair and reuse initiatives, which increase the usefulness of apparel and cut down on waste (Choi & Cheng, 2015; Nayak, Houshyar, et al., 2020).

The fashion businesses must prioritize ethical labour practices in addition to alleviating negative environmental practices, which are covered under CSR (Nayak, Van Thang, et al., 2021). Some of the ethical practices include giving employees fair treatment, paying fair compensation and providing safe and healthy working environments. Because it fosters employee trust and loyalty and guarantees that the business works in a socially responsible manner, ensuring that workers are treated ethically and fairly is crucial for the long-term viability of the fashion industry. The other aspects of ethical practices are free from child labour, forced labour and discrimination in any form. CSR can be defined as "the voluntary integration, by companies, of social and environmental concerns in their commercial operations and in their relationships with interested parties (Employment, 2001)".

Transparency, which refers to the practice of releasing information about the production and supply chain of fashion products, is another crucial element (Garcia-Torres, Rey-Garcia, Sáenz, & Seuring, 2022). The usage of traceability systems, which allow customers to trace the history of their garments, as well as social and environmental impact measurements, which allow businesses to assess their sustainability performance, can help in the sustainability journey (Nayak, George, Haq, & Pham, 2022). In order to ensure that the fashion industry runs in a way that is not only profitable but also sustainable over the long term, sustainable principles in the garment supply chain often involve the use of ecofriendly materials, ethical labour practices and responsible production procedures. Fashion companies can enhance their social and ethical policies, lessen their environmental effect and contribute to a more sustainable future by implementing these principles.

## 4.3 APPROACHES TOWARDS SUSTAINABILITY

### 4.3.1 Concept of Fast Fashion and Associated Demerits

There has been a rapid increase in the products from fast fashion brands, which are cheaper and trendy. The products representing the latest trends are manufactured with a very short lead time for the mass market. The concept of fast fashion is to provide consumers with the latest trends in fashion that are inexpensive and move rapidly from runway to retail (Hayes & Jones, 2006). It takes 8–12 weeks from concept to consumer compared to a minimum of about 44 weeks for the traditional fashion brands. Global fast fashion brands introduce new trends almost on a weekly basis to stay on-trend and meet the customer demand. There are several environmental and social impacts associated with the fast fashion products as mentioned below.

- **Excessive Purchase and Rejection:** As the fast fashion is trendy, consumers purchase inexpensive clothes in excess and reject them more often than the traditional clothing. Even some clothes are rejected after one wear (Gupta & Gentry, 2016). This leads to the problem of landfill, waste management, GHG generation and resource depletion.
- **Lack of Transparency:** Several brands do not disclose information relating to the ethical practices during the manufacturing and supply chain processes (Fraser & van der Ven, 2022). Customers may reject the brands if they become transparent and provide the information to them. The unethical practices come into highlight only after some information are available on a mass scale or some incidents such as Rana Plaza Disaster happen (Sinkovics, Hoque, & Sinkovics, 2016).
- **Unsafe Working Conditions and Sweatshop:** As the lead time for fast fashion is too short and the products are cheap, some manufacturers don't take care of the working conditions and employee's safety (Bhattacharjee, 2019). Cheap fashion uses cheap raw materials in addition to using cheap labour in many developing countries. The employees work longer working hours with very low wages; often the wages do not meet the minimum wage-level set for the country. Hence, the workers can't lead a normal life and have to do lots of adjustments.
- **Greenwashing:** Fast fashion brands are often found to be greenwashing the consumers, which relates to providing misleading information or

bury the information related to unethical practices in the supply chain process (Adamkiewicz, Kochanska, Adamkiewicz, & Łukasik, 2022). Greenwashing can be defined as "unjustified and misleading claims by fashion brands about their products being more ecofriendly or ethical than they really are". The greenwashing can include false information, lack of information, lack of transparency and over and ambitious claims that are not real (Adamkiewicz et al., 2022). Greenwashing is a major concern in fashion as many brands take advantage of growing demand for trendy clothes that are sustainable. The consumers, however, are misled by greenwashing and buy the product, assuming it as sustainable.

- **Environmental Pollution:** It is already mentioned in the first point that the rejection of clothes more often leads to environmental pollutions (Niinimäki et al., 2020). Further, these clothes were produced by the processes that also had negative environmental and social impacts. For example, dyeing in fast fashion results in water pollution or the cultivation of conventional cotton uses excessive water and synthetic fertilizers. On a similar note, many cheap clothes are made from synthetic fibres such as polyester or nylon that take several hundred years to biodegrade in the landfill. Many often, the blended fabrics used in apparel or cheap fibres in the clothing are hard to recycle.

## 4.4 SUSTAINABLE FASHION SOURCING

The fashion sector, one of the biggest and most powerful sectors in the world, significantly affects both the environment and society (Ting & Stagner, 2021). It takes a lot of natural resources, including water, land and energy, as well as chemicals and auxiliaries to make garments, footwear and accessories. In addition, the fashion sector has a substantial carbon footprint, with some estimates placing its share of world carbon emissions at approximately 10% (Jones & Wright, 2018). Sustainable fashion sourcing is the practice of obtaining goods and materials for the fashion industry in an ethical and responsible manner. This entails taking into account the environmental effects of the materials and production methods as well as the social and ethical repercussions of the resources' sourcing and fair treatment of the labour force involved in the manufacturing (Chen & Slotnick, 2015). When it comes to sustainable sourcing, the fashion sector faces several significant obstacles, which need to be addressed to become sustainable.

- **Use of Plastics:** The raw materials used to make clothing, footwear and accessories have a negative impact on the environment due to the use of huge quantities of plastics. For instance, the manufacturing of synthetic fabrics like nylon and polyester can result in large amounts of GHGs. Additionally, the EOL clothing of these plastics are not biodegradable; hence, they lead to contamination of land and nearby water sources (Barra & González, 2018).
- **Social and Ethical Concerns:** The fashion industry has a track record of using child labour, forced labour and exploiting employees in developing nations, in addition to having unsafe working conditions and labour standards. Additionally, there are frequently serious problems associated with

the supply chain, making it challenging to guarantee that employees are treated fairly and ethically. These problems are associated with several levels of subcontracting, complexity of the supply chain network and lack of transparency (Birtwistle & Shearer, 2001).
- **Excessive Waste Generations:** Fashion sourcing process has been associated with excessive solid waste generation and increased carbon emissions (Choi, 2013). Firstly, the packaging materials used during transportation are discarded to landfill by many fashion businesses. As these materials are primarily made from plastics, it leads to GHG emissions and waste management problems (Nayak, Nguyen, Patnaik, & Khandual, 2021). Secondly, the sourced inventory associated with fashion manufacturing travels miles before being converted into final products. The final products also need to travel more miles to reach consumers at various global destinations. The transportation activities are associated with huge amount of carbon emissions that pollute the planet (McDonald, Hung, & Akbari, 2020).

The current model in many countries is in a linear fashion, which includes a take-make-dispose approach or linear economy model as shown in Figure 4.1a (Dissanayake & Weerasinghe, 2021). In order to become sustainable and save resources, it is essential to become circular. Fashion products, which include apparel, footwear and accessories, that are designed to reduce waste and enhance resource efficiency are referred to as circular fashion (Figure 4.1b). This may entail creating easily repairable and recyclable products as well as putting in place mechanisms to gather and reuse or recycle used goods (Mishra, Jain, & Malhotra, 2021). There are several approaches to become circular, such as 3R (reduce, recycle and reuse), upcycle and composting. It depends on the product type, the brand's choice and availability of facilities to select a specific route for circularity. Sustainable fashion sourcing, which seeks to lessen the negative effects of the fashion industry on society and the environment, is a developing movement that strives to address these issues. Sustainable fashion sourcing can be done in a variety of ways, which include:

- Sustainable raw materials for apparel production,
- Sustainable logistics and warehouse management in garment production,
- Waste management in the supply chain,
- Modular designing and durable products,
- Ethically made fashion and
- Approaches by consumers during purchase, usage and disposal phases.

### 4.4.1 Sustainable Raw Materials for Apparel Production

Textile fibres are the building blocks for manufacturing of fashion and textiles. The first step for a specific garment style is the selection of appropriate raw material (i.e., fibres and yarns), which is finalized in the conceptualization stage by designers, and it greatly influences the total environmental impact (Islam, Perry, & Gill, 2021). Further, the process routes to manufacture a specific garment are selected based on

(a)

(b)

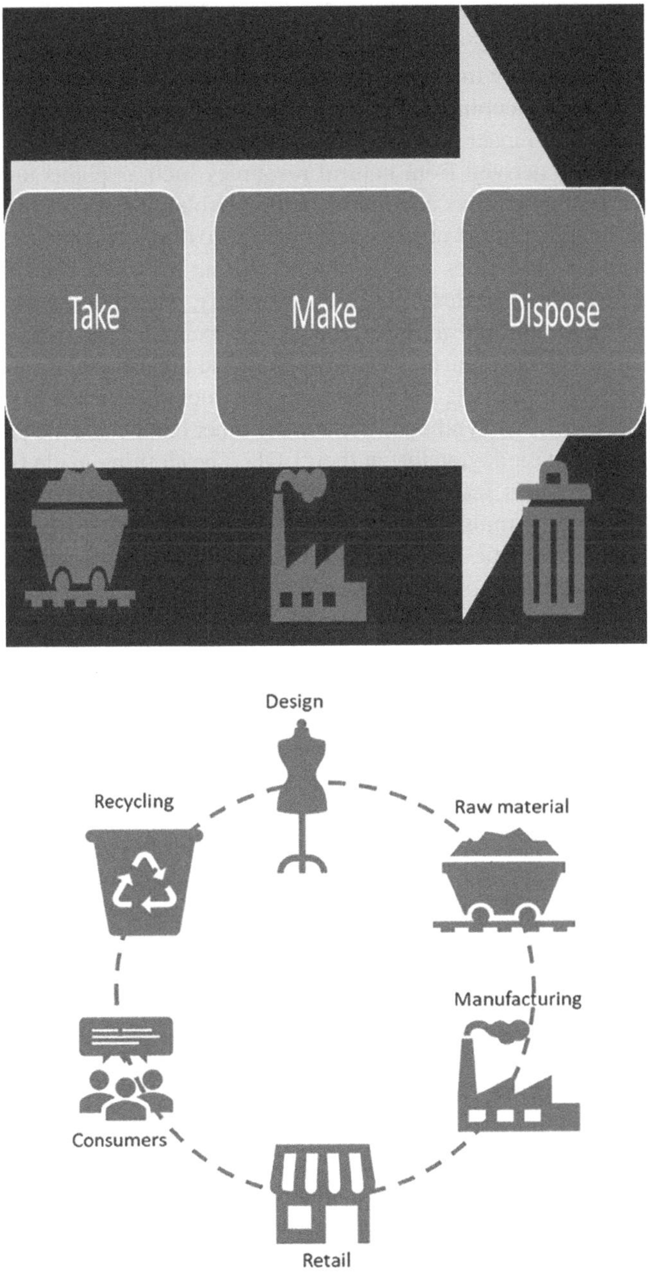


**FIGURE 4.1** Models followed in fashion: (a) linear model and (b) circular model.

the type of raw materials. For example, selection of 100% cotton and polyester/cotton blends needs different processing routes. Hence, to avoid the environmentally damaging processes especially in chemical processing, raw materials should be carefully selected. Hence, the conceptualization stage along with the designers play a vital role in the environmental impact of garment manufacturing (Nayak, 2020).

Textile fibres are derived from natural resources such as plants and animals in addition to the synthetic fibres derived from the petroleum resources. Natural fibres like cotton use a large amount of fresh water (about 20,000 L/kg) and toxic chemicals (such as herbicides, pesticides and defoliants) during the cultivation stage (Tausif, Ahmad, Hussain, Basit, & Hussain, 2015). Similarly, wool fibre is associated with environmental pollution and animal cruelty. For example, scouring of wool that cleans greases and other impurities by using soap and alkaline solutions cause water pollution (Hasanbeigi & Price, 2015). Shearing of wool fibres leads to injuries to the sheep due to carelessness. Synthetic fibres on the other hand consume a large amount of energy and are not biodegradable at their EOL. The clothing made from polyester fibres shreds microfibres into the water courses, which then enter the food chain through the aquatic animals and can cause harm to the human digestive system. Hence, the use of synthetic fibres is also associated with environmental problems when the environmental impact is considered.

Some of the synthetic fibres such as Lyocell and Tencel are more ecofriendly than polyester, nylon and acrylic, due to their renewability and biodegradability nature. In the last two decades, the technological developments in polymer science have resulted in the commercial production of new fibres such as soybean protein, Lenpur, SeaCell and PLA or corn fibres, which are considered to be more sustainable (Maduna & Patnaik, 2023). Using these sustainable fibres in the manufacturing of clothing, footwear and accessories is one of the most effective strategies to lessen the environmental impact of the fashion industry. Other fibres include organic cotton (which is farmed without the use of synthetic fertilizers or pesticides), as well as recycled materials (like recycled polyester and recycled nylon), which lessen the demand for fresh resources. Hence, the use of these new fibres provides a solution to the problems associated with the use of traditional fibres such as cotton, wool, polyester and nylon. Various new fibres that are more environmentally sustainable for garment manufacturing are discussed in the following sections.

#### 4.4.1.1 Organic Cotton

Organic cotton is produced without the use of any hazardous chemicals, including pesticides, herbicides, fertilizers, defoliants or growth regulators, and is derived from seeds that have not undergone genetic modification (Gardetti & Muthu, 2019). In addition to its excellent qualities, such as being free from allergic responses and having a softer feel than conventional cotton, organic cotton minimizes soil and water pollution. When compared to conventional cotton, organic cotton cultivation reduces product toxicity to practically nil and reduces it overall by 93%. Organic cotton improves the environment, fosters a real economy and safeguards our wellbeing. In order to produce organic cotton fibre, not only are synthetic fertilizers substituted with organic ones, but also a variety of systematic agricultural practices are used (Maduna & Patnaik, 2023). Farms are prepared two years prior to planting for the conversion from conventional to organic cotton, and more ecologically friendly manual and mechanical weed control techniques are used. The soil, climate and level of disease and insect resistance are all taken into consideration while choosing the organic cotton variety.

#### 4.4.1.2 Organic Wool

Sheep raised in humane and decent farm circumstances without exposure to pesticides or hormones are considered as organic wool (Fletcher, 2013). When compared to conventional wool, organic wool production promotes animal ethics and is free of harmful chemicals and hormones, despite the fact that there is not much of a difference in the fibre qualities. Wool produced organically has a lower environmental impact, making it more sustainable. Similar to organic cotton, which also has to be verified by outside organizations, there is an increasing demand for organic wool on a global scale. The following requirements must be met in order for wool to be considered organic: (a) the feed and forage used by the sheep must be organic, and the grazing land cannot have been treated with synthetic pyrethroids or pesticides; (b) no synthetic hormones or genetic engineering should be used to enhance sheep growth; (c) only approved chemicals should be injected into the sheep to prevent sheep-scab and (d) the sheep cannot be treated with any type of pesticide that is harmful to the sheep (Poole, Church, & Huson, 2009).

#### 4.4.1.3 Wild Silk

Wild silk is produced without the use of dangerous chemicals and is grown in areas of open woodland, which is an easy supply of food for silkworms. Due to advantages like: (a) the wild silk cultivation can promote forest preservation and provide income for the tribes; (b) the wild silk cocoons are collected after the moth has emerged as a butterfly as opposed to killing the moth while in-situ for the cultivated silk and (c) as the wild silk cocoons are broken, the continuous silk filament is produced. Although wild silk is extremely difficult to bleach or dye, it has a beautiful gold lustre by nature (Karthik & Rathinamoorthy, 2017).

#### 4.4.1.4 Hemp

Hemp, one of the fastest growing natural fibres, was initially utilized around 10,000 years ago. Hemp, which may be used to make textiles and garments, has a texture that is comparable to flax and has excellent drape. Hemp may be used as a stand-alone fibre or combined with other fibres like spandex, polyester and cotton (Ahmed, Islam, Mahmud, Sarker, & Islam, 2022). Hemp fibre has the best level of durability among all natural fibres. Other benefits of hemp include its strength, which is around 300% more than that of cotton, as well as its high breathability and absorbency, good abrasion resistance, resistance to mold and mildew, UV resistance and antibacterial qualities (Zimniewska, 2022). Hemp materials can be coloured easily, can be washed or dry cleaned and become softer after washing. Since hemp takes less water to grow and doesn't require the use of pesticides, herbicides or seeds that have been genetically modified, hemp cultivation is seen to be environmentally friendly. According to Table 4.1, hemp fibre has the highest yield of any plant fibre. Hemp farming provides vital nutrients to the soil and stops soil erosion. Only organic materials, such

**TABLE 4.1**
**Average production of various natural fibres per hectare of land**

| Fibre name | Average fibre production (kg/ha) |
|---|---|
| Hemp | 1,600 |
| Flax | 1,000 |
| Cotton | 800 |
| Organic cotton | 500 |
| Wool | 65 |

as compost and natural manure, may be used to grow hemp. Due to the hemp plant's similarity to the marijuana plant, hemp farming is prohibited in many nations. When compared to the manufacturing of organic cotton, hemp fibre can lower the environmental impact by 50%. The only environmental issue with hemp fibre manufacturing is the water pollution caused by the retting process needed to remove the fibre. By using procedures like dew retting, steam explosion and enzymatic retting, this can be prevented. Table 4.1 shows the average amount of different natural fibres produced per hectare of land.

#### 4.4.1.5 Flax

The cultivation of flax requires no pesticides or herbicides and uses less fresh water than cotton, making the production of flax fibre more environmentally friendly than cotton (Debnath, 2019). Additionally, because flax seeds don't require as much deep ploughing, less energy is used and there is less soil erosion on sloped fields. Biodegradability, ease of waste disposal, less water use, protection of non-renewable resources and decreased carbon dioxide ($CO_2$) emission are just a few of flax's environmental advantages. Flax plants absorb a significant quantity of $CO_2$ over the course of their lifetimes. Flax fibre outperforms cotton in terms of strength, abrasion resistance, shape retention, absorbency and breathability. About 20% less energy than glass fibre and 10% less energy than carbon fibre are used in the production of flax fibre. The materials from EOL flax may be utilized to create composites for usage in the building, automotive and thermal insulation industries (Rahman, 2021).

#### 4.4.1.6 PLA Fibre

Made from renewable resources such as sugar, beetroot, maize starch and sugarcane, PLA fibre is a linear thermoplastic aliphatic polyester fibre. Since PLA fibre degrades quickly, it can be utilized as compost after the EOL. In terms of energy usage, $CO_2$ emissions and EOL choices, PLA is preferable over many polymers made from petroleum (Tarique et al., 2021). The primary virtue of PLA's short life cycle because of its compostability made it the subject of initial promotion for uses in single-use packaging. The advantages of using and producing PLA fibres over synthetic fibres like polyester are listed below. There are certain unfavourable effects of PLA synthesis, which have been listed as well (Nayak, 2022). The commercialization of PLA fibre has not yet been accomplished since there are more drawbacks than advantages.

Advantages:

- The limited supply of synthetic raw materials is not a challenge for PLA as it won't run out since PLA fibres are made from naturally replenishable resources.
- Compared to the creation of synthetic fibres, the PLA fibre process uses less energy and emits less gases into the atmosphere.
- Corn fibre is renewable as it can be renewed when needed. Hence, unlike synthetic fibres there is no threat of PLA fibre running out or getting exhausted.
- PLA fibre is biodegradable at the end of its life. Hence, it is free from the problems of land pollution due to waste problems.

Disadvantages:

- Large amounts of land, which are now being utilized to produce food, are required for the commercial production of the raw materials for PLA fibres.
- Foods (corn, sugarcane or sugar beetroot) will be redirected away from human consumption and into the production of fibre.
- Despite the biodegradability of PLA garments at the EOL, their disposal in landfills will cause the production of GHG like methane, hazardous land and water owing to colour leaching and the creation of human carcinogens.
- The price is greater compared to other fibres like polyester because PLA fibre is more expensive and less widely available.
- Although PLA has many characteristics with polyester, its lower melting point limits its use in high-temperature applications and makes transfer printing with PLA challenging.
- As the PLA fibres degrade during chemical processing with the use of water and temperature, the characteristics of the fabric will alter.

#### 4.4.1.7 Lyocell Fibre

Made from eucalyptus trees (*Eucalyptus regnans*), which can grow very quickly and with little to no pesticide use, Lyocell is a rayon fibre (Prakash & Kumar, 2023). It is a rayon fibre that was made utilizing dry-jet wet spinning as an environmentally friendly textile fibre. In the same family of rayon fibres as viscose, Lyocell is similar to that fibre as well as other cellulosic fibres like cotton and flax. Lyocell, however, has the lowest environmental impact of all the rayon fibres during production. Up to 98% of the amine oxide solvent used to dissolve the cellulose for fibre spinning can be recovered and utilized again (Jiang, Bai, Chen, & Liu, 2020). The solvent is also harmless and safe to use. Other advantages of Lyocell include wrinkle resistance, softness, biodegradability, use of renewable raw materials (such as Eucalyptus trees), bleach-free production (since the fibre is white and pure) and reduced water and energy usage. The wetting of the fibre, which causes a modest reduction in mechanical strength and fibrillation, is the sole issue with Lyocell (i.e., small fibrils separating from the main fibre or filament)

#### 4.4.1.8 Soybean Protein Fibre

Soybean protein fibre is a by-product made from the hulls of soybeans. It may be grown in large quantities at a low price and is a renewable plant source. Despite the fact that the fibre was created in the 1950s, there has been little progress in its commercialization. The fibre is becoming more significant as people become more conscious of sustainable materials made from renewable resources (Choudhury, 2023). Using a cutting-edge bioengineering technique, fibre is produced from the soybean cake. Although clothing made from soybean fibres has an elastic grip and a delicate hand akin to cashmere, it is less durable than cotton and hemp. In addition to having a sheen akin to silk, soy fibre also offers outstanding comfort qualities like cotton. The fibre is superior to cotton in moisture transport and comparable to cotton in moisture absorption, which is the primary determinant in comfort. It is also antimicrobial and UV resistant.

#### 4.4.1.9 Casein or Milk Protein Fibre

Casein, sometimes referred to as milk protein fibre, is made from milk protein. The fibre made of milk protein has advantages over both natural and manufactured fibres. They feature a silk-like sheen and long-lasting antimicrobial qualities that are natural. The fibres are highly pleasant since they have great air permeability, breathability and water transport qualities. The pH of the fibre is comparable to the skin and includes roughly 15 different types of amino acids, which contributes to healthy skin maintenance. Casein fibres are widely used in women's clothes, children's clothing, athletic wear, uniforms and intimate apparel. Casein fibre is one of the greatest sustainable fibres since it is biodegradable, environmentally benign and made entirely from renewable resources (Thangavelu & Subramani, 2016).

#### 4.4.1.10 Lenpur Fibre

Lenpur fibre is a cellulosic fibre made from the wood pulp of certain species, including white pine, and is a renewable resource. Lenpur fibre has advantages over other cellulosic fibres, including softness, increased moisture absorption, increased heat range and inherent antibacterial and antiodour qualities (Nayak, Nguyen, Panwar, George, & Ulhaq, 2020). Lenpur fibre may be mixed easily since it is compatible with many different textile fibres. Ring spinning creates smooth yarns that may be used right away in knitting machines. Lenpur can be used in place of cotton fibre because of its comfort-enhancing qualities. Additionally, it may be utilized to create designer premium clothing.

#### 4.4.1.11 SeaCell Fibre

SeaCell fibre is made from seaweed, which is always thought to have positive effects on health and medicine. By actively exchanging materials with the skin, SeaCell fibre may feed the skin to keep healthy (Maduna & Patnaik, 2023). Iron is present in the fibre, and it is a fantastic supply of iodine, which is crucial for thyroid function. Another vitamin found in seaweed is vitamin B12, a shortage of which can cause a specific kind of anaemia. It contains both anti-inflammatory qualities and a wealth of minerals with antibacterial effects. These characteristics make the SeaCell fibre an intrinsically healthy material for clothing that is kind to the skin.

## 4.5 SUSTAINABLE LOGISTICS AND WAREHOUSE MANAGEMENT IN GARMENT PRODUCTION

Sustainability of the fashion industry depends on efficient logistics and warehouse management in the manufacturing and distribution process of clothing. Fashion firms have come under growing pressure in recent years to lessen their environmental impact and enhance the sustainability of their supply chains. Utilizing sustainable logistics and warehouse management techniques during the garment production process is one method businesses can achieve. The vital aspect is the utilization of environmentally friendly transportation strategies for sustainable logistics in the fashion industry (Garza-Reyes, Villarreal, Kumar, & Molina Ruiz, 2016). Utilizing electric or hybrid vehicles and putting carbon offset programs in place are two examples of how to combat the problem. Additionally, fashion enterprises can lessen their carbon footprint by choosing suppliers local to their factories and warehouses, which can help them avoid the need for expensive long-distance shipping (Bektaş & Laporte, 2011).

Additionally, it is essential to guaranteeing the sustainability of the clothing production process is effective warehouse management. Employing automation and advanced technologies is one method that businesses can use to enhance their warehouse management procedures. By eliminating the need for manual labour, automated material handling systems, for instance, can increase productivity and lower the warehouse's carbon impact. The usage of real-time tracking technologies can also aid in inventory optimization and lower the danger of waste brought on by overstocking or surplus inventory. The use of technologies such as RFID and internet of things (IoT) can help to improve the transparency and accuracy of information in fashion supply chain (Nayak, Singh, Padhye, & Wang, 2015).

Utilizing energy-efficient lighting and heating systems is a crucial component of sustainable warehouse management (George, Ulhaq, & Nayak, 2020). This may aid in lowering the warehouse's energy use and carbon footprint. Companies can further lessen their environmental effect by implementing water conservation measures like using low-flow faucets and water-saving equipment. A sustainable fashion sector must, in general, have sustainable logistics and warehouse management. Fashion firms may drastically lessen their environmental impact and increase the sustainability of their supply chain by employing ecofriendly transportation strategies, reusable packaging materials, automation and technology and energy-efficient systems.

This chapter also discusses some of the findings from the RFID project, which involved preliminary qualitative in-depth interviews with four business owners of Vietnamese Small and Medium-Sized Enterprises (SMEs) involved in fashion and textile manufacturing. The interviews were conducted inspired by the framework of TBL to measure sustainability. A standardized interview protocol was used to ensure comprehensiveness among the interviewees (Nayak, George, Haq, & Pham, 2022). The interview questions focused on getting insights about sustainability by the use of various technologies such as RFID in Vietnamese fashion supply chain. The interview questions focused on getting insights about sustainability using various technologies such as RFID. It allowed us to compare different views and extract meaningful insights from the interviews. The interviews are then transcribed and coded by using the qualitative data analysis software NVivo. Some of the findings are given in the extract form in the following section.

- Extract 1: The RFID technology helps fashion businesses in reducing waste, as one of the businesses observed from its business operation.

  > I believe RFID technology can help to reduce waste because one of the problems we're facing is that when we have many stocks, and our business is slowing down, we cannot sell many products at the same time, so we have excessive stock. Fortunately, because we're selling rackets, which can be in the warehouse for a long time, some products will decrease in value when they're in the warehouse for a long time. With such stocks, we cannot sell anymore, and we must dispose of them by burning or burying them. That is a huge waste, so I believe with RFID technology, stock shrinkage will be reduced, and we can sell more products to reduce the waste.

- Extract 2: RFID technology can help to reduce the use of electricity and manpower. As mentioned by one of the participants:

  > RFID technology can reduce electricity consumption as it is much faster than barcode. What we have seen is we are doing barcode scanning one by one and it takes a lot of time. I think with RFID we can do that faster as it can scan many tags at the same time, and I believe it can decrease our electricity usage. Further, the use of RFID technology can reduce the labour cost as it can scan several tags simultaneously.

- Extract 3: RFID can help the customers to track the right type of product in a retail store, increasing customer satisfaction relating to sustainability. As mentioned by one of the interviewees:

  > Any product with the RFID tag is highly visible compared to barcode. The products have information inside an RFID tag so the customer can track the information inside the tag to see the supply chain process of producing that product whether it's environmentally friendly. Not only for the customer, but it is also helpful for the company to track the information for improvements and convenience of sustainability. The customers get right information to take sustainability decisions.

- Extract 4: Utilizing RFID technology can improve the sustainability of the supply chain due to increased efficiency and accuracy. As explained by one of the participants:

  > RFID technology in the supply chain helps our organization to reduce waste by increasing the accuracy of demand forecasting, decreasing the

risk of overproduction, and decreasing the need for product returns. It reduces the lead time due to accuracy of information that can be readily available when needed.

- Extract 5: RFID technology helps to reduce overtime in apparel manufacturing units. As explained by one of the participants:

> RFID technology helps our organization to reduce the overtime problems, which was a common problem before the implementation of RFID. The manual process of record keeping takes more time, is tiresome and inaccurate. RFID avoids waste and improves the quality of garments. Hence, the problem of overtime is reduced with the use of RFID technology.

### 4.5.1 Waste Management in the Supply Chain

This section focuses on the increased use of recycled and upcycled fashion and textile products. At the EOL of fashion and textiles, several of the products have a potential to be recycled or upcycled into new products (Nayak, Houshyar, et al., 2020). Similarly, packaging materials such as reusable plastic containers and pallets can be reused in the making of new clothing, which is another crucial part of sustainable logistics. This could lessen waste generation and the fashion industry's overall environmental impact as many of the fashion and textiles end up in the landfill. To further lessen their influence on the environment, businesses can also establish waste-reduction programmes, such as recycling and composting initiatives (Hustvedt, Meier, & Waliczek, 2016).

Planet earth is at risk due to the rise in the pollution level due to manufacturing as well as excessive waste produced from the EOL products. The amount of fashion waste is increasing at an alarming rate due to the popularity of fast fashion. Plastics (bottles, nets and packaging bags) are one of the main reasons of increasing pollution levels, as it takes more than 200 years for their biodegradation (Rhodes, 2018). Several fashion and textiles are manufactured from synthetic materials such as nylon and polyester, which are considered as plastics (Moorhouse, 2020). Fishing nets made from nylon represents 10% of the debris in the ocean and totals up to 600,000 tonnes per annum that is dumped into the ocean. It is crucial to find the suitable way to deal with old nets that no longer can be used.

The nylon nets and gears can be recycled into new products such as wheels for chairs, computer parts, toothbrushes and clothing. For example, pro-surfer Kelly Slater has launched a clothing line made from ocean trash, including the fishing nets and other ocean wastes. This is just an example of recycling fishing gears into new, usable objects rather than dumping them into the landfills or causing harm to the marine life (Moorhouse & Moorhouse, 2017). Like Nylon, the plastics from polyester are also causing serious environmental pollution. Polyesters in the form of soda bottles, industrial waste and worn-out garments can be converted into polyester fibres to produce clothing. The use of recycled polyester helps to reduce the use of virgin material derived from petroleum resources, which is limited. The use of recycled nylon and polyester helps in reducing the pollutions caused from incineration or landfill (Nayak, Nguyen, Panwar, & Jajpura, 2020).

In addition to recycling, upcycling is also being widely used by many fashion brands. In upcycling process, discarded clothes are converted into new clothes, which are often beautiful (Cassidy & Han, 2017). Upcycling is not the process of breaking a product and remaking new products but designing a new product with newer design concepts. There is a tremendous rise in the use of recycled and upcycled fashion and textile products. The main difference between upcycling and recycling is that with upcycled clothes, waste is looked at creatively and transformed into something new. On the other hand, material is broken down first and then converted into usable clothes in recycling process (Pensupa, 2020).

## 4.6 MODULAR FASHION AND DURABLE PRODUCTS AS A MEANS OF SUSTAINABILITY

Modular fashion refers to clothing items that have detachable pieces, so that one can easily alter the clothing item to suit changing needs and tastes over time. Modular fashion can incorporate and combine all or some of the construction methods used for traditional garment manufacturing (Chen & Lapolla, 2021). Modular clothing is a transformable clothing that aligns with a few different aspects of environmentally ethical fashion design. For one, a garment made in a manner that can change appearances easily may encourage consumers to buy fewer clothing items, given its versatility. Additionally, modular clothing enables users to affect the design of a garment that may cause them to feel more emotionally attached to the garment, delaying its psychological obsolescence.

Modular design is an ideal concept as it reduces the number of garments and can solve multiple purposes. Modular design refers to designing fashion through the universal module subsystem by the arrangement of subassemblies and components to generate a variety of specific or similar product features and performance (Chen & Lapolla, 2021). It not only satisfies consumer demand for products with the latest trends and cheap prices but also tackles environmental issues caused by fast fashion. The modularity concept first appeared in architecture in the 1920–1930. However, it was not widely accepted in fashion design until the 1980s, when pioneer MULTIPLES started to sell modular garments that gave clients more options to express their styles. Modular designs provide flexibility for the wearer to select a range of combinations only from few garments. Modular garments also save resources and waste generation at their EOL. One example of modular design has been shown in Figure 4.2.

Generally fast fashion products are associated with short product life cycles that lead to a large-scale rejection of clothing. Hence, rapid disposal linked with overconsumption of fashion products is an obstacle to sustainability. Many fast fashion products are produced with low quality, so that they can be sold at cheaper prices. As a result, consumers are buying these cheap and low-quality products in excessive amounts, which have a short shelf life. Durable products can provide potential sustainable solutions to the overconsumption of fast fashion apparel. Producing products that are physically durable and have functional design features that can last for a long time can save the amount of resources used in a garment. Further, durable

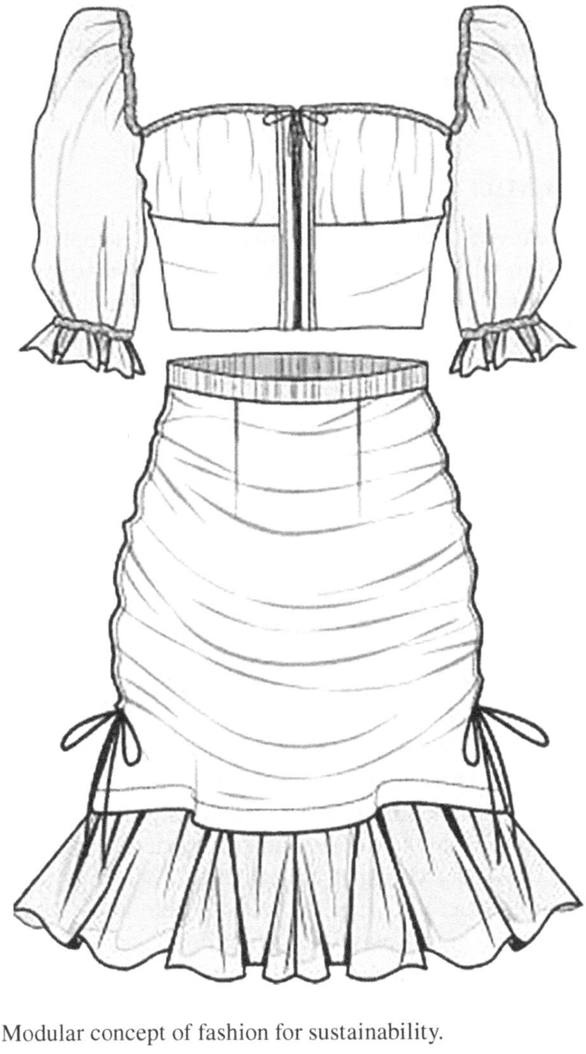

**FIGURE 4.2** Modular concept of fashion for sustainability.

fashion apparel with emotionally enduring designs embrace meaning and create an emotional connection for the consumer. Hence, the consumers will be wearing the product for a longer time.

The designers should consider functional design of apparels for longevity, and that functional longevity is ultimately the first step towards sustainable fashion. Physical durability alone may not be the ultimate solution to fashion apparel overconsumption. It has been shown that sustainable consumption requires emotionally durable designs, but products are worthless if they are not physically durable. Hence, both functional and emotional coherence are necessary to accelerate progress towards greater sustainability (Niinimäki, 2017). Fibre content plays a significant role in product durability. For example, polyester is more durable than cotton; hence, they are blended to enhance the durability of the garment. It has been shown that luxury goods possess the unique feature of being more durable than ordinary products, hence, sustainable (Sun, Bellezza, & Paharia, 2021). Due to improved awareness on sustainability, some consumers are willing to buy durable products; however, the number is quite small.

### 4.6.1 Ethically Made

There have been many violations of ethics in the fashion supply chain process. The workers are forced to work overtime, paid less than the standard wage, work overtime with no additional pay and work in unsafe conditions, which are some of the negligence in the ethical fashion drive (Nayak, 2020). Similarly, not caring for the environmental impacts, releasing the effluents directly to the water courses without filtration, releasing GHG and waste generation are some of the problems that go against fashion ethics. One example where fashion ethics was not followed is the Rana Plaza disaster that killed more than 1,100 people and injured 2,500 (Barua & Ansary, 2017). There are several other examples where fashion ethics are violated in developing as well as developed countries.

A crucial component of sustainable fashion sourcing is to make sure that the materials and products are sourced ethically while considering the social and environmental effects of the production process (Ha-Brookshire, 2017). To guarantee that materials and products are procured responsibly, this can entail collaborating with suppliers and manufacturers who have strict environmental and social policies and putting traceability systems in place (Birtwistle & Moore, 2007). Although the fashion supply chain is rather complex, the traceability system can help to identify the breach of code of conduct at any stage. The use of various technologies such as RFID can help to improve the traceability in the supply chain process.

Ethical fashion considers the environmental impacts of fashion production as well as wellbeing of society (Haug & Busch, 2016). Selecting new materials with lesser environmental impact (e.g., organic cotton, linen, lotus and hemp), repurposing materials at EOL, reducing plastics in the product, caring for the labour force, taking care of the society and avoiding child labour and forced labour are typical examples of ethical fashion practices. Ethical fashion also considers animal rights where raw materials are produced from animals such as leather goods. The raw materials should be prepared by avoiding animal cruelty and eliminating pollution of the environment.

## 4.7 APPROACHES BY CONSUMERS DURING PURCHASE, USAGE AND DISPOSAL PHASES

Consumers can take care of fashion sustainability by being responsible while making fashion purchases, while using the fashion products and at the EOL of fashion. For example, they can reject the products from a brand if they are manufactured by not following ethical practices. The brands need to be transparent by providing product information to consumers (Fraser & van der Ven, 2022). Here are some of the steps consumers can take to become sustainable.

- **Reduce Clothing Amount and Buy Local:** Rather than buying large amounts of fast fashion products, the consumers should focus on durable products. The less we buy, the less we put the stress on the global supply chain. Consumers should avoid impulse buying and buy only the product when it is needed. Buying locally may help to change the way the global clothing supply chain functions. It will reduce the carbon emissions involved with the global sourcing and logistics operations (McNeill & Moore, 2015).
- **Buy Ethical and Durable Products:** Consumers have the right to select or reject a brand based on ethical practices. If the product does not take care of the fashion ethics during manufacturing, they can reject the product. Also, they should make a habit of buying good-quality products that are durable and last longer (Sun et al., 2021). Durable products help to reduce the amount of EOL clothing ending in the landfill and reduce the GHG emissions.
- **Check Clothing Labels:** Labels tell us a lot about where the clothing comes from, what the composition is and how to take care of the garment (Nayak & Padhye, 2015a, 2015b). This can help the consumers to purchase sustainable products and take the right care of the garments. Some manufacturers have also been making a move towards sustainable eco-fashion, with a low impact on the environment throughout the garment life cycle, and fashion brands are taking notice. Consumers should also buy recycled clothing and clothing made from new natural fibres (biodegradable and renewable) to reduce the environmental impacts.
- **Buy Secondhand Clothing:** Buying secondhand clothing and reusing clothing is a great way to reduce waste and the impact on the environment. Demand for fast fashion is increasing clothing wastage. Buying secondhand clothing can reduce the amount of waste and save environmental pollution in addition to saving money.
- **Wash Less and Iron Less:** It is a well-known fact that most of the energy used in the life cycle of a fashion product is related to post-purchase or consumer usage stage. The energy consumption during usage of garments is about 67% compared to 33% during garment manufacturing (Nayak & Ratnapandian, 2018). Further, the laundering of synthetic textiles such as polyester releases a large quantity of microplastics that end up in oceans and enter the human food chain. We can reduce the environmental concerns with "wash less and iron less concept". Whenever possible, air dry can be used to dry clothes on a sunny day, which can save energy consumption by reducing the environmental impacts.
- **Right Approach at the EOL:** Until now the global average value for recycling of fashion and textiles is about 14%. These figures are as low as 2% in developing countries and can be close to 20% in developed countries. Hence, it is evident that 86% of the EOL fashion and textiles end up in the landfill due to lack of recycling facilities, lack of consumer awareness and low demand for secondhand clothing in many countries. Lack of recycling facilities is a major problem in developing countries as there are not many initiatives by the government or private sector. Similarly, many consumers

are unaware of what to do at the EOL clothing and they dump into the landfill. Further, there is no or very little demand for recycled clothing in some countries, which prevents recycling. Improving the recycling facilities could help to improve the amount of EOL clothing getting new life.

## 4.8 LCA OF FASHION AND TEXTILES

Any fashion product has a specific life cycle, which refers to the time period from the raw material stage till the disposal stage (Zamani, Sandin, & Peters, 2017). Garment LCA is the process of evaluating the environmental impacts of a garment, from fibre manufacturing, through spinning, weaving, finishing, garment manufacturing, distribution, usage to the disposal or recycling stage (Zamani et al., 2017). The total carbon emissions will change based on the product type, processes selected and the distance it travels to reach to the consumers. The conceptualization stage plays a vital role in the total environmental impacts of a particular garment. For example, using a 100% cotton garment will have different amount of carbon emissions compared to a polyester-cotton blend. Hence, designers play a major role in deciding the total environmental impacts of a product. The total carbon footprint (cradle to grave) of different fashion products based on fibre type has been given in Table 4.2.

Recently, LCA has been given increasing importance due to significant rise in the environmental awareness during garment manufacturing (Gupta, Kushwaha, Dave, & Mahanta, 2022). Sustainability of a product is directly related to its life cycle. Hence, it is essential to know the environmental impact during a garment's entire life cycle stages to become sustainable. In order to reduce the environmental impact during the life cycle of a garment, the concepts of recycling and upcycling

**TABLE 4.2**
**Carbon footprint (cradle to grave) of different fashion products based on fibre type**

| Product name | $CO_2$ equivalent (kg) |
|---|---|
| Polyester T-shirt (130 g) | 15.0 |
| Wool sweater (265 g) | 1.95 |
| Cotton T-shirt (110 g) | 3.0 |
| Stretch denim (cotton & elastane) (477 g) | 14.0 |
| Polyester dress (478 g) | 10.0 |
| Polyester nylon jacket (444) | 18.0 |
| Underwear and hosiery | 1.9 |
| Gloves | 1.1 |
| Swimwear | 1.8 |
| Shawls and ties | 1.8 |
| Cotton long shirt white (222 g) | 10.8 |
| Sweat jacket average | 13.4 |

*Source:* Roos et al. (2015).

are being adopted by many fashion brands. Instead of a single life, the produces are being designed to be used multiple times. Hence, the raw materials used in these products are never going to be discarded to the landfill and have an infinite life. However, there has not enough research to compare the total environmental impacts of the products designed for the single use and multiple use. This can be possible if the researchers investigate the LCA of each type of product and compare the results. The popularity of fast fashion is creating ecological imbalance as large quantities of unused clothes are dumped by the consumers into the landfill (Joung, 2014). The fast fashion brand as well as the consumers should find ways such as 3R (reduce, reuse, recycle), upcycling and donations to reduce the environmental impacts of fashion.

## 4.9 CASE STUDIES OF SUSTAINABLE APPAREL BRANDS

### 4.9.1 Everlane

Everlane is an American fashion brand that caters to both men and women, which primarily uses ecommerce for its business (Zhao, Lee, Li, & Sun, 2022). This brand sells a wide range of products, including apparel, footwear and bags. In addition to its headquarters in San Francisco (California), Everlane operates retail locations in New York City, Boston, Los Angeles, Austin and Palo Alto. It is the company's mission to sell clothing at a reasonable price. Currently, about 97% of Everlane's polyester and nylon apparel materials are made from certified recycled fibres. The fabrics for these apparels are created from recycled plastic bottles, discarded fishing lines, fabric scraps from pattern cutting and other materials that would otherwise end up in landfills (Leonas, 2017). In the footwear sector, approximately 45% of the polymer components are made from recyclable materials. Hence, Everlane follows the circular economy (CE) concept while selecting raw materials for its products.

The denim industry is well known for its negative impacts on environments during the manufacturing process (Nayak, George, Jajpura, Khandual, & Panwar, 2022). When Everlane revised denim manufacturing processes in 2017, they promised to do everything differently to reduce the environmental impacts. Some of the most environmentally conscious manufacturers in the world produce Everlane's denim. These companies surpass the environmental standards set for the industry, as they recycle their water, use renewable energy and repurpose waste to produce waste-free superior jeans and denim (Periyasamy & Periyasami, 2023). Generally, denim is typically made from traditional cotton grown with harmful chemicals and frequently contains elastane yarn, which pollutes the air and water. However, Everlane is converting all of its jeans and denims to natural (organic cotton) and introducing Roica-V550 yarn, the first chemical-free stretch fabric in the world. Due to these innovations and the use of microplastic-free dyes and recycled trims, Everlane will be one of the caring companies to produce non-polluting denims.

Everlane's raw materials have always prioritized sustainability starting from farms to factories, and the company has doubled its efforts to drastically reduce its environmental impact by innovative materials and processes. Further, the company prefers sustainable natural fibres such as organic cotton and linen in addition to utilizing recycled synthetic fibres due to their durability (Maduna & Patnaik, 2023). Producing functional alternatives to recycled materials, such as elastane-recycled flexible fabrics, places them in direct competition with their suppliers. Geodis, Everlane's distribution partner in Kutztown (Pennsylvania), plays a vital role in the company's operations to reduce its environmental impact, beginning with its waste stream management and recycling.

There are more than 8 billion tonnes of plastic on planet earth, which is a major problem in many countries (Rhodes, 2018). Out of the total global consumption of plastics, less than 10% are recycled on a global scale (Zheng & Suh, 2019). In 2018, Everlane made its first environmental commitment: "By 2021, all virgin plastic will be eliminated from their supply chain, including products, packaging, poly bags, offices, and stores". Currently, all of Everlane's packaging bags are made either from 100% recycled plastics or Forest Stewardship Council-certified paper. To meet the demand, it is necessary to develop a significant quantity of recycled plastic, which is currently a challenge for many fashion brands. It is vital to achieve landfill-free status in a facility that handles so many shipping materials and the majority of Everlane's physical products. The packaging materials are recycled and reused; waste-to-energy partnerships are utilized for the remainder that can't be recycled. In addition to using reusable shipping packages, they have partnered with Happy Returns to reduce the amount of recycled cardboard.

Everlane purchases renewable energy credits (RECs) for all the electricity used in their stores and offices. Using RECs, they invest in vast quantities of renewable energy derived from the air, the sun and hydropower for the national grid. This strategy reduces the nation's reliance on fossil fuels and electricity-related emissions. For in-store events, all Everlane locations use glassware and recyclable flatware instead of single-use plastic. All new employees receive a package containing reusable cutlery, a MiiR water bottle and Stasher bags on their first day. Fare Resources, a nutrition programme that favours organic and/or locally produced foods, is implemented in their workplaces, and they have significantly decreased the amount of single-use glass and food waste. Their employee-led sustainability group evaluates employee education programmes on topics such as local ballot initiatives and composting.

### 4.9.2 Hugo Boss

Hugo Boss, based in Metzingen (Baden-Württemberg, Germany), is a luxury fashion brand. Hugo Ferdinand Boss founded the company to manufacture versatile clothing in 1924. Hugo Boss has evaluated the environmental impact of its products using the Natural Capital Protocol of the Capitals Coalition (previously the Natural Capital Coalition) since 2016. The method assigns monetary values to significant environmental effects such as water consumption, ecosystem quality and $CO_2$ emissions, which helps the analysis more comparable for reducing environmental impacts (Zentes et al., 2017). Hugo Boss works closely with animal protection organizations including People for the Ethical Treatment of Animals. In addition, it will increase its communication with the animal protection organization FOUR PAWS in 2020. The Company's Animal Welfare Policy, which was published at the start of 2021 and applies to the entire Group, was formulated (AL-Issa, Thanasi-Boçe, & Ali, 2022).

Utilizing specialized product labels, Hugo Boss consistently identifies exceptionally sustainable items as RESPONSIBLE designs. According to the RESPONSIBLE Product Policy, RESPONSIBLE models use 60% or higher sustainable raw materials and exceed additional environmental and social standards for production, shipping and packaging. Hugo Boss prioritizes the use of water-saving methods in manufacturing and colouring techniques that use less water (Pandey, Pandit, Pandey, & Mishra, 2020). Hugo Boss has planned that 30% of the company's products will contribute to a specific water preservation footprint by 2025. This indicates that certain dyeing, tanning, washing and chemical treatment processes will reduce water usage by at least 30%.

Hugo Boss has collaborated with several ecofriendly material initiatives for using Less Water collection. For example, Recot2 cotton yarn, SeaCell fibre made from dried algae and recycled wood are some of the fibres that use less water. Similarly, Lenzing AG's Tencel cellulose fibres are produced with less water and using sustainable approaches (Arana, Franco, Joshi, & Sedhai, 2020). Other products manufactured from linen require less supplemental watering and fewer chemical fertilizers during fibre production (Debnath, 2019). These innovative approaches reduce the usage of water and save resources.

Hugo Boss has been recognized by the Carbon Disclosure Project (CDP) for the second time for its commitment to climate conservation in the supply chain. The CDP evaluates the extent to which companies engage their suppliers on climate change and recognizes the leaders as Supplier Engagement Leaders. This places Hugo Boss in the top 8% of the 13,000 businesses that responded to the CDP. Most of our emissions are caused by the upstream supply chain, which is the focus of our brand's initiatives. Hugo Boss has committed to carbon neutrality by 2045 for this reason.

A statement that our suppliers will be instructed and educated on energy efficiency and environmental protection. Hugo Boss created the resource efficiency module (REM) with ZDHC and other parties as part of their current commitment. The ZDHC Supplier Platform's REM tool assists textile and footwear manufacturers in reducing their environmental impact. Suppliers can monitor resource consumption, identify opportunities for improvement and devise strategies to conserve resources and boost output. The company began giving away the REM tool to 50 of our vendors last year. This year, there will be 100 more added. The brand chooses suppliers based on their share of our sourcing revenue and manufacturing $CO_2$ emissions.

## 4.10 CASE STUDIES ON THE USE OF TECHNOLOGY

The use of various technologies in supply chain management enables organizations to respond rapidly and effectively to fluctuations in demand, such as those caused by the Covid-19 pandemic. According to a study by the Boston Consulting Group, "Digitalization has become increasingly important for businesses to maintain agility and respond quickly to rapidly changing market conditions" (Nobile, Noris, Kalbaska, & Cantoni, 2021). One of the most significant advantages of technology in the supply chain is its capacity to enhance visibility and transparency. According to the World Economic Forum, "Digital solutions can enhance supply chain visibility, allowing businesses to respond more quickly to disruptions, increase collaboration, and reduce the risk of stock shortages".

The increased visibility by the use of technologies enables businesses to respond to changes in demand more quickly and to better manage their inventory levels, thereby reducing the risk of stock shortages and excess inventory. Dubey et al. (2020) showed that big data and artificial intelligence (AI) play a significant role in the sustainable fashion and textile industry. The adoption of AI can reduce the environmental impact of the fashion industry by 59% while simultaneously increasing profits by 60%. Companies such as Zalando and Lenzing have implemented AI-based solutions to optimize inventory management, reduce waste and minimize energy consumption (Hindolia & Arya, 2022).

The use of big data and AI allows for more efficient supply chain management and informed decision-making about the sustainability of products. In addition, AI can be used to forecast customer demand, leading to more personalized and sustainable product offerings (Lee, 2021). The resulting integration of AI and sustainability in the fashion and textile industry is expected to continue to grow in the coming years. Digital twin technology based on the AI, in which virtual replicas of physical products are used for sustainability analysis, is also being adopted in the fashion industry (Lim, Zheng, & Chen, 2020).

The IoT has been increasingly utilized in the fashion and textile industry to streamline supply chain processes and improve efficiency (Manavalan & Jayakrishna, 2019). IoT implementation in the fashion industry is done in combination of the use of RFID technology by fashion retailer Zara. RFID tags are attached to each garment, allowing them to be tracked throughout the supply chain, from production to store shelves. This technology has helped Zara improve inventory management and reduce the amount of unsold or excess inventory, leading to cost savings and increased profitability. IoT has also been implemented in the textile supply chain to improve efficiency and reduce waste. One example is the use of smart fabric sensors by textile manufacturer Alvanon. These sensors can track the real-time movement and wear of garments, allowing for more accurate sizing and fit. This technology has helped Alvanon reduce the amount of unsold or excess inventory, as well as improve the fit and comfort of garments for customers.

However, implementing IoT in the fashion industry also comes with challenges. For example, fashion trends change quickly and there is often a need for a quick turnaround in production. This can lead to pressure on supply chain partners to meet tight deadlines, which can be difficult if they are not equipped with the necessary technology. Additionally, there is a risk of data privacy and security breaches if RFID technology is not properly managed (Nayak et al., 2015). Further, there is a need for robust and reliable wireless networks to ensure that data is properly transmitted (Garrido Azevedo & Carvalho, 2012).

The quick marriage of textiles and electronics now makes it possible to produce conductive yarn as well as integrate sensors into clothing easily and widely. Smart textiles have the ability to usher in a new age in retail by integrating with smartphones to analyze physiological data such as temperature, respiration, heart rate, stress, activity or even hormone levels (Chae & Hinestroza, 2020). Smart clothing is created by incorporating smart wearables into clothes, and it offers a promising alternative to or extension of smartphones and other portable linked devices as the future interface between the physical and digital worlds.

Smart textiles consist of discrete devices fabricated from or incorporated into textiles fibres. Here we report the realization of a fully operational 46-inch smart textile lighting/display system. RGB fibrous LEDs are coupled with multifunctional fibre devices that are capable of wireless power transmission, touch sensing, photo detection, environmental/bio-signal monitoring and energy storage. A flexible and foldable high-performance hydrogen ($H_2$) sensor on all textile substrate is fabricated by inkjet-printing of reduced graphene oxide. The inkjet-printing process provides the advantages of compatibility with various substrates, non-contact patterning and cost-effectiveness.

Textile dual-band logo antenna is integrated with a reflector for application in IoT wearable devices. Numerical and experimental investigations of the antenna performance in two environments—(1) in free space and (2) on models of a human body, notebook and laptop—reveal that the proposed fully textile logo antenna achieved radiation efficiency between 25%–30% (in free space) and 20%–38% (on the human body).

Design is an important tool in the CE globe for converting waste to wealth. Industry 4.0 technology and digital advances have had a favourable influence on CE transition, by analyzing product lifetime data. To achieve a more efficient and successful CE solution, strong coordination among the many players of the textile circular ecosystem is essential. In this regard, IoT plays a critical role by disseminating information and data to all parties. Industrial 4.0 is expected to give several benefits to a firm or organization but putting it into effect will not be easy. The two industrial sectors have never been considered concurrently in Indonesia. Many of the problems are associated with internal and external communication. At all levels, there are frequent misconceptions between the industry 3.0 and industry 4.0 models.

In the Vietnamese fashion and textile industry, RFID technology may change economic dynamics and provide a more sustainable future (Nayak, George, Haq, & Pham, 2022). RFID offers speed, accuracy and on-time information availability. This paper sought to analyze the long-term benefits of using RFID technology by utilizing empirical research. The results indicate that the fashion sector should switch to RFID because of the advantages of sustainability and supply chain transparency. The study suggested that the fashion industry should integrate RFID technology due to the benefits of sustainability and transparency in the supply chain.

## 4.11 CONCLUSIONS

Fashion and textile manufacturing is a major activity in many developing countries such as Vietnam, Bangladesh, India and China. To keep the product cost low and survive in the global competitive market, the garment manufacturers around the world are involved with practices that are harmful to the planet and people. From a sustainability standpoint, various processes involved in garment manufacturing are inherently at odds as designers and product developers face several challenges to streamline a garment style. Globalization has led to the shifting of garment manufacturing to the developing countries, which neglect the three aspects of sustainability: environmental, social and economic. The developing countries are taking the advantages of lack of stricter legislations to violate the global standards set for planet and people while making profit. This chapter has discussed some of the environmental problems related to garment manufacturing, and the various approaches to become sustainable during the entire garment supply chain process.

The sustainable practices in garment manufacturing starts from conceptualization, where the designers play an influential role by selecting sustainable raw materials. Sustainable sourcing is the major task while suitable fashion is taken into consideration. Selection of ecofriendly and biodegradable fibres such as organic cotton, flax, hemp, PLA and Lyocell can reduce the environmental impacts of fashion manufacturing. Sustainable logistics and warehouse management also play a major role to reduce the environmental impacts of transporting, storing and distributing goods. LCA is a method used by many fashion brands to find the total environmental impact of a product. The process of recycling of fashion wastes, packaging materials and treatment of effluent that is discharged to the ecosystem is also vital in the sustainability journey. Using recycled raw materials mixed with new raw materials can help to preserve the resources, which will otherwise deplete towards the end of the century. The case studies from two fashion brands, Everlance and Hugo Boss, are explained how they are becoming sustainable. Finally, the findings of a survey conducted in Vietnamese fashion and textile industries are also discussed in this chapter.

## REFERENCES

Adamkiewicz, J., Kochanska, E., Adamkiewicz, I., & Łukasik, R. M. (2022). Greenwashing and sustainable fashion industry. *Current Opinion in Green and Sustainable Chemistry, 38*, 1–8. DOI: https://doi.org/10.1016/j.cogsc.2022.100710

Ahmed, A. F., Islam, M. Z., Mahmud, M. S., Sarker, M. E., & Islam, M. R. (2022). Hemp as a potential raw material toward a sustainable world: A review. *Heliyon, 8*(1) 1–15. https://doi.org/10.1016/j.heliyon.2022.e08753

AL-Issa, N., Thanasi-Boçe, M., & Ali, O. (2022). Boosting luxury sustainability through blockchain technology. In *Blockchain Technologies in the Textile and Fashion Industry* (pp. 17–46): Springer.

Arana, C., Franco, I. B., Joshi, A., & Sedhai, J. (2020). SDG 15 life on land: A review of sustainable fashion design processes: Upcycling waste organic yarns. In Isabel B. Franco, Tathagata Chatterji, Ellen Derbyshire, and James Tracey (Eds.). *Actioning the Global Goals for Local Impact: Towards Sustainability Science, Policy, Education and Practice* (pp. 247–264), Springer.

Barra, R., & González, P. (2018). Sustainable chemistry challenges from a developing country perspective: Education, plastic pollution, and beyond. *Current Opinion in Green and Sustainable Chemistry, 9*, 40–44.

Barua, U., & Ansary, M. A. (2017). Workplace safety in Bangladesh ready-made garment sector: 3 Years after the Rana Plaza collapse. *International Journal of Occupational Safety and Ergonomics, 23*(4), 578–583.

Bektaş, T., & Laporte, G. (2011). The pollution-routing problem. *Transportation Research Part B: Methodological, 45*(8), 1232–1250.

Bhattacharjee, S. S. (2019). Fast fashion, production targets, and gender-based violence in Asian garment supply chains. In *Labor, Global Supply Chains, and the Garment Industry in South Asia* (pp. 207–227): Routledge.

Birtwistle, G., & Moore, C. M. (2007). Fashion clothing–where does it all end up? *International Journal of Retail & Distribution Management, 35*(3), 210–216.

Birtwistle, G., & Shearer, L. (2001). Consumer perception of five UK fashion retailers. *Journal of Fashion Marketing and Management: An International Journal, 5*(1), 9–18.

Brun, A., Caniato, F., Caridi, M., Castelli, C., Miragliotta, G., Ronchi, S.,..., & Spina, G. (2008). Logistics and supply chain management in luxury fashion retail: Empirical investigation of Italian firms. *International Journal of Production Economics, 114*(2), 554–570.

Cassidy, T. D., & Han, S. L.-C. (2017). Upcycling fashion for mass production. In *Sustainability in Fashion and Textiles* (pp. 148–163): Routledge.

Cervellon, M. C., & Wernerfelt, A. S. (2012). Knowledge sharing among green fashion communities online: Lessons for the sustainable supply chain. *Journal of Fashion Marketing and Management, 16*(2), 176–192.

Chae, Y., & Hinestroza, J. (2020). Building circular economy for smart textiles, smart clothing, and future wearables. *Materials Circular Economy, 2*(1), 1–4.

Chen, C., & Lapolla, K. (2021). The exploration of the modular system in textile and apparel design. *Clothing and Textiles Research Journal, 39*(1), 39–54.

Chen, J.-Y., & Slotnick, S. A. (2015). Supply chain disclosure and ethical sourcing. *International Journal of Production Economics, 161*, 17–30.

Choi, T.-M. (2013). Local sourcing and fashion quick response system: The impacts of carbon footprint tax. *Transportation Research Part E: Logistics and Transportation Review, 55*, 43–54.

Choi, T.-M., & Cheng, T. E. (2015). *Sustainable Fashion Supply Chain Management*: Springer.

Choudhury, A. K. R. (2023). Sustainable protein fibres. In *Sustainable Fibres for Fashion and Textile Manufacturing* (pp. 181–226): Elsevier.

Choudhury, A. R. (2014). Environmental impacts of the textile industry and its assessment through life cycle assessment. In *Roadmap to Sustainable Textiles and Clothing* (pp. 1–39): Springer.

De Brito, M. P., Carbone, V., & Blanquart, C. M. (2008). Towards a sustainable fashion retail supply chain in Europe: Organisation and performance. *International Journal of Production Economics, 114*(2), 534–553.

Debnath, S. (2019). Flax fibre extraction to fashion products leading towards sustainable goals. In *The UN Sustainable Development Goals for the Textile and Fashion Industry* (pp. 47–57): Springer.

Di Benedetto, C. A. (2017). Corporate social responsibility as an emerging business model in fashion marketing. *Journal of Global Fashion Marketing, 8*(4), 251–265.

Dissanayake, D., & Weerasinghe, D. (2021). Towards circular economy in fashion: Review of strategies, barriers and enablers. *Circular Economy and Sustainability, 2*(1), 1–21.

Dubey, R., Gunasekaran, A., Childe, S. J., Bryde, D. J., Giannakis, M., Foropon, C., & Hazen, B. T. (2020). Big data analytics and artificial intelligence pathway to operational performance under the effects of entrepreneurial orientation and environmental dynamism: A study of manufacturing organisations. *International Journal of Production Economics, 226*, 1–12.

Employment, E. C. D.-G. f. (2001). *Promoting a European Framework for Corporate Social Responsibility: Green Paper:* Office for Official Publications of the European Communities.

Fletcher, K. (2012). Durability, fashion, sustainability: The processes and practices of use. *Fashion Practice, 4*(2), 221–238.

Fletcher, K. (2013). *Sustainable Fashion and Textiles: Design Journeys*: Routledge.

Fraser, E., & van der Ven, H. (2022). Increasing transparency in global supply chains: The case of the fast fashion industry. *Sustainability, 14*(18), 1–24.

Garcia-Torres, S., Rey-Garcia, M., Sáenz, J., & Seuring, S. (2022). Traceability and transparency for sustainable fashion-apparel supply chains. *Journal of Fashion Marketing and Management: An International Journal, 26*(2), 344–364.

Gardetti, M. A., & Muthu, S. S. (2019). *Organic cotton, Is it a Sustainable Solution. Springer.*

Garrido Azevedo, S., & Carvalho, H. (2012). Contribution of RFID technology to better management of fashion supply chains. *International Journal of Retail & Distribution Management, 40*(2), 128–156.

Garza-Reyes, J. A., Villarreal, B., Kumar, V., & Molina Ruiz, P. (2016). Lean and green in the transport and logistics sector—A case study of simultaneous deployment. *Production Planning & Control, 27*(15), 1221–1232.

George, M., Ulhaq, I., & Nayak, R. (2020). Sustainable approaches in warehousing and inventory management in the fashion industry. In *Supply Chain Management and Logistics in the Global Fashion Sector* (pp. 138–158): Routledge.

Goyal, A., & Nayak, R. (2020). Sustainability in yarn manufacturing. In *Sustainable Technologies for Fashion and Textiles* (pp. 33–55): Elsevier.

Gregoratti, C., & Miller, D. (2011). International framework agreements for workers' rights? Insights from River Rich Cambodia. *Global Labour Journal, 2*(2), 1–22.

Gupta, R., Kushwaha, A., Dave, D., & Mahanta, N. R. (2022). Waste management in fashion and textile industry: Recent advances and trends, life-cycle assessment, and circular economy. In Sunpreet Singh, Lalit Goswami, and Chaudhery Mustansar Hussain (Eds.), *Emerging Trends to Approaching Zero Waste* (pp. 215–242), Elsevier.

Gupta, S., & Gentry, J. W. (2016). The behavioral responses to perceived scarcity—The case of fast fashion. *The International Review of Retail, Distribution and Consumer Research, 26*(3), 260–271.

Ha-Brookshire, J. (2017). *Global Sourcing in the Textile and Apparel Industry*: Bloomsbury Publishing USA.

Hacking, T., & Guthrie, P. (2008). A framework for clarifying the meaning of Triple Bottom-Line, Integrated, and Sustainability Assessment. *Environmental Impact Assessment Review, 28*(2), 73–89.

Hasanbeigi, A., & Price, L. (2015). A technical review of emerging technologies for energy and water efficiency and pollution reduction in the textile industry. *Journal of Cleaner Production, 95*, 30–44.

Haug, A., & Busch, J. (2016). Towards an ethical fashion framework. *Fashion Theory, 20*(3), 317–339.

Hayes, S., & Jones, N. (2006). Fast fashion: A financial snapshot. *Journal of Fashion Marketing and Management: An International Journal, 10*(3), 282–300.

Hindolia, A., & Arya, J. (2022). Hermeneutic approach in understanding the role of artificial intelligence: Fashion & apparel industry/luxury brand retailing. *Academy of Marketing Studies Journal, 26*(2), 1–14.

Hustvedt, G., Meier, E., & Waliczek, T. (2016). The feasibility of large-scale composting of waste wool. In Subramanian Senthilkannan Muthu and Miguel Angel Gardetti (Eds.), *Green Fashion: Volume 1* (pp. 95–107), Springer.

Islam, M. M., Perry, P., & Gill, S. (2021). Mapping environmentally sustainable practices in textiles, apparel and fashion industries: A systematic literature review. *Journal of Fashion Marketing and Management: An International Journal, 25*(2), 331–353.

Jang, J., Ko, E., Chun, E., & Lee, E. (2012). A study of a social content model for sustainable development in the fast fashion industry. *Journal of Global Fashion Marketing, 3*(2), 61–70.

Jiang, X., Bai, Y., Chen, X., & Liu, W. (2020). A review on raw materials, commercial production and properties of lyocell fiber. *Journal of Bioresources and Bioproducts, 5*(1), 16–25.

Jones, S., & Wright, C. (2018). Fashion or future: Does creating shared value pay? *Accounting & Finance, 58*(4), 1111–1139.

Joung, H.-M. (2014). Fast-fashion consumers' post-purchase behaviours. *International Journal of Retail & Distribution Management, 42*(8), 688–697.

Karthik, T., & Rathinamoorthy, R. (2017). Sustainable silk production. In *Sustainable Fibres and Textiles* (pp. 135–170): Elsevier.

Keeble, J. J., Topiol, S., & Berkeley, S. (2003). Using indicators to measure sustainability performance at a corporate and project level. *Journal of Business Ethics, 44*(2), 149–158.

Khurana, K., & Ricchetti, M. (2016). Two decades of sustainable supply chain management in the fashion business, an appraisal. *Journal of Fashion Marketing and Management, 20*(1), 89–104.

Lee, Y. K. (2021). Transformation of the innovative and sustainable supply chain with upcoming real-time fashion systems. *Sustainability, 13*(3), 1–14.

Leonas, K. K. (2017). The use of recycled fibers in fashion and home products. In *Textiles and Clothing Sustainability: Recycled and Upcycled Textiles and Fashion* (pp. 55–77), Springer.

Li, Y., Zhao, X., Shi, D., & Li, X. (2014). Governance of sustainable supply chains in the fast fashion industry. *European Management Journal, 32*(5), 823–836.

Lim, K. Y. H., Zheng, P., & Chen, C.-H. (2020). A state-of-the-art survey of Digital Twin: Techniques, engineering product lifecycle management and business innovation perspectives. *Journal of Intelligent Manufacturing, 31*, 1313–1337.

Maduna, L., & Patnaik, A. (2023). Introduction to sustainable fibres for fashion and textiles. In R. Nayak (Ed.), *Sustainable Fibres for Fashion and Textile Manufacturing* (pp. 27–48): Elsevier.

Manavalan, E., & Jayakrishna, K. (2019). A review of Internet of Things (IoT) embedded sustainable supply chain for industry 4.0 requirements. *Computers & Industrial Engineering, 127*, 925–953.

Marcuccio, M., & Steccolini, I. (2005). Social and environmental reporting in local authorities: A new Italian fashion? *Public Management Review, 7*(2), 155–176.

McDonald, S. D., Hung, N. M., & Akbari, M. (2020). Transportation and logistics for a sustainable fashion sector. In *Supply Chain Management and Logistics in the Global Fashion Sector* (pp. 159–182): Routledge.

McNeill, L., & Moore, R. (2015). Sustainable fashion consumption and the fast fashion conundrum: Fashionable consumers and attitudes to sustainability in clothing choice. *International Journal of Consumer Studies, 39*(3), 212–222.

Mishra, S., Jain, S., & Malhotra, G. (2021). The anatomy of circular economy transition in the fashion industry. *Social Responsibility Journal, 17*(4), 524–542.

Moon, K. K.-L., Youn, C., Chang, J. M., & Yeung, A. W.-h. (2013). Product design scenarios for energy saving: A case study of fashion apparel. *International Journal of Production Economics, 146*(2), 392–401.

Moorhouse, D. (2020). Making fashion sustainable: Waste and collective responsibility. *One Earth, 3*(1), 17–19.

Moorhouse, D., & Moorhouse, D. (2017). Sustainable design: Circular economy in fashion and textiles. *The Design Journal, 20*(sup1), S1948–S1959. https://doi.org/10.1080/14606925.2017.1352713

Nayak, R. (2020). *Supply Chain Management and Logistics in the Global Fashion Sector: The Sustainability Challenge*: Routledge.

Nayak, R. (2022). *Sustainable Fibres for Fashion and Textile Manufacturing*: Elsevier

Nayak, R., Akbari, M., & Far, S. M. (2019). Recent sustainable trends in Vietnam's fashion supply chain. *Journal of Cleaner Production, 225*, 291–303.

Nayak, R., George, M., Haq, I. U., & Pham, H. C. (2022). Sustainability benefits of RFID technology in Vietnamese fashion supply chain. *Cleaner Logistics and Supply Chain, 5*, 1–12.

Nayak, R., George, M., Jajpura, L., Khandual, A., & Panwar, T. (2022). Laser and ozone applications for circularity journey in denim manufacturing—A developing country perspective. *Current Opinion in Green and Sustainable Chemistry, 38*, 1–8.

Nayak, R., Houshyar, S., Patnaik, A., Nguyen, L. T., Shanks, R. A., Padhye, R., & Fegusson, M. (2020). Sustainable reuse of fashion waste as flame-retardant mattress filing with ecofriendly chemicals. *Journal of Cleaner Production, 251*, 1–9.

Nayak, R., Jajpura, L., & Khandual, A. (2023). Traditional fibres for fashion and textiles: Associated problems and future sustainable fibres. In *Sustainable Fibres for Fashion and Textile Manufacturing* (pp. 3–25): Elsevier.

Nayak, R., Nguyen, L., Patnaik, A., & Khandual, A. (2021). Fashion waste management problem and sustainability: A developing country perspective. In *Waste Management in the Fashion and Textile Industries* (pp. 3–29): Elsevier.

Nayak, R., Nguyen, L. T., Panwar, T., George, M., & Ulhaq, I. (2020). Sustainable supply chain management: Challenges from a fashion perspective. In *Supply Chain Management and Logistics in the Global Fashion Sector: The Sustainability Challenge* (pp. 1-30), Routledge.

Nayak, R., Nguyen, L. V. T., Panwar, T., & Jajpura, L. (2020). Sustainable technologies and processes adapted by fashion brands. In R. Nayak (Ed.), *Sustainable Technologies for Fashion and Textiles* (pp. 233–248): Elsevier.

Nayak, R., & Padhye, R. (2015a). The care of apparel products. In *Textiles and Fashion* (pp. 799–822): Elsevier.

Nayak, R., & Padhye, R. (2015b). *Garment Manufacturing Technology*: Elsevier.

Nayak, R., & Padhye, R. (2017). *Automation in Garment Manufacturing:* Elsevier.

Nayak, R., & Padhye, R. (2018). Introduction to automation in garment manufacturing. In *Automation in Garment Manufacturing* (pp. 1–27): Elsevier.

Nayak, R., Panwar, T., & Nguyen, L. V. T. (2020). Sustainability in fashion and textiles: A survey from developing country. In *Sustainable Technologies for Fashion and Textiles* (pp. 3–30): Elsevier.

Nayak, R., & Patnaik, A. (2021). *Waste Management in the Fashion and Textile Industries*: Woodhead Publishing.

Nayak, R., & Ratnapandian, S. (2018). *Care and Maintenance of Textile Products Including Apparel and Protective Clothing*: CRC Press.

Nayak, R., Singh, A., Padhye, R., & Wang, L. (2015). RFID in textile and clothing manufacturing: Technology and challenges. *Fashion and Textiles, 2*(1), 1-9.

Nayak, R., Van Thang, L. N., Nguyen, T., Gaimster, J., Morris, R., & George, M. (2021). Sustainable developments and corporate social responsibility in Vietnamese fashion enterprises. *Journal of Fashion Marketing and Management: An International Journal, 26*(2), 307–327.

Nguyen, H. T., Le, D. M. D., Ho, T. T. M., & Nguyen, P. M. (2021). Enhancing sustainability in the contemporary model of CSR: A case of fast fashion industry in developing countries. *Social Responsibility Journal, 17*(4), 578–591.

Niinimäki, K. (2017). Sustainable consumer satisfaction in the context of clothing. In *Product-Service System Design for Sustainability* (pp. 218–237): Routledge.

Niinimäki, K., Peters, G., Dahlbo, H., Perry, P., Rissanen, T., & Gwilt, A. (2020). The environmental price of fast fashion. *Nature Reviews Earth & Environment, 1*(4), 189–200.

Nobile, T. H., Noris, A., Kalbaska, N., & Cantoni, L. (2021). A review of digital fashion research: Before and beyond communication and marketing. *International Journal of Fashion Design, Technology and Education, 14*(3), 293–301.

Pandey, R., Pandit, P., Pandey, S., & Mishra, S. (2020). Solutions for sustainable fashion and textile industry. In P. Pandit, S. Ahmed, K. Singha, & S. Shrivastava (Eds.), *Recycling from Waste in Fashion and Textiles: A Sustainable and Circular Economic Approach* (pp. 33–72). Wiley.

Pensupa, N. (2020). Recycling of end-of-life clothes. In *Sustainable Technologies for Fashion and Textiles* (pp. 251–309): Elsevier.

Periyasamy, A. P., & Periyasami, S. (2023). Critical review on sustainability in denim: A step toward sustainable production and consumption of denim. *ACS Omega, 8*(5), 4472–4490.

Poole, A. J., Church, J. S., & Huson, M. G. (2009). Environmentally sustainable fibers from regenerated protein. *Biomacromolecules, 10*(1), 1–8.

Prakash, C., & Kumar, S. K. S. (2023). Regenerated synthetic fibres: Bamboo and lyocell. In *Sustainable Fibres for Fashion and Textile Manufacturing* (pp. 229–246): Elsevier.

Rahman, M. Z. (2021). Mechanical and damping performances of flax fibre composites—A review. *Composites Part C: Open Access, 4*, 1–18.

Rhodes, C. J. (2018). Plastic pollution and potential solutions. *Science progress, 101*(3), 207–260.

Roos, S., Sandin, G., Zamani, B., & Peters, G. (2015). Environmental assessment of Swedish fashion consumption. In *Five Garments–Sustainable Futures*: Mistra Future Fashion 1–142.

Shen, B. (2014). Sustainable fashion supply chain: Lessons from H&M. *Sustainability, 6*(9), 6236–6249.

Shen, B., Zheng, J.-H., Chow, P.-S., & Chow, K.-Y. (2014). Perception of fashion sustainability in online community. *The Journal of the Textile Institute, 105*(9), 971–979.

Sinkovics, N., Hoque, S. F., & Sinkovics, R. R. (2016). Rana Plaza collapse aftermath: Are CSR compliance and auditing pressures effective? *Accounting, Auditing & Accountability Journal, 29*(4), 617–649.

Sun, J. J., Bellezza, S., & Paharia, N. (2021). Buy less, buy luxury: Understanding and overcoming product durability neglect for sustainable consumption. *Journal of Marketing, 85*(3), 28–43.

Tarique, J., Sapuan, S., Khalina, A., Sherwani, S., Yusuf, J., & Ilyas, R. (2021). Recent developments in sustainable arrowroot (*Maranta arundinacea* Linn) starch biopolymers, fibres, biopolymer composites and their potential industrial applications: A review. *Journal of Materials Research and Technology, 13*, 1191–1219.

Tausif, M., Ahmad, F., Hussain, U., Basit, A., & Hussain, T. (2015). A comparative study of mechanical and comfort properties of bamboo viscose as an eco-friendly alternative to conventional cotton fibre in polyester blended knitted fabrics. *Journal of Cleaner Production, 89*, 110–115.

Thangavelu, K., & Subramani, K. B. (2016). Sustainable biopolymer fibers—Production, properties and applications. In Subramanian Senthilkannan Muthu and Miguel Angel Gardetti (Eds.), *Sustainable Fibres for Fashion Industry: Volume 1* (pp. 109–140), Springer.

Thee, K. W. (2009). The development of labour-intensive garment manufacturing in Indonesia. *Journal of Contemporary Asia, 39*(4), 562–578.

Ting, T. Z., & Stagner, J. (2021). Fast fashion-wearing out the planet. *International Journal of Environmental Studies*, *80*(4), 1–11.

Vachon, S., & Klassen, R. D. (2008). Environmental management and manufacturing performance: The role of collaboration in the supply chain. *International Journal of Production Economics, 111*(2), 299–315.

Zamani, B., Sandin, G., & Peters, G. M. (2017). Life cycle assessment of clothing libraries: Can collaborative consumption reduce the environmental impact of fast fashion? *Journal of Cleaner Production, 162*, 1368–1375.

Zentes, J., Morschett, D., Schramm-Klein, H., Zentes, J., Morschett, D., & Schramm-Klein, H. (2017). Vertical players—Manufacturers and verticals. In Joachim Zentes, Dirk Morschett, and Hanna Schramm-Klein (Eds.), *Strategic Retail Management: Text and International Cases* (pp. 115–137), Springer.

Zhang, M., Kong, X. X., & Ramu, S. C. (2016). The transformation of the clothing industry in China. *Asia Pacific Business Review, 22*(1), 86–109.

Zhao, L., Lee, S. H., Li, M., & Sun, P. (2022). The use of social media to promote sustainable fashion and benefit communications: A data-mining approach. *Sustainability, 14*(3), 1178–1192.

Zheng, J., & Suh, S. (2019). Strategies to reduce the global carbon footprint of plastics. *Nature Climate Change, 9*(5), 374–378.

Zimniewska, M. (2022). Hemp fibre properties and processing target textile: A review. *Materials, 15*(5), 1–29. https://doi.org/10.3390/ma15051901

# 5 Textile Surface Ornamentation Techniques

*Ruby Kashyap Sood and Anu Sharma*

## 5.1 INTRODUCTION TO PRINTING TECHNIQUES

Printing can be trailed and associated to painting on the cave's walls. It is defined as localized dyeing or patterned dyeing on the fabric. Textile printing can be traced back as an ancient art from 4th and 5th century B.C. The first technique used for printing of fabric was block printing from China around 4,500 years ago. In the 17th century, Europe also started printing the fabric and by the 18th century, rollers were invented for printing of the fabric. In the 18th century, "Toile de Jouy" from France was introduced by Christophe Philippe Oberkamph, a specialized print with one colour on a light background fabric inspired from Rococo art and showcasing illustrious sketches inspired from natural elements and documentation of day-to-day life of that period. In 20th century, screen printing with rotary screens took over the market for fast and mass productions. Currently in the 21st century, digital printing, which is a quick and exclusive printing technique, is in demand for printing of the fabric.

The colouring material used in ancient times for both dyeing and printing was mostly obtained from the natural plant and resources. However later to obtain bright and fast colours, synthetic dyes were invented and came into use. There are four fundamental styles of printing like direct printing, resist printing, discharge printing and mordant printing.

### 5.1.1 Direct Printing

Direct printing is the application of dye directly on a light-coloured or dyed fabric and is the oldest form of printing style. Block printing, roller printing and screen printing are examples of techniques of direct printing. Direct printing that is obtained using pigment colours is called as pigment printing and the process is called as dry printing. To fix the pigment paste on the fabric, binder and fixers are mixed with the colour to achieve the result.

### 5.1.2 Resist Printing

Resist printing as the name suggests is a style of printing wherein a certain part of the fabric is resisted by use of the molten wax or by tying of the fabric. The first style of resist printing is called as Batik and the latter one is defined as the tie and dye

 DOI: 10.1201/9781003432739-5

technique. This printing style is done by dyeing the fabric in the dye bath after the resist material/technique has been applied. The use of direct dyes to acid dyes and reactive dyes in the dye bath depends upon the material of the fabric and kind of the colour required.

### 5.1.3 Discharge Printing

This is a printing style done on dark backgrounds, where the colours are discharged in the design area and the light colour is left on the fabric. The sodium sulphoxylate formaldehye is one of the chemicals used to remove or discharge the colour from the dyed fabric. Sometimes other colours are printed over the light area with contrasting colours to form new designs.

### 5.1.4 Mordant Printing

Mordant printing is a style of printing where a mordant is required to make the dyestuff attach to the fabric. All the natural dye printing processes require metallic mordants like alum, potassium dichromate, stannous chloride, copper sulphate and stannic chloride to form different colours. These can be added with the dye or applied before or after the dyeing or printing process to get the desired colours. Every natural dye reacts separately with all different kinds of mordants; hence, many experiments and explorations are conducted for achieving different colours and its variations.

## 5.2 PRINTING AND DYEING TECHNIQUES

Printing is the most common and highly implemented technique of value addition on textile fabrics. The cost of the printing technique depends upon the kind of printing and the machine used in the printing of the fabric. The prints can be created and printed on the fabrics using various machines like roller, hand block and screen printing.

### 5.2.1 Roller Printing

Roller printing is mostly used in the textile mills where in an hour 5,500 m (approximately) fabric is printed. It is the costliest process of printing as for every colour there is a roller engraved with the design. The roller is made up of copper and is engraved using laser machines to create designs on the rollers. The size of each roller is 25 inches to 32 inches in circumference and length of rollers is equal to the fabric width to be printed. In India 25 inches is preferred for creating designs as repeat; however, in international market 32-inch repeat designs are preferred. Around five to six colours are preferred for printing using this technique. The rollers are engraved nowadays using laser beams that make the process very fast and effective for preparing the rollers. As this process requires costly copper cylinders for printing, this technique is used only in mills or places where the production of printed fabric is in very high quantity.

Technique of Roller Printing: Two cylinders are used between which the textile material is pressed for printing. One of these cylinders is the engraved design roller, while the other one is to apply pressure on the fabric to get pressed against the engraved roller. Two blades are used in this process. One of them is the doctors blade, which removes colour from the unengraved area of the roller and works exactly like squeegee in the screen-printing process. The other one is the lint blade that removes lint deposited on the roller from the fabric during the printing process. The fabric to be printed is backed with a back grey cloth that absorbs the excessive dye along with a woollen blanket that provides padding to the fabric when it is moving through the rollers while printing (Figure 5.1).

The rollers can stay for a long time if taken care properly. At times the colours are pigment colours while sometimes dyes are used for printing the fabric; based on the kind of colours used for the printing, the process is called as wet printing for dyes and dry printing for the pigment printing.

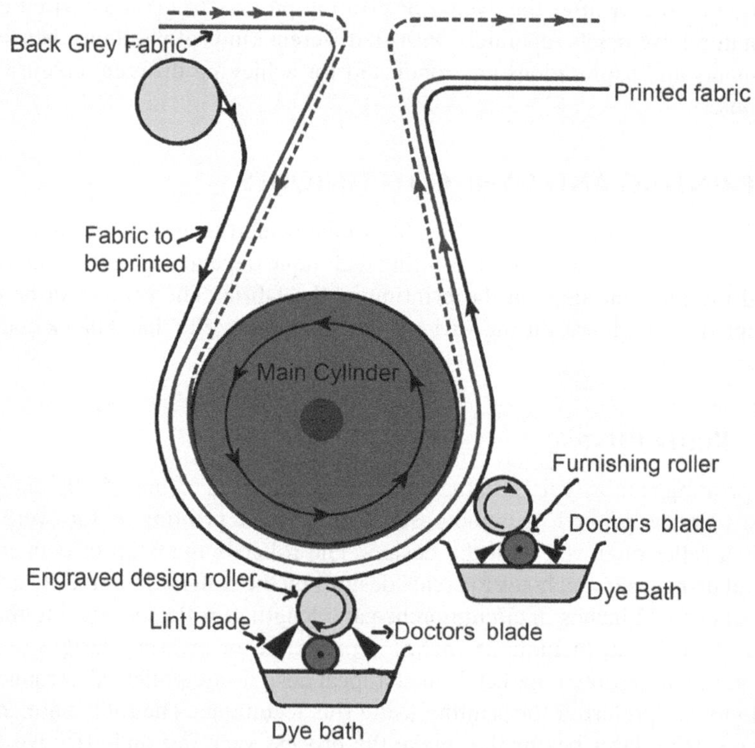


FIGURE 5.1 Roller printing process.

### 5.2.2 Block Printing

Block printing is an age-old technique done manually, where the carved wooden and metal blocks are used to transfer print on the textile material. This printing technique is a slow process and requires lot of manual labour for creating a complete design on any material. The minimum block size is from 2 inches and goes up to 14 inches as per the design required. The preferred size of the blocks is 6 inches by 6 inches, which is square in shape. This size is manageable by a block printer to apply equal pressure on the whole block, which gives a smooth print. However, bigger blocks from this size sometimes lead to patchy prints due to uneven application of the hand pressure by a printer. The blocks are 2–3 inches in height and have a raised platform to handle them. The design is carved on a wooden block and the rest of the space is chiselled out, so that the protruding design gets the colour from the colouring tray and prints it onto the fabric. The blocks are carved in three forms: *gad, rekh* and *datta*. The *rekh* block creates the outline of the designs, *gad* blocks fill the designs and *datta* complements the designs created by the *gad* and *rekh* blocks.

For colour, pigment dyes are used with gum arabica. The dye to be applied is spread in a tray having foam and squeegee (a thin wooden plank with a handle having a rubber blade at the end) and every colour has a separate tray like this with a squeegee. The block to be printed with one colour is first printed on a fabric pinned on a padded table and then the second colour is applied with the second block; this process is repeated till all the colours are printed and the design is completed. Mostly three to five colours are preferred in a block print design. The pigment prints are fixed using dry heat or by drying the fabrics under the sun.

In India, hand block printing is famous from Rajasthan's Bagru and Sanganer districts, where in early years natural dyes were used to create block prints on the fabrics.

### 5.2.3 Screen Printing

Screen printing is a faster process than the hand block printing technique and is used as the process where large quantity of printing is required. This technique is a good solution for mass production of printing on textiles.

The screen is a wooden or metal framed device with nylon or polyester mesh having closed and open pores according to the design. The first step for preparing a screen is to create an acetate layer with the design. Then, screen is prepared with coating of light reactive emulsion that gets hardened on exposure to the light. The acetate layer is then put on the screen mesh and is exposed to bright light, wherever the light passes through the screen gets hardened, and mesh gets blocked. The screen is then washed off and is ready to print a fabric. This process is used for every single screen for every single colour in the design. The mesh size is selected based on the kind of dye being used and the texture of the fabric to be printed (Figure 5.2).

The pigment dyes are mixed with binder and thickener to form the printing colour paste that has a thinner consistency for garments as compared to the home furnishing products. For garments, high polymer binder is used with the pigment dye. Disperse dyes, reactive dyes and discharge dyes are also used for different types of fabrics; however, the use of pigment dyes is the most common for screen printing. Here the

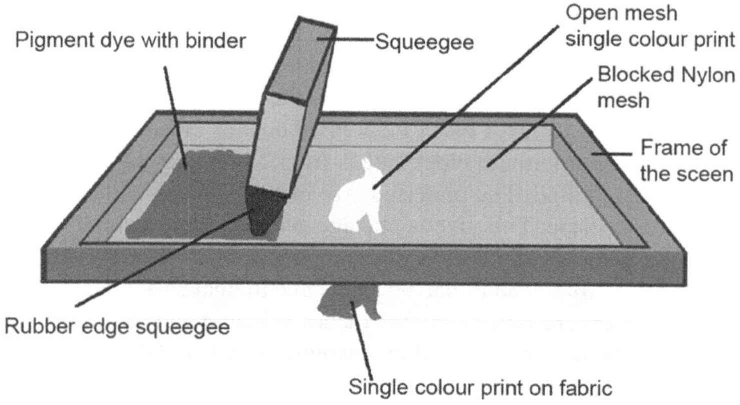


**FIGURE 5.2** Screen-printing process.

**FIGURE 5.3** Single colour print and its exposed screen artwork. Artwork by Amit Sharma.

screen repeat size can be set within 4–72 inches in width as per the end product requirement.

The colour dyes are applied on these frames and are pushed through the open pores of the mesh allowing the dye to pass through the pores leaving behind the imprint of the design on the fabric. The squeegee (a thin wooden plank with a handle having a rubber blade at the end) is used to press the colouring dye through the screen, and its size is equal to the width of the screen size.

For every single colour in the design there is one single screen that means if there are four colours in a design there would be four screens excluding the background colour of the fabric. There are different methods of screen printing depending upon the quantity of the fabric to be printed. Hand screen printing, flatbed printing and rotary screen printing are the methods by which screen printing is achieved on any textile material (Figure 5.3).

#### 5.2.3.1 Hand Screen Printing

It's a manual method wherein the printing is done by hands. In the first step, the fabric is laid on a padded table with a screen and the printing paste of the first colour to be printed. Two men stand on either side of the printing table and pass the squeegee from one end of the fabric width to the other end, which presses the printing dye through the open mesh of the screen resulting in printing of the fabric. Similar process is repeated for all the colours till the whole design is printed. For every single screen/single colour, two humans are involved for one complete length of the fabric laid on the table. Once the first colour is printed, the second colour screen is laid on the fabric on the exact place where the first colour screen was placed and the second colour is pressed through the screen on to the fabric. This process is repeated in a continuous manner for all the colours till all the colours are printed. By this process about 80–90 m of fabric is printed in one hour. The printing of garments or parts of garments can also be done using hand screen-printing technique; hence, this is a versatile printing technique that can be used for a smaller quantity as well as for a specialized or engineered print as well.

#### 5.2.3.2 Flatbed Printing

The flatbed printing is a similar process as the hand screen printing wherein the colours are placed in a similar format, however here, the placement of screens on the fabric takes place with the help of a conveyor belt instead of using any manual labour. The fabric moves on a rubberized conveyer belt and reaches every screen one by one as per the design and gets printed through the squeegee that presses the printing colour through the open mesh of every screen.

Any kind of screen printing can be done in various sizes from 4 to 72 inches in width depending on the size of the repeat of the design. In India six to eight colours are mostly used for screen printing on the textiles while in international market up to 16 colours of screen printing is also done. This technique is majorly used for printing long width of fabrics with overall prints and is a cheaper version of printing in mass production.

Screen printing is mostly done on the white-coloured backgrounds of fabrics; however, if the background colour is any other colour other than the white for printing motifs on top of it, it's called as blotch printing. There is one more style of screen printing called as khadi printing, wherein white colour pigment print is used for printing motifs or designs on any dark-coloured fabrics.

Nowadays foil (silver, gold or copper colour) printing is also being achieved using screen-printing technique, wherein the glue is applied using a single-coloured screen of the design and then foil is pasted on the glue by applying heat. In place of foil if fibres are placed on the glue, it is called as flock printing.

#### 5.2.3.3 Rotary Screen Printing

The rotary screen printing is used in huge mass productions in mills, as it's the fastest process of printing textile material than the above two mentioned processes. This process is very similar to roller printing, except in place of engraved rollers here rollers with films having design repeats are used. The size of these rollers varies between

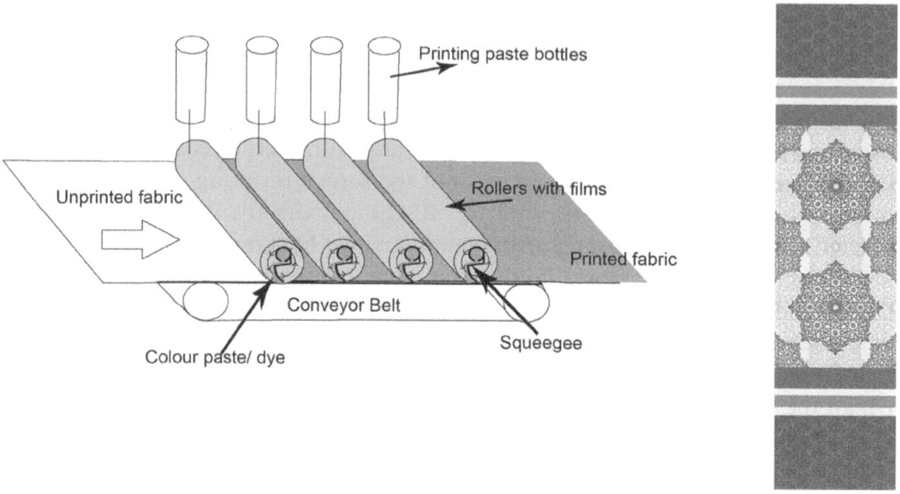


**FIGURE 5.4** Rotary screen-printing process and its artwork in repeat in the size of the roller of 25 inches across the full width of the fabric. Diagram by Amit Sharma.

25 inches and 32 inches in circumference having the film with exposed design to be printed on the fabric. As in screen print, here also for every colour there is one roller. The fabric moves on a conveyer belt and the fixed rotating rollers prints the fabric. This process is preferred for 12–16-colour printing in a design (Figure 5.4).

Once the fabrics are printed by any method, for final treatment they are dried at high temperature with dry heat and the process is called as curing.

### 5.2.4 Sublimation and Transfer Printing

Transfer print as the name suggests, it is the transfer of print from one medium to the other i.e. from paper to the fabric. This process takes place by application of heat. The disperse dyes are used to create artwork on the paper and with the application of heat, the artwork gets transferred as a layer on the fabric. Special quality of papers is used for transferring the prints on the fabric.

In sublimation process the dyes get transferred onto the fabric. Sublimation means turning the dyes from solid state to gaseous state. This printing technique can only be applied on fabrics that have an affinity for the disperse dyes, like nylon, acetate and polyesters. All quality of prints, vis-à-vis water coloured, sketch and solid coloured any kind of print, can be achieved using this technique on different materials having polyester coating like ceramics, mouse pads, etc. This kind of printing can be done in small units as well (Figure 5.5).

Sublimation can be best achieved if it's done on a white or light-coloured backgrounds. Sublimation printers work on MCYK printing and have no combination to produce white colour; hence, for any sublimation print having white in the artwork, a white base background of the fabric/product is required. These prints are long lasting and have good washing fastness.

**FIGURE 5.5** Artwork for sublimation print with water colour effect. Artwork by Amit Sharma.

### 5.2.5 Tie and Dye

Tie and dye is a technique of value addition that involves the process of resisting pre-planned areas on the fabric in various ways, followed by dyeing and unravelling of the tied areas to reveal a fascinating pattern on the cloth. Tie and dye is categorized as a resist dyeing technique and is practised in different parts of the world, largely in India, Japan, Africa and Indonesia. There are different names to tie and dye like Bandhani or Bandhej from Gujarat and Rajasthan, Shibori in Japan, Adire Alabere or Adire Oniko in Africa and Planji in Indonesia.

There are different methods of tying or resisting the fabric prior to colouration. The technique of tying the fabric will influence the overall pattern achieved once the ties are removed after the process of dyeing. There is an array of mechanical methods of resisting the fabric, ranging from binding, folding, gathering, knotting, clamping and stitching along with their variations that create innumerable patterns on the fabric. Different materials like string, rubber bands, wire and other unconventional mediums are used for the purpose of tying the cloth. Since tie-dye is a manual technique, each time the outcome is different, even if the fabric is tied in a similar fashion. Tie-dye produces a one-of-a-kind piece that makes it unique and more valued in comparison to mass-produced, machine-made products.

The uniqueness of tie-dye lies in its unusual textures created by resisting and penetration of dye around the protected areas that diffuses to create tonal colour effects. There is possibility of developing assorted tie-dye patterns by combining methods of

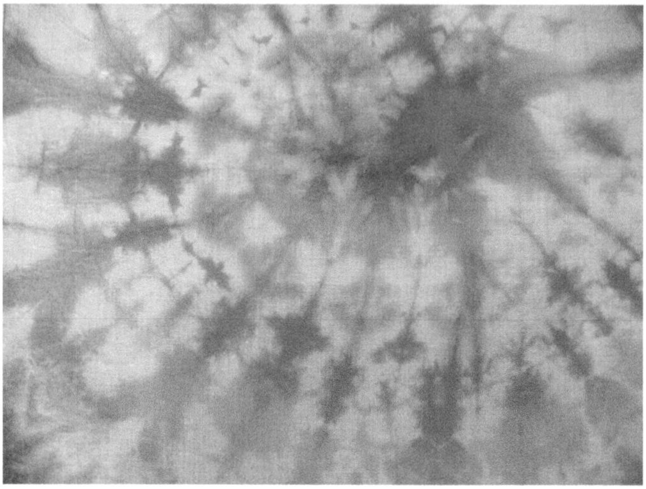

**FIGURE 5.6** Multi-coloured tie-dye. Sample by Deepika Sah.

resisting and multi-coloured effects by repeating the process of tie-dye to introduce a new colour each time. No tie-dye fabric after opening the ties is completely smooth and flat. There is an inherent texture created by the taut resisting that can be retained or else ironed out to smoothen the cloth (Figure 5.6).

The process of tie and dye begins with the pre-preparation of the fabric. Commonly light weight and fine cottons and silks are used for tie and dye for better dye penetration through the layers of the resisted fabric. Pre-treatment of the fabric involves desizing the fabric by soaking it in water for six to eight hours that facilitates better dye absorption. The next step entails folding the fabric into layers for ease of tying and also assisting in achieving repetitive patterns. If the resisting pattern is complex, the design is traced on the fabric using the prick and pounce method, that is pricking along the lines of the design on a butter paper to create tiny holes and pouncing a powder mixture combining a fugitive colour like *neel* or *geru* and kerosene, through the holes to leave a dotted impression on the fabric. Various mechanical methods can be employed to resist the fabric prior to dyeing. After tying the fabric in a particular manner, the resisted fabric is dyed in the lightest colour of the planned colour scheme. Following dyeing, the fabric is washed to get rid of the excess dye. The dyed fabric is further tied to protect the coloured areas in order to add the next colour. The process of renewing ties and dyeing would be repeated depending on the number of colours planned for the design. The dyed fabric is washed and rinsed before opening the tied areas.

### 5.2.6 Shibori

Shibori is the Japanese tie-dye technique practised since ancient times. Traditional Shibori samples from Japan are distinctively blue and white, resisting a pattern on white fabric and dyeing with natural indigo. There are several resisting methods followed in traditional Japanese Shibori, from binding, clamping or wrapping around a pole to hand stitching. The techniques are known by different names like Itajime

Shibori (clamping), Kumo Shibori (bound resist), Arashi Shibori (pole wrapping) and Mokume Shibori (stitch resist), to name a few.

Itajime Shibori is achieved by fan folding a fabric and placing a block of wood on the top and bottom of the folded bundle. To hold the blocks in place, the fabric with the blocks is clamped using C-clamps or other types of clamps to achieve uniform compression of the fabric. Different patterns can be achieved by varying the folding methods and the shape and size of the blocks clamped on the fabric.

Kumo Shibori or bound resist is another traditional Japanese Shibori technique that creates a spider web effect, a circular shape with a darker centre with radiating lines. The pattern is achieved by pinching the cloth at a point and binding with thread. The size of the circle would depend on the extent of thread tied from the centre point. Smaller bound resists, Kanako Shibori, that are similar to fine Bandhani dots are also practised.

Arashi Shibori or pole wrapping is a popular Shibori technique that produces different striped patterns or pleated textures. The technique involves wrapping the fabric around a pole and tying thread on top of the fabric to hold it in place. Traditionally solid wooden poles were used, which are nowadays replaced with lightweight, hollow PVC pipes or glass, stainless steel poles and even glass bottles. Pole-wrapped textiles are usually heat set to retain the beautiful pleated texture. A variation of pole wrapping is called Bomaki, where the fabric is stitched to create a cylindrical structure that can be slipped over a pole and gathered to compress and thus resist the dye uptake to create interesting organic line patterns.

Mokume Shibori employs needle and thread to hand sew the design. Hand stitching is a versatile technique as small running stitches can be carried out along any motif and then the thread is pulled to gather the stitches tightly and knotted to form a dotted resist line. It is important to use a stronger thread for hand sewing to avoid thread breakage and obtain a better resist. Hand-stitched Shibori is best achieved on tightly woven fabrics. Another hand-stitched technique, Ori-Nui, is carried out by sewing just below a fold to achieve double dotted lines (Figure 5.7).

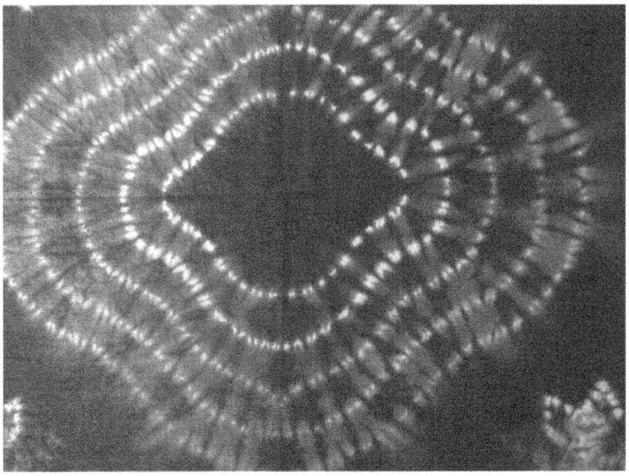

**FIGURE 5.7** Hand-stitched Shibori to achieve dotted resisted lines. Sample by Deepika Sah.

The texture created due to resisting and compressing the fabric can be retained by steaming the tie-dyed fabric and applying mechanical pressure. After the fabric is completely dried, the resists are removed to reveal the Shibori patterns and textures. The Shibori texture can also be heat set permanently if carried out on thermoplastic fabrics like polyester.

## 5.3 EMBROIDERY AND NON-LOOM TECHNIQUES

Embroidery, the art of needlework, is a technique that adorns a plain fabric. It is a value-addition technique that utilizes a needle and thread. Embroidery as a means of embellishment can be planned on different products, smaller to bigger items, ranging from apparel, home textiles to fashion accessories. The needlework can be planned in specific parts of a product or as an all over embellishment. The technique is possible on a wide variety of fabrics including natural, synthetic, blends, non-wovens as well as leather. There are different embroidery threads ranging from cotton, silk, wool to metallic and synthetic threads that are used to create beautiful embroidered products. For a more decorative effect, embroidery also integrates other materials like sequins, beads, pearls, coins, mirrors, quills, precious and semi-precious stones, etc.

Embroidery has been practised since antiquity. There are different embroidery styles prevalent across the globe with each region having its own distinct designs. Primarily women embroidered their clothing and household products, a means to express their creative energies. The main purpose of embroidery was to decorate textiles. It is believed that the basis of embroidery was embedded in primitive beliefs. As a measure to protect them from unknown powers, man tattooed patterns on his body or embellished his clothing with symbolic motifs.

This section will cover the different styles of embroidery classified on the basis of the method of stitching.

### 5.3.1 Hand Embroidery

Hand embroidery, as the name suggests, is done by hand using a needle or a special hook needle that is known as an awl, *aari* or a tambour. Embroidery done with a hook needle is also called Aari embroidery or Tambour embroidery. It is recorded that the Aari embroidery originated in India during the Mughal period and later, in the 18th century, was brought to France and Britain, where it was referred to as Tambour embroidery. Aari or Tambour embroidery is done on a fabric that is stretched on a frame. The hook needle is used to create a chain stitch and apply different trims like beads, sequins etc.

Hand embroidery is a versatile technique that makes use of a variety of stitches, threads and fabrics. Each hand-embroidered piece is unique and thus, valuable. Two hand-embroidered pieces with the same design will never appear the same due to the handcrafted quality. Hand embroidery is a time consuming process that involves skilful hands to create a personalized and artistic creation that are regarded as prized heirlooms by the holder.

There are two styles of hand embroidery, free style embroidery and counted thread embroidery. Free style embroidery is flexible, allowing usage of different stitches to embroider any kind of intricate design. It is possible on any kind of fabric. As the name suggests, counted thread embroidery is carried out by counting the warp and weft of the ground fabric, thus creating more geometric designs. Counted thread embroidery would require a more open-weave fabric in order to count the yarns for stitching. Fabrics like Aida or even-weave cotton or linen are most suitable for this style of embroidery.

Embroidery stitches are broadly classified into three major categories: flat, looped and knotted or raised stitches. There are innumerable variations under each category that makes hand embroidery a very creative art. Flat stitches are two-dimensional that lie on the fabric surface. Examples of flat stitches include running stitch, satin stitch and herringbone stitch. Looped stitches are stitches that are linked to one another. Chain stitch is a typical type of a looped stitch where a series of loops form a chain. Knotted or raised stitches are three-dimensional stitches raised from the textile surface. Some examples are French knots, bullion knots, coral stitch, etc.

Hand embroidery uses a variety of threads ranging from stranded cotton embroidery floss, pearl cotton, wool, twisted and untwisted silk and metallic varieties. The selection of needles for needlework is dependent on the type of embroidery style and fabric type. The most commonly used needle is Crewel needle with medium long eye and sharp point that is ideal for free style embroidery. A tapestry needle with a long eye and blunt tip is preferred for counted thread work. A milliner needle that is a thin needle with a round eye and sharp point is used for knotted stitches while a chenille needle with a long eye and sharp point is selected for ribbon embroidery.

Before starting embroidery, the design is transferred on the fabric. The design can be traced using a light-box or transferred by using carbon papers or heat transfer methods. The design can also be printed on a water-soluble stabilizer and placed on the fabric to carry out the embroidery. The stabilizer can be dissolved in warm water after the embroidery. The commonly used method of transferring embroidery design on fabric is prick and pounce method, as discussed earlier in Tie and Dye section.

For ease of embroidering, the fabric to be embroidered is stretched on a frame. Depending on the extent of area to be embroidered, either hoops or bigger frames called slate frames or adda are used.

Embroidery stitches are selected according to the different effects desired in a design. To achieve outline, stitches like running stitch, stem stitch or back stitch are appropriate. For filling in a shape, filling stitches such as satin or herringbone are suitable. Certain stitches like running or chain stitch can be used for both outline and filling effects. To introduce texture and three-dimensional appearance, knotted stitches, trims like beads, sequins and other embellishments can be applied or threads of different thicknesses can be couched on the fabric.

### 5.3.2 Machine Embroidery

Machine embroidery is the method of creating embroidered designs on fabric using a sewing machine or specialized embroidery machines. The first hand embroidery machine was discovered in the beginning of the 19th century by French inventor, Joshua Heilmann. The machine comprised a frame for setting up the fabric, a needle assembly and a handle for movement of the needle. This invention transformed the embroidery industry and led to further advancements, namely the Schiffli embroidery machine that used a combination of two threads, one in the needle and the other bobbin thread encased in a shuttle. The Schiffli machine was invented by Isaak Groebli in Switzerland. Post World War II, the production of embroidery machines was on the rise. A major advancement in machine embroidery technology occurred in the 1980s with the success of the first computerized embroidery machine. Prior to the advent of computers, embroidery designs were punched on paper and ran on the embroidery machines. The computerized embroidery machine led to further advanced versions that increased the design possibilities, a more convenient and efficient method of embroidering on fabric that is suitable for commercial purposes. Considering machine embroidery is cost-effective and production friendly, if finds wide usage in the fashion industry to carry out embroidery on various products.

Machine embroidery is a faster and convenient method in comparison to hand embroidery. It is advantageous for businesses producing large quantities as output is quick and embroidery is precise and identical across pieces. However, machine embroidery also has its own drawbacks as there is limited scope to experiment with stitches and achieve a more three-dimensional look that is possible with hand embroidery. Machine embroidery is ideal for cost-effective, mass-produced items.

Different types of threads can be used to execute machine embroidery. Primarily rayon and polyester threads are used as they are economical, available in a wide range of colours and can endure high-speed machine stitching without breakage or fraying. Rayon embroidery threads are softer and have a higher sheen in comparison to polyester threads. On the other hand, polyester embroidery threads are stronger, colour fast and shrinkage resistant.

Machine embroidery can be done on different varieties of fabric. It is important to stabilize the fabric that is to add a layer of fusing on the back side of the fabric prior to machine embroidery. The function of a stabilizer is to support and stiffen the fabric and avoid puckering and distortion during the process of embroidery. Different stabilizers are used that are selected according to the fabric to be embroidered. For example, a cut away fusing is used on knitted fabrics and a tear away fusing is used on woven fabrics. There are also water-soluble stabilizers that are backed on sheer fabrics or applied on top of pile fabrics like loop terry to flatten the nap to carry out the embroidery smoothly. Water-soluble fusible interlinings are also helpful for lace-like effects and cutwork. Besides stabilizing the fabric, it is pertinent to hoop the fabric so that it stays taut and in the same place while the machine is running to embroider the design.

Embroidery machines can be broadly classified into manual free-motion embroidery machines and computerized embroidery machines. The selection of an embroidery machine would depend on the scale of work and requirement by an individual embroidery designer, small-scale or a large-scale business that is producing more volumes. Further, embroidery machines can also be categorized into single-head embroidery machine and multi-head embroidery machine. As the name suggests, a single-head embroidery machine holds one needle and a multi-head embroidery machine has multiple needles ranging from 2 to 12 to even more number of heads. A multi-head embroidery machine executes embroidery simultaneously on multiple products, thus ideal for mass production.

#### 5.3.2.1 Free-Motion Manual Embroidery Machine

The free-motion manual embroidery machine is a single-needle, single-head machine that is operated manually, allowing free-hand stitching in different styles within any shape. Different stitches are possible on the manual embroidery machine using a darning foot, from outline stitches to filling stitches, chain stitch, boucle stitch, *dori* work etc. along with appliqué, cutwork and quilting. Since it is a single-needle machine, needle has to be re-threaded each time to introduce a new colour in a multi-coloured embroidery design. Thus the process is labour dependent and takes more time in comparison to a fully automatic computerized embroidery machine.

#### 5.3.2.2 Computerized Embroidery Machine

The introduction of the computerized embroidery machine completely revolutionized the process of embroidery. There are different types of computerized embroidery machines ranging from single head to multi-head, multi-needle to semi-automatic and fully automatic versions. A computerized embroidery machine is computer controlled with in-built digitized design that is embroidered on the fabric without an operator's assistance. A fully automatic computerized embroidery machine only requires a supervisor to begin the machine and keep a check on its functioning while it is in use. The multi-needle computerized embroidery machine yields faster results as all needles are threaded in the beginning itself and can run continuously as it does not require re-threading for colour change.

Embroidery on a computerized machine entails different steps, starting from creating a digital embroidery design. There are specific embroidery softwares that assist in digitizing design files, editing designs and selecting suitable stitches and colours. The next step involves uploading of the final digitized file on the embroidery machine. The fabric to be embroidered is stabilized and hooped to avoid puckering during the embroidery process. After threading the needles and positioning the stretched fabric on the machine, the embroidery on the intended design begins. The operation is monitored to address any problems like needle or thread breakage, skipped stitches, imbalanced stitching or puckering.

### 5.3.3 Non-Loom Techniques

It is believed that the earliest of textiles were created by manipulating fibres with hands. Perhaps, basketry was the precursor to textiles and techniques like intertwining, interlacing, interloping and knotting were adopted to construct fabric. Over a period of time, different non-loom techniques evolved and are still used in different parts of the world. Non-loom techniques, as the name suggests, do not require a loom, but very minimal tools, thus making it highly handy and convenient for the maker. This may be the reason for certain non-loom techniques like crocheting and hand knitting to flourish as domestic handmade crafts as it was much easier for women to crochet or hand knit in between household chores.

Non-loom techniques can be categorized into weft-based techniques and warp-based techniques. When the textile structure builds horizontally, it is weft-based, such as knitting, crochet, netting and tatting. When the textile grows vertically, it is warp-based, like macramé, braiding, sprang or ply-splitting.

Textiles constructed off the loom find wide application, from making smaller to bigger articles, and endless possibilities of creating decorative edges and trims to accessorize different products. The following section describes crochet, tatting and macramé.

#### 5.3.3.1 Crochet

Crochet is a method of creating a textile using a simple tool, a hook needle and yarn. The term "Crochet" is derived from a French word meaning hook. The technique involves inter-looping of a single yarn using a long-handle hook needle to build a structure. Different yarns made of cotton, wool, silk and other fibres can be used to crochet. Nowadays, unconventional materials like plastic, wire, ribbon and fabric scraps are also used to develop innovative crochet products. The crochet hook made of metal, wood or plastic is available in different sizes from 2.25 to 30 mm thickness, in order to crochet with the finest to the thicker, chunky yarns respectively.

The basic foundation stitch in crochet is a chain stitch that further grows row-wise to create a textile. To create a chain, a slip knot is used to secure the yarn on the crochet hook followed by inter-looping to make a chain of the requisite length. The basic crochet stitches include single, double, half double, treble, double treble and triple treble along with texture stitches like bobble, puff and popcorn. There is also scope to combine different stitches to achieve interesting patterns. Crochet is a versatile technique as there is scope to increase or decrease loops in a row for the purpose of shaping into specific products.

A wide array of products are possible with Crochet, from baby blankets, throws, cushion covers to dresses, shawls, hats, bags and soft jewellery. Crochet trims and edgings are other possibilities that add value to a product.

#### 5.3.3.2 Tatting

Tatting is a technique used to create intricate lace work including lace edgings and ornamental mats or doilies. It is believed that tatting was derived from knotting technique and eventually in the 19th century emerged as an intricate-lace-making technique using a combination of knots and loops to add value to clothes by way of decorative edgings and lace collars.

The most common method of tatting uses a small shuttle, tapering at both ends. The thread is wound on the shuttle and the knots are created over the fingers. The typical knots are single stitch and double knots to create rings, chains and picots, which are combined to create intricate motifs and designs. The other method of tatting uses a needle in place of a shuttle.

Tatting is an intricate craft that can be used to add decorative lace edges to clothing and home textile products. Nowadays, the technique is also used to create delicate jewellery and small fashion accessories.

#### 5.3.3.3 Macramé

Macramé is a method of constructing a textile using different knots. It is recorded that the word Macramé is derived from an Arabic word, "migramah" meaning fringe. The origin of Macramé dates back to the 13th century, a technique used by Arabic weavers to knot the loose ends of handwoven textiles like shawls and towels.

FIGURE 5.8 Alternating square knots. Photography by Ashima Tiwari.

The primary Macramé knots are half knot, square knot and half hitch. There are many variations possible with these basic knots. Macramé is a simple method to create a textile as it just requires yarn for knotting and an element like a ring, rod or yarn itself that can be stretched between two fixed points for mounting the working yarns for making knots. Prior to knotting, it is important to determine the length of the yarn required to create a knotted piece. The single yarn calculation is usually four to six times the ready knotted textile. Macramé can create both narrow and wider textiles. The width of the Macramé textile will influence the number of yarns that need to be mounted for knotting. A wider textile is created by connecting adjacent yarns with knots alternating in rows (Figure 5.8).

Macramé is a versatile technique and has gained popularity over a period of time. Macramé finds wide application from small fashion accessories like soft jewellery, waist belts, decorative hangings to pot-holders, wall arts and clothing. Since Macramé is a handmade method, it is valued and holds greater relevance in context of sustainability. Macramé has been explored by both international and national designers and is observed on the fashion runways as well.

## 5.4 FABRIC MANIPULATION TECHNIQUES AS MEANS OF SURFACE ORNAMENTATION

Fabric manipulation covers a gamut of techniques that assist in imparting interesting textures on a plain material. Fabric manipulation is a widely used technique to embellish a fabric with the help of a thread and needle. Sewing with hand or using a machine, these techniques create varied effects such as gathers, pleats, puffs, ridges and bulges to completely restructure a two-dimensional material into a three-dimensional surface.

The fabric manipulation techniques were practised since ancient times. Over time, some of the techniques that have functional application in garments, like gathers, pleats or tucks, are effectively used to impart a beautiful texture on the fabric. The major categories of fabric manipulation are gathering, shirring, pleating, smocking, tucking, quilting and filled-in effects.

### 5.4.1 Gathers

Gathering is a technique that contracts the fabric into very tiny folds by stitching along a fabric edge and drawing the thread to bunch up the fabric. Gathers can be achieved by hand or machine stitching. In the case of hand gathering, two parallel rows of running stitches are executed followed by pulling the threads to achieve the gathers. While using a lockstitch machine, few adjustments such as increasing the stitch length to 4 mm and slight loosening of the upper tension are required to pull the bobbin thread without breaking. Gathering can be conveniently created using special attachments like a gathering foot or a ruffler that automatically gathers the fabric while stitching. Elastic gathering for enhanced stretchability can be achieved by winding elastic thread in the bobbin or by stitching an elastic cord or band.

Since the fabric compresses on gathering, it is important to estimate the fabric requirement. It is suggested that for medium-weight fabrics, the fabric length should be 2.5 times the ready gathered edge, and for lightweight fabrics, at least three times of the finished length should be calculated.

Different gathered effects can be achieved by planning the gathering along a single edge, adjacent edges, opposite edges or all side gathering on a piece of fabric. Gathered ruffles are created by gathering on a fabric strip. Single edged gathered ruffles are possible by gathering on one edge of the strip and double edged gathers are accomplished by gathering through the centre of the strip. Gathering all around a shape and drawing the thread creates smaller inflated gathered elements like yo-yos and puffs. There are a myriad of possibilities with the gathering technique to achieve beautiful textures.

### 5.4.2 Pleats

Pleats are calculated folds planned along the fabric edge that are held in place with a line of stitching. Commonly pleats are used to create fullness in garments, allowing for easier movement to the wearer. Folded pleats can be ironed to obtain sharp creases or they can be more free flowing and fluid. Over a period of time, pleats like folds on fabric are also used to develop creative surfaces.

There are broadly two types of pleats, namely flat pleats and projecting pleats. Flat pleats are two-dimensional folds on fabric that are fixed in place by stitching along the fabric edge. The three basic flat pleats are knife pleats, box pleats and inverted box pleats. Since pleating involves folding, thus contracting the fabric, the amount of fabric required for forming a number of pleats needs to be calculated. A typical flat pleat comprises pleat depth that is the measurement of the width of the pleat and the underfold, the area on the fabric where a pleat is placed. The underfold is twice the pleat depth. Thus the extra fabric requirement for pleating is calculated as one pleat underfold times the total number of pleats desired.

Projecting pleats are three-dimensional pleats that are raised from the surface of the fabric. Some examples are double-box pleat, accordion pleats, three pinch fold or four pinch fold pleats. Projecting pleats are ideal for fabric manipulation with a wide scope for variations.

### 5.4.3 Shirring

Shirring is similar to gathering as fabric is compressed into soft puckers by doing multiple parallel rows of gathering at intervals. The requirement of fabric for shirring is similar to gathering that ranges from 1.5 to 3.5 times depending on the fabric weight and the amount of fullness desired. Shirring can be done by hand or machine, as explained for gathering.

There are many variations possible with the shirring technique. Rows of gathering executed both parallel to the warp or weft of a fabric generate cross shirring or waffle effect. This type of shirring produces an overall puffed appearance. Pattern shirring is another method to achieve interesting textures. In pattern shirring, a gathering foot is most handy to automatically gather along the repetitive pattern lines that can be curved, zig-zag, spiral and many other line variations. There is scope to explore shirring as a technique for textured surfaces.

### 5.4.4 Smocking

Smocking is both a decorative and functional technique that entails drawing a fabric into very fine pleats with hand stitching and securing the folds with embroidery. In the process, the technique imparts elasticity to the fabric. For ease of stitching parallel rows to create fine pleats, a smocking pleater is used in the industry. The smocking pleater machine has multiple needles to create rows of stitches simultaneously on the fabric that can be later pulled to pleat up the fabric. For smocking, the fabric calculation to three to four times the ready width of the smocking planned.

There are mainly two types of smocking, English smocking and direct smocking. The English smocking is a two-step process. Firstly, pleats are created on the fabric followed by embroidery to hold the mini pleats together. Later, the threads stitched for drawing the fabric are removed. Different embroidery stitches like cable stitch, wave stitch, feather stitch and variations are used in smocking. The embroidery stitches are executed to create patterns like diamond and honeycomb.

Direct smocking is a one-step process that works on a dotted pattern grid transferred on the fabric. Stitching by picking adjacent dots and pulling the thread creates a fold. The pattern followed for stitching the dots can vary, thus leading to different smocked designs. Another style of direct smocking is North American smocking that is based on a dotted pattern grid and using a combination of pulled stitches and slack stitches to create finer folds and a textured pattern. There are different North American smocking patterns such as lattice, lozenge, flower and variations that can be executed based on the grid dots picked to make a pulled or slack stitch (Figure 5.9).

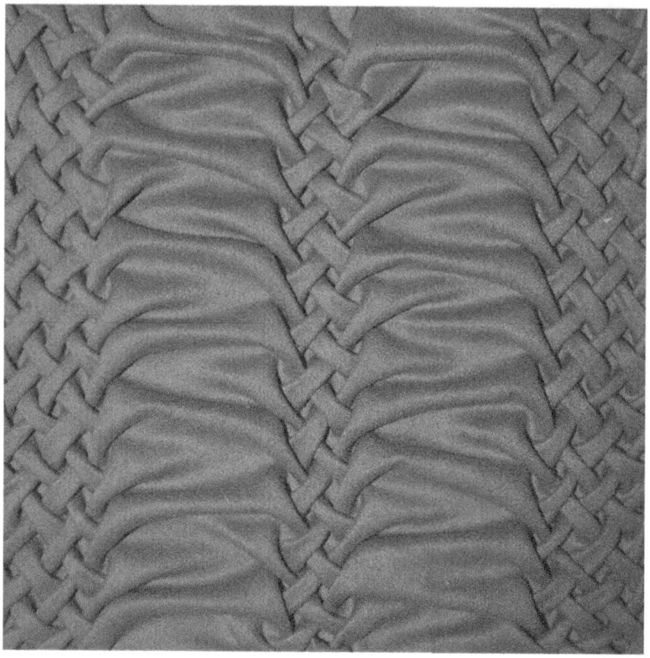

FIGURE 5.9 North American smocking. Sample by Prashant Manjhi.

## 5.5 MISCELLANEOUS SURFACE ORNAMENTATION TECHNIQUES

Along with the surface ornamentation techniques mentioned above, there are many others that are used based on the requirement and development of customized products. These techniques are either Indian or adapted from other countries for their aesthetic values and designs.

### 5.5.1 Hand Painting

It's a printing technique done by hand using brushes and acrylic-based fabric paints or dyes. This technique is the oldest and an indigenous technique of creating prints on the fabric. Kalamkari and Leather Puppetry from Andhra Pradesh are illustrious examples of hand painting on different materials in India. Hand painting is mostly done by pen or brushes using natural as well as synthetic dyes. Nowadays traditional arts like Madhubani from Bihar and Gond art from Madhya Pradesh are also being practised on fabrics for craft popularity and reachability. These traditional crafts are inspirational in terms of their motifs, designs and painting techniques for many designers.

### 5.5.2 Marbling

Marbling is a printing technique that is done using paints, adhesives or any dispersants in water. The paint floats on thickened water and is then transferred onto the fabric. This is a Japanese technique of printing from 12th century and is called as *suminagashi*, meaning "floating ink", used for printing papers for the document. The marbling prints resemble the textures on the stone. In 15th century, the same technique was introduced in Central Asia in Turkey.

For marbling any fabric, marbling size, fabric mordant (mostly alum), acrylic paints and any open and broad tray are required. The fabric is first treated with alum so that it's affinity towards dyes can be increased specifically in cottons. The dye bath is prepared having marbling size and water in the tray. The acrylic colours in liquid consistency are poured on the thickened water as a layer (colours are suspended) and abstract patterns are created in the dye bath with these colours. The dried mordant-treated fabric is laid gently in a stretched form on the coloured abstract pattern that gets printed onto the fabric. The fabric is then dried and washed thoroughly thereafter.

### 5.5.3 Acrylic Pouring

Painting on fabrics is made easy by the use of acrylic paints that are easily available in the market. Acrylic paints when applied on fabrics get hard and generate cracks after drying, hence a fabric medium is mixed along with water to make the paint flexible even after drying. These paints can be applied on different fabrics like cotton, polyester, silk, etc. The fabrics are first washed to remove any starch or co-polymer before application of the acrylic paint. The paint is diluted in required amount using water and fabric medium and is then applied on the fabric as per design requirement. In acrylic pouring, the colours are mixed as per designer's choice in a pouring liquid consistency and then poured over the fabric to create abstract patterns. The process concludes with heat setting of the paint using dry heat by iron or drying under the sunlight.

### 5.5.4 Inclusion of Sundries

In today's era when all the landfills are full and the environment is a major concern for everyone, sustainability becomes the core concept to implement in all the fields of work. Based on the same theory many fashion and textile brands have developed collections, putting a halt to fast fashion and promoting increased life cycle span of a garment or a textile product.

Many brands and designers create haute couture garments and accessories with intricate hand embroideries, using sustainable fabrics and sustainable raw materials like wooden beads, linen, wool and silk threads for embroidery, buttons made up mother of pearl, coconut shell, wood, fabrics, horns, shells, metal, bones, etc. These collections are sometimes created from biodegradable fibre that has been upcycled or recycled with high value addition, giving the whole collection a new life and value.

## 5.6 CASE STUDIES OF APPAREL BRANDS UTILIZING SURFACE ORNAMENTATION TECHNIQUES

There are many fashion and apparel brands that develop intricate surface ornamentation for their collections and have made their mark in the national as well as international market for their design aesthetics and high values.

### 5.6.1 Shades of India

Shades of India is a company in Noida run by Ms. Mandeep Singh Negi and Mr. David; Ms. Mandeep's passion for Indian textiles is reflected in her brand through her collections. The company creates collections of apparel, home and accessories using interesting fabric textures with prints and surface ornamentation techniques.

The brand is established for people who love and value craft and textiles. Its positioning is in the higher-end market targeting clients who are ready to pay for the value of the textiles and craftsmanship. The collections have Indian garments with interesting details like sarees, blouses, kurtas, plazzos, pyjamas, trousers, dupattas, stoles and shirts.

Every ensemble has unique fabric texture and prints that is created based on the inspiration given by Ms. Mandeep and her team. The collections are created using fabrics like cotton mulmul, blend of silk cotton like Chanderi and Maheshwari, different weight silks, organzas, organdies, velvets, voiles and all natural fibre fabrics of high quality. These fine woven fabrics are given beautiful textures using smocking techniques, pin tucks, gathering, couching and different kinds of hand embroideries using beads, sequences, laces and crochet techniques.

The highlight of each garment is its finishing and detailing. The kurta and blouses' necklines and sleeve ends are interestingly finished with handwork to give them exquisite look in tone and tone colour palette. The edges of the lower garments are finished with hand or machine embroidery to give them an appealing look. Special laces and buttons used in the garments are created with handwork and beads, which are exclusive highlights of the brands.

The prints used in the collections are also bold and mostly created through digital print or screen print techniques. From engineered to all over prints are the highlights of the brand with intricate details and amusing colour palettes.

The accessories collection of the brand consists of the hair bands, bags and potlis and jewellery. All the accessories have intensive hand work on preferred dark colour backgrounds. The hand embroideries are done using various kinds of stitches like French knots, couching, running stitches and many more with beads, sequences and zari threads. The jewellery is created by threads and textiles of various varieties. The beads of the necklaces are created from fabrics and are then stringed together using interesting yarn craft techniques.

The home collections of the brand consist of bedding, cushions, table collection, curtains and tie backs. The brand also does interior design projects customized for the clients. The textures, colours and surface applications are all based on the space for which the collection is being designed. Here also as in apparel and accessories, interesting textures are created over a basic fabric or printed fabric using quilting, pin tucks, gathering, tufting and various hand embroidery techniques. The important fabrics used in the home collections are cotton voile, cotton velvet, silk, cotton silk, linen and cotton satins.

The textures are created on the fabrics using machine embroidery, layering of fabrics, sandwiching yarns between two layers of fabrics and many more interesting and innovative ways to create interesting designs and generate value for the product.

This brand is known nationally and internationally for creating timeless pieces using Indian traditional embroideries and textiles and giving customers to own heirloom in their wardrobes and homes.

### 5.6.2 Mura Collective

Mura Collective is renowned for its signature and ingenious Shibori designs and techniques. It has been awarded four times the significant UNESCO Seal of Excellence for its handcrafted and sustainable method of tie and dye.

Mura creates designs using tie and dye, hand and machine Shibori techniques. This brand's USP lies in the way the techniques are used to create designs with natural dyes. The brand targets the clients who are inspired from handcrafted textiles, values and promotes them.

The product list of the brand includes fabrics and Indian garments like sarees, kurtas, jackets, shrugs under womenswear and shirts for men with stoles, shawls and dupattas. Home linen like bedsheets and cushion covers are also produced in Mura. The fabrics used are fine medium-weight cottons, silks and merino wool in apparels.

As the brand supports and values sustainability theory, the brand launched upcycling collection from the leftover fabrics, created masks with bags and launched them during the COVID period.

The products showcase bright colour palette with few traditional colours like indigo and mehroon. The design patterns consist of abstract patterns, floral patterns, geometric patterns and patterns inspired from the natural elements, obtained from various Shibori techniques done by hand and by the machine. The garments are decorated with more than one tie and dye technique having a mixture of textures and forms. The design of any product is a combination of patterns with repeat motifs and engineered patterns.

Thus, Mura is a brand that works with natural fibre fabrics and natural dyes and hence has made its own mark in the customer's eye using traditional surface ornamentation technique in creating high-end valued products.

### 5.6.3 Arvind Limited

Arvind Mills located in Santej, near Ahmedabad, Gujarat, is one among the largest Denim fabric producer in India with 22 global patents for environmental solutions. It is a Lalbhai Group company, which serves to apparel brands like Aeropostale, Arrow, US Polo, Tommy Hilfiger, CK and Sephora. Denims, Wovens, Voiles and Knits are the main four branches of the company.

The mill produces woven fabrics and has inhouse printing facility as well. At Arvind mills both hand screen and rotary screen-printing techniques are used; however, for large scale of fabric production rotary screen printing is preferred. The rotary screens are made of nickel and are coated with a photo emulsion ink. The mesh size of the screen depends upon the kind of design to be printed, if the design is patchy then mesh size of 25, 40, 60 or 80 is used while for intricate and sharper designs mesh size of 155, 165 or 195 is used.

In Arvind Mills, there is a production of about 1,000 m of printed fabric in a day using this technique. Disperse, pigment, reactive and discharge dyes are used for printing different kinds of fabric produced in the mill. There are different temperatures required for application of each dye. The fabrics after printing are washed and cured using dry heat/steam depending upon the kind of dyes used. Fixing agents are also used to fix the dyes onto the fabric while printing.

The designs produced at Arvind fashions are according to the client's requirement from national as well as international market. The production is mass produced as it is a fabric-manufacturing mill. Designs created for prints depend upon the requirement of the client and its product. Geometrical, floral, abstract and all kinds of designs are created by the design team based on the trend research and client requirements. The designs have fixed repeats as per the rotary size, but different layouts are set in computer-aided design for the printing unit. The colours also vary according to the requirement of the client. The colours are documented and printed as per the Pantone shade card. More the number of colours the costlier is the print and hence limited colours are preferred in rotary screen printing.

Arvind fashion is a mill where the products are created and innovated with sustainable approach using sustainable fibres, finishes and in value-addition techniques. The brand also works in other spheres attached with their mills, for example energy and water conservation, CSR projects for betterment of human life and many more.

In all, the company works towards environmentally friendly and sustainable approach as mentioned in every work related to the brand.

## BIBLIOGRAPHY


1. Arvind Fashions, *Arvind Fashion*, 2023. Retrieved February 27, 2023, from https://www.arvindfashions.com.
2. Barden, B., *The Crochet Stitch Bible*, Midtown Manhattan, New York City: Chartwell Books, 2000.
3. Campbell, J. and Bakewell, A. M., *The Complete Guide to Embroidery Stitches: Photographs, Diagrams, and Instructions for over 260 Illustrated Stitche*, Midtown Manhattan, New York: Plesantville, 2006.
4. Corbman, B.P., *Textile Fibre to Fabric*, McGraw-Hill, New York, 1983.
5. Dunnewold, J., *Complex Cloth: A Comprehensive Guide to Surface Design*, Washington: Bothell Publisher, 1996.
6. Splash By Lo, *History of Textile Printing*, 2023. Retrieved February 27, 2023, from https://splashbylo.com/en/blog/historia-estampado-textil.
7. BrainKart, *Methods of Printing*, n.d. Retrieved February 25, 2023, from https://www.brainkart.com/article/Methods-of-Printing_37451/.
8. Mura Collective, *Collections*, n.d. Retrieved February 16, 2023, from https://muracollective.com/collections/all.
9. Pizzuto, J.J., Price, A., and Cohen, A.C., *J.J. Pizzuto's Fabric Science*, London: Fairchild Books, Bloomsbury Publishing Plc., 1994.
10. Sang, S.M. 14-Process control in printing of textiles. In Shang S.M. (Ed.), *Textile Fibre to Fabric*, 339–362. New Delhi: Woodhead Publishing Series in Textile, 2013.
11. Sekhri, S., *Textbook of Fabric Science, Fundamentals to Finishing*, PHI Learning Private Limited, Delhi, 2016.

12. Shades of India, *Product Category*, n.d. Retrieved February 12, 2023, from https://www.shadesofindia.com/product-category.
13. FLD POS, *What Is Screen Printing?*, n.d. Retrieved February 15, 2023 from https://www.fldpos.com/info/what-is-screen-printing-23622597.html.
14. Inkwell Nation, *What Is Sublimation?*, n.d. Retrieved February 15, 2023, from https://inkwellnation.com/printing/what-is-sublimation.
15. Wolff, C., *The Art of Manipulating the Fabric*, Iola, Wisconsin: Krause Publications, 1996.

# 6 Size Charts, Body Measurements, and 3D Body Scanners

*Harpreet Kaur and Kulbir Kaur*

## 6.1 INTRODUCTION

The term "Anthropometry" is derived from a combination of Greek words *anthrop* (meaning human) and *metricos* (meaning measurement) (Gupta, 2014). Anthropometry is the field of study related to the systematic measurement of the human body. The term branches from *anthropology* (*the* "science of humanity", *the study of human societies and cultures and their evolution*).

The discipline, which is now one of the primary techniques of physical anthropology, began in the 19th century when early studies of human biological and cultural evolution sparked an interest in the systematic description of both living and extinct populations.

The physical attributes are dimensions of limbs and other body parts; reach of limbs in standing, sitting, and various postures; weight, etc. The tools used for this purpose are stadiometers for height, bicondylar callipers for bone diameter, skinfold callipers for the thickness of skin and subcutaneous fat, scales for weight, etc., and the handling of all these tools requires great precision.

Today, the majority of this data is used by **ergonomists** to design physical facilities that are user-friendly because personnel are, in essence, interchangeable components of any system. *Ergonomics is the study of how equipment and furniture can be arranged so that people can do work or other activities more efficiently and comfortably* (*Health Benefits of Ergonomics | Portable Laptop Stand*, 2018).

Anthropometric measurements were largely employed for cultural purposes (e.g., artwork) to portray beauty, power, and other desired aspects of the human form by the ancient civilisations of Rome, Greece, and Egypt. The "width of a human hand" or "length of a human foot" was popular units of measurement.

Artists during the Renaissance used human proportions to apply anthropometric measures to their creations. One of the most famous instances is the work of the legendary artist Leonardo da Vinci (the iconic Vitruvian Man), who used cadavers to gather measurements of the human body. To gain correct anthropometric measurements, other painters relied on live models and historical achievements.


DOI: 10.1201/9781003432739-6

### 6.1.1 History of Anthropometrics

- Although techniques of taking human measurements find their origin in ancient times, Alphonse Bertillon is considered to have made a major contribution to its present-day practices.
- Based on his classification method known as the "anthropometric system" or "judicial anthropometry", Alphonse Bertillon is regarded as the father of anthropometrics.
- To tackle the issue at hand, Bertillon devised a system of classification based on anthropomorphic measurements assuming that bone density is fixed past the age of 20 years and human dimensions are inherently variable.
- Bertillon measured the height, breadth, foot size, length and width of the head, length of the middle finger, and length of the left forearm of criminals in prison, as well as other morphological and differentiating traits.
- The use of this anthropometric system was called "Bertillonage" and became popular around the world during the late 1800s and early 1900s.

## 6.2 SIZE CHARTS

Today, the apparel industry is one of the largest industries in the world generating a revenue of some 1.53 trillion US dollars in 2022, a little decline from the previous year. However, income is expected to rise to more than 1.7 trillion dollars by 2023 (Smith, 2023). An industry catering to consumers globally of different races, ethnicities, and sizes requires a huge database for standardisation of size for ready-to-wear (readymade) garments. Currently, the apparel industry follows systems of size standardisation that are different for different countries, for example, the US, UK, and EU systems of standard sizes. But there is no single universally accepted standardisation of size because the human body doesn't come in a standard size. With that being said, we have established that people from different lands have body types unique to their demography.

This brings us to the question of whether the prevalent size charts can serve the consumer who falls outside the sample size studied for the standardisation of human body measurements. Well, the answer is no. As we now know, people have different body types that do not align with the standard size charts followed today, especially the Asian consumer pool. While the majority of the manufacturing for the apparel industry takes place in Asian countries, there does not exist a standard size chart specific to the Asian population. Asia is a huge continent; even among Asians, there would be a requirement for at least three different studies, dividing the continent into three demographic regions based on the similarity among the human figures in those regions.

## 6.3 SIZE CHART FORMULATION

Clothing sizing was one of the first commercial applications of anthropometry. All historical large-scale surveys were conducted to aid with the sizing of mass-produced, ready-to-wear military clothes and equipment. The problem of garment sizing is extremely complex, in addition to being interdisciplinary in nature, making the entire process of developing a sizing system both costly and time-consuming.

- Size UK cost 1.2 billion dollars in 2000
- Size Sweden took five years to finish

The amount of time and money needed to collect and analyse data for garment sizing has been a major deterrent to tackling the problem. Furthermore, the development of sizing systems required input from:

a. Anthropologists (planning and conducting an anthropometric survey)
b. Statisticians and mathematicians (population sampling plan, statistical and data analysis)
c. Pattern makers (critical measurement selection)
d. Garment technologists and designers (production-related issues – size plan and fit).

The data analysis procedure has been by far the weakest link. That is, how to process the huge amount of data gathered to arrive at a suitable range of sizes that may serve as the basis for mass garment manufacturing. Because the problem is inherently non-linear, simple statistics cannot be used to analyse it.

The sizing problem is ubiquitous in **dress forms,** also. The dress forms (*an adjustable dummy used in dressmaking that can be made to conform to a person's figure*) or **mannequins** (*a mannequin is a life-sized model of a person that is used to display clothes, especially in shop windows*) used by the designers to create and showcase their designs come in fixed sizes, which again are based on the research and data analyses of a group of population that doesn't always do justice to the unique fitting needs of the consumer. The most common sizes used for dress forms in India are 8, 10, and 12. These sizes are not at all fitting to the population and do not have the same bust-to-hip ratio or waist-to-hip ratio as those of Indian consumers.

The Bureau of Indian Standards, in their study, has given in detail standards for the standing heights and anthropometric dimensions of Indian children between the age group of 5 and 17 years: graphical representations of body measurements, reach dimensions, eye levels, and ratios of body segments with standing height.

## 6.4 COMPONENTS OF DRESS FORMS AND TAKING MEASUREMENTS

In the business of apparel production, measurements are crucial. If the measurements are not correct, the whole process, no matter how precisely and carefully executed, would yield undesirable outputs. Hence, the need to consider different body types in the sample for readymade garments becomes essential. While taking measurements, there are some balance lines running perpendicular to the axis of the body, often referred to as the "centre front" or "centre back".

### 6.4.1 Limitations Associated with Anthropometry

It is ironic that the application of anthropometry to apparel fitting remains the faultiest to this day. They are based on simple averages or a percentile system, presuming that human bodies have mathematical precision and increments in their shapes and sizes. It is outrageous that some companies test fit their garments on professional models that are not the figure that resonates with the masses, who are the actual consumers of those products.

According to the current study's analysis of Indian population statistics, 20% of the population matches the dimensions of an average figure. Given this insight, it may be preferable to adopt a system in which the bodies are not required to suit the sizes, but rather the sizes are designed to fit the actual shapes of the bodies. To put it another way, create a sizing system that covers 80% of the population. Outliers, or those who are notably different in size or proportion from standard dimensions, make for 10%–20% of the population that is not covered. These will need to be addressed as distinct categories (e.g., plus-size category).

Traditional instruments like callipers and measuring tapes are being phased out in favour of digital devices. Among digital gadgets, the most recent technology is around low-cost 3D optical imaging techniques, which find applications in healthcare, research, and even everyday life. Contactless anthropometry has several advantages over more expensive or invasive systems. Technically, relevant information from 3D scans or sensors is obtained in three steps: data gathering, procedure, and measurement. Data acquisition for scanners can be accomplished using a variety of approaches, including structured light scanners and time of flight scanners. Machine learning, which is a combination of artificial intelligence and data mining, has the potential to improve data gathering and analysis in kinanthropometry applications. Kinanthropometry is the study of size, shape, proportionality, composition, biological maturation, and body function to understand the process of growth, exercise, sports performance, and nutrition. Far from being a replacement for traditional methods, digital anthropometry represents a great opportunity to improve accuracy, validity, practicability, and the use of self-monitoring processes under professional supervision.

### 6.4.2 One Size Doesn't Fit All

Style and variety have become so essential in our lives that today we have dedicated brands for toddlers, kids, and adults, catering to their needs and wants and designing the products accordingly. These brands develop products keeping the consumer in mind. But today most countries are ageing countries, which means that the major population of consumers is above 60 years of age or approaching 60. The physique of an ageing person is different from that of a young adult. Their body structure changes with the shoulders drooping, legs becoming thin, and body fat accumulating around the belly, so the clothes of sizes that have been standardised based on the measurements of young adults don't always fit them right. Hence, the standard size charts followed for adult clothing are not appropriate for an elderly consumer.

Even the designs of the garments available in the market are not in synergy with the needs and wants of the elderly. In the later years of life, comfort becomes the utmost priority and the aesthetics are secondary. It doesn't mean that aesthetic design can be entirely avoided but a design that looks good but isn't comfortable will not appeal to this consumer pool. Hence, designing products has to be done keeping in mind the comfort and feel-good factor of the consumer while keeping the aesthetic beauty intact, according to the likings of the consumer.

## 6.5 3D BODY SCANNERS

3D scanners play a revolutionary role in the garment industry by enabling them to take quick measurements and, by extracting this data, customised digital fits can be generated for the customer. 3D scanners also help in representation of relationship between garment and the body. The technology is upgrading the industry on many levels by speeding up the measurement process, shortening the time to collect a data pool for sizing, sorting, and creating various categories based on sizes, shapes, and angles. The 3D representation allows immediate evaluation of the scan and could be used recurrently eliminating the need for taking repeated measurements. Whereas the technology's future seems promising in the industry, there is still copious scope for improvement. One such example is while scanning people with amputations, contemporary scanners don't show satisfactory results. They show a black shadow in the place of the missing body part and the measurement around the area is also not accurate. Part of the reason for this obstacle is that the technology used for 3D body scanning uses light rays for the purpose that is unable to give accurate results on figures with amputations.

Following are a few 3D body scanning technologies currently available in the market.

Laser Scanning: This type of scanning relies on a triangulation and light-plane technique to operate. A couple charged device (CCD) scans the field of view while a laser serves as the light source. The movement of the light on a body is detected by the CCD. Laser-based body scanners have a scanning speed of roughly 60,000 points per second.

White or structured light scanning: Projectors are used with body scanners that use structured light technology. They cast a sequence of white light stripes onto the target, which are photographed. The curvature of the stripes covering the subject helps to describe the 3D contour of the body. Studies have demonstrated that light projection systems are faster than laser-based systems during the scanning sequence. These types of scanners do, however, last a little longer when it comes to measurement extraction (Istook and Hwang, 2001).

Human Solutions (formally TechMath): Vitus SmartXXL, a 3D body scanner by Human Solutions, uses laser technology. In 12 seconds, it gathers roughly 140 body dimensions in a 3D image of the body using 8 sensor heads and an optical triangulation technique. Data can be exported in ASCII, OBJ, STL, and DXF formats (Human Solutions, 2009).

Structured light is used by the Telmat (SYMCAD) turbo flash/3D System for Measuring and Creating Anthropometric Database. For data collection, a set of white light stripes are projected onto the subject. As a result of errors caused by natural body sway movements and undergarment reflectance, its new "ST" (Special Tracking) version applies 3D digitising to its data capture (Telmat Industrie, 2009).

## 6.6 CONCLUSION

As a result of the huge disruption created by the COVID-19 epidemic, digital anthropometry has emerged as an active field of study. Practitioners will almost certainly adopt hybrid processes (human-machine-based methods) to obtain anthropometric data in the future. Skinfolds are those metrics that may be kept in future kinanthropometry protocols unless portable ultrasound equipment drops in price to make them more accessible, which may replace manual processes.

## REFERENCES

GuptaD (2014). Anthropometry and the design and production of apparel: an overview. In *Elsevier eBooks* (pp. 34–66). https://doi.org/10.1533/9780857096890.1.34

*Health Benefits of Ergonomics | Portable Laptop Stand.* (2018, October 10). Coffee shop hero. https://coffeeshophero.com/benefits-of-ergonomics/#:~:text=Ergonomics.,activities%20more%20efficiently%20and%20comfortably

Human Solutions. (2009). [online]. https://www.human-solutions.com

Istook CL, Hwang S (2001). 3D Body scanning systems with application to the apparel industry. *Journal of Fashion Marketing and Management*, *5*(2), 120–132.

Telmat Industrie. (2009). [online]Ultra-Fast 3d Body Scanning Booth, Symcad, 3D body Scanner by Telmat. https://www.symcad.com

# 7 Pattern Making Techniques

*Sudha Subramanian*

## 7.1 INTRODUCTION

The apparel development is an artistic process of incorporating style in the regular attire of the individual. The apparel designing is the first phase of the garment development in which the design will be created by sketching the design and then converted into a garment. In between this process, the phase required to achieve the proper fit of the garment is a technique that develops a flat pattern out of the body measurements of the clients. The two major such techniques are,

1. Pattern Drafting
2. Draping

## 7.2 PATTERN DRAFTING

Garment industry is one of the major contributors for the total revenue of the country and it faces challenges with the increasing demand. In this manufacturing process, pattern making is one of the crafts that needs more skill for the development of customised to standardised garment styles [1]. Pattern making is an art of manipulating and shaping the flat fabric in order to conform to the curves of the human body. Thus it acts as a link between the design development and garment production by rendering the garment design in the form of garment components. Fischer [2] defines that pattern helps the designer to render a flat sketch that can be transformed into three-dimensional designs. Shoben and Ward [3] state that pattern is the foundation for the fit and design that provides sufficient ease for the comfort and ease of the garment. According to MacDonald, flat pattern is a design process in which the basic pattern is developed in order to fit the standard dress form [4]. Pattern is the foundation block, which includes style lines, gathers, drapes, pleats etc., that supports in conformability and fit of the garment. In general, pattern is made out of measurements of the human body from which toile is developed to analyse the fit, proportion and shape of the garment. The measurements used for the pattern development are based on the personalised measurement or standard size through which custom-fit garment or ready-to-wear apparel can be developed. The patterns are also mentioned as slopers, blocks, master pattern or foundation that is developed by printing on paper containing instructions required for sewing operation. These developed patterns can be graded to different sizes from smaller to larger to fit the varying body figures. The fit of these patterns are analysed for its correctness by cutting the pattern in


DOI: 10.1201/9781003432739-7

the sample fabric and the fit of the developed garment are checked. Through this productivity of the style will be increased and the rejections will be reduced. The productivity will also be increased by arranging the pattern on to the marker by which the cutting waste can be reduced and time will be saved. The effective use of this technique helps in saving the time, reducing the fabric wastage, increasing the productivity and able to reach the customer satisfaction. Thus, fit is one of the major elements that can be achieved by proper development of pattern.

Some of the advantages of the pattern drafting process are [5],

- Development of pattern for wide variety of design is possible with drafting
- The fitting problem of the individual can be overcome through pattern drafting for their measurement for a specific style of garment
- Cheaper method compared to draping process.

### 7.2.1 Body Shapes and Pattern Drafting

The industries generally consider the ideal body shapes for taking the dimensions, pattern development and analysis of fit for varying style requirements. Similarly, they consider the ideal figure for showing their collections to the market. But, in market based on the body size of the women, they are classified into various body shapes. According to a study, women body shapes are classified as follows [6],

1. Pear type – chest girth is at least 8 cm less than their hip girth
2. Apple type – chest girth is at least 9 cm more than their hip girth
3. Rectangle type – difference of 7 cm or less between their chest and hip girths
4. Hourglass type – chest girth is at least 37 cm larger than their waist girth.

Another classification on female body shapes is shown in Figure 7.1 [7].

Hong Cai [8] analysed the influence of body shape on pattern making process and highlights the significance of physical characteristics of the human body on fashion design development [8]. In this research, it was stated that the neck is the intermediate link between the head and body structure. The neck structure of the human body influences the collar design of the garment at three different phases, front neck point, rear neck point and inclined neck point. These three phases form the basic neck line shape of wide, high and flat rear side and narrow, low and curved front side. The variation in these phases of neck line affects the fit of the collar band of the garment style without affecting the basic movement of the neck. With respect to the shoulder position, the tilting degree of the shoulder, shoulder thickness and off the shoulder difference are to be considered in the design development process. These factors influence the development of upper clothing on considering the broadening of the proportion with the increase in chest circumference. Thus, it is highly important in considering the externally curved front shoulder line and the internally curved rear shoulder line that adopt to the fashionable pattern development. In the human body, chest is the plumpest area and basic parameter in the design development process. Thus, in the development of pattern for women, exact positions and protuberant of breasts are to

| Body Shape | Definitions | Shape |
|---|---|---|
| Hourglass | A very small difference between bust and hip circumferences<br>The ratios of bust-to-waist and hips-to-waist are about equal and significant | 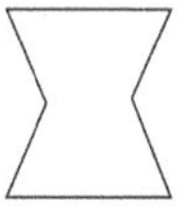 |
| Top hourglass | A larger bust circumference than hip circumference<br>The ratios of bust-to-waist and hips-to-waist measurements are significant enough to produce a definite waistline | 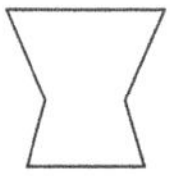 |
| Bottom hourglass | A larger hip circumference than the bust circumference<br>The ratios of bust-to-waist and hips-to waist are significant enough to produce a definite waistline | 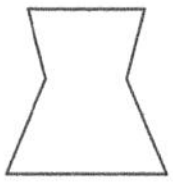 |
| Spoon | A larger circumferential difference in hips and bust<br>The bust-to-waist ratio is lower than the hourglass shape<br>The hip-to-waist ratio is high |  |
| Triangle | A larger hip circumference than that of the bust<br>The ratio of hip-to-waist was small<br>Larger in the hips than in the bust without having a defined waistline | 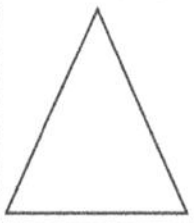 |
| Rectangle | Bust and hip measurement fairly equal<br>Bust-to-waist and hip-to-waist ratios are low<br>There is not a clearly discernible waistline |  |
| Oval | The average of the stomach, waist and abdomen measurement is less than the bust measurement | 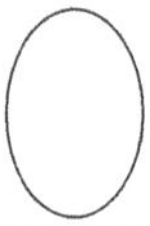 |
| Diamond | The average of the stomach, waist, and abdomen measurements is more than the bust measurement | 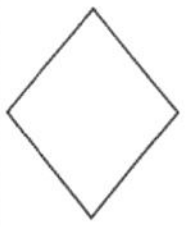 |
| Inverted Triangle | A larger bust circumference than that of the hips<br>A small bust-to-waist ratio |  |

**FIGURE 7.1** Classification of female body shapes.

be considered for proper fit of the garment. This can be achieved by means of incorporating a pleat or darts in the breast area to conform to the concave-convex changes of the chest profile. While considering the waist position, it is the thinnest position of the female body profile. It is one of the parts that supports the activities of the human body by flexing and considered as a part to be shape in pattern making. Thus, while making the pattern it is to be taken into consideration the high rear side and low front side of the waist in fluctuating manner. It is also important to consider the specific activity and seasonal characteristics on designing the waist portion of the pattern.

### 7.2.2 Measurements and Pattern Drafting

Pattern development by considering the measurements of different body parts provides basic foundation for knowing the relationship between the body and the pattern [9]. This relationship in turn is implemented in the tailoring technique that leads to the perfect fit of the garment. In the development of pattern, very few measurements are used in the drafting and other measurements are applied in relation to the other. Thus it is highly important to understand the link between each and every measurements of the body. Some of the measurements considered in the drafting of pattern are given in Table 7.1 [10]. In addition to this, there are some of the measurements that are highly useful but it is left unattended many of the times [11, 12]. Due to this, the pattern fit is not achieved properly and alterations are carried out to solve

**TABLE 7.1**
**Body measurements**

| Measurements | How to take body measurements | Individual's measurements + | Ease* (inches) = | Total measurements |
|---|---|---|---|---|
| | Removing clothing or wearing only intimate wear, gently wrap tape around body keeping tape level | | | |
| Bust | Measurement at fullest point | | 3–4 | |
| High bust | Measurement just above breasts, under the arms and across fullest part of the back | | 3–4 | |
| Low bust | Measurement just under breasts and around neck | | 3–4 | |
| Neckline | Measurement of neck circumference 1 inch above base of neck | | 0.5–1 | |
| Shoulders | Measurement from base of neck to top of shoulder bone | | 0.5 | |
| Back | Measurement from shoulder to shoulder across back | | 1–1.5 | |
| Shoulder cap (depending on sleeve length, four measurements may be needed) | Measurement with arm slightly bent (like slipping hand into front pants pocket) | | | |
| a. Shoulder cap | Top of shoulder to 0.5 inch below wrist bone | | N/A | |
| b. Upper arm length | Measurement of top of shoulder to outside centre of elbow | | N/A | |

*(Continued)*

**TABLE 7.1 (*Continued*)**
**Body measurements**

| Measurements | How to take body measurements | Individual's measurements + | Ease* (inches) = | Total measurements |
|---|---|---|---|---|
| c. Upper arm circumference | Measurement around upper arm at armpit | | 0.5–2 | |
| d. Wrist | Measurement around widest part of hand | | 0.5–1 | |
| Derriere | Measurement of widest part of hips | | 2–4 | |
| Back | Measurement down back from prominent bone at base of neck down to waistline | | 1–2 | |
| Thighs | Measurement 9 inches down from waistline, around heaviest part of single thigh | | 1–3 | |
| Inseam | Measurement from crotch to ankle bone on inside of leg | | 0.5–1 | |

*Source:* Ref. [10].

* The chart provides a range of ease measurements for constructing a fitted garment. Ease amount depends on the pattern instructions, type of fabric used, fit and style preference.

the problem through experimentation. Figure 7.1 [9] shows the measurements of distinction with respect to the angle of the body. In Figure 7.2, measurement A describes the distribution of the bust suppression around the bust area. This also will give the under bust depression and equality of suppression of both the busts. This will give an understanding to the measurer on the bust shaping of the customer. Similarly, measurement B in Figure 7.2 shows the projection of the scapula and the shoulder blade. This measurement will help in rendering the curvature of the back running from the shoulder blade over the shoulder. This curvature assists in calculating the application of dart in the pattern in addition to the shaping requirements. Measurement C provides the information on the distribution of the curvature running between back of waist and scapula. This helps in shaping around and to analyse the strength required at the back for a garment. Finally, measurement D gives the understanding of the waist shaping in relation to the shoulder. Through these measurements one could be able to draft a pattern with required fit considering the various suppressions and depressions in the body profile.

In addition to the type of measurements, the method of taking the measurement also has the influence in the fit problem in the developed pattern. The measurement to be taken should be chosen based on the style and fit of the garment. For example,

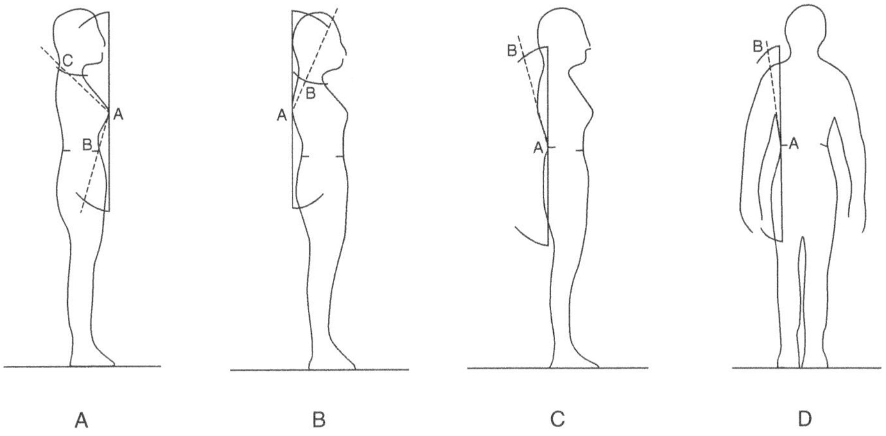


**FIGURE 7.2** Body proportions.

**Source: Ref. [9].**

if the garment consists of the strap in the closely fitted design, the measurer must measure the strap position of the wearer. Otherwise, it results in the imbalance strap that affects the fit and appeal of the garment on wearer. Similarly, if the measurer did not consider the measurement from waist to hip, it will affect the shape of the hip and in turn leads to looseness or tightness in that region for the wearer. This is because, the measurements to be considered not only depend on the style but also on the age and sex of the wearer [13]. It is also evident from the studies that identifying the landmarks in the body is highly crucial in drafting the pattern to the requirement of the customer. The inconsistency in identifying the landmarks results in the variation in pattern drafting that leads to fit problems to the wearer [14]. So it is important in developing the structured manner of identifying the landmarks and recording it in the efficient and effective way to the pattern developer that helps in improving the fit of the garment [15].

## 7.3 COMPUTER-AIDED DESIGN AND PATTERN DRAFTING

Computer-aided design (CAD) technology is one of the sources to develop the fashion clothing automatically in a short period of time. The technology uses a computer with the specific software that allows the designers and engineers to plan, model and evaluate the design before being manufactured. CAD technology helps as a time saving process that helps in easy storage of the digital form of pattern for easy retrieval on requirement. It helps in efficient way of producing the pattern, marker and alterations in faster, easier and efficient manner [16]. CAD is one of the easier and advanced techniques for making a pattern and simulating it. Through CAD software, we can create patterns or scanning it to the CAD system and can be modified according to the style variation on customer demand. It can be used for various applications such as design sketching, technical drawing development, fabric print developments and so on [1]. Pattern making software is used to create a pattern by providing input on measurements and data required to make it. Using the pre-programmed software, the image of the pattern will be developed for the personalised measurement. It also stores the information related to the pattern such as flat sketch, specification sheet, trim requirements, size grade charts and costing. It also supports in developing the marker for the style, i.e. arranging the patterns on to the fabric for cutting with maximum fabric utilisation and minimum wastage. With the use of CAD, matching of the stripes and plaids can be done automatically and the amount of wastage can be estimated. CAD also supports in mapping the texture on the pattern and could able to simulate the texture by draping on to the dummies.

Computer-assisted designing helps in incorporating the changes easily and effectively on the screen for further processes of grading and marker planning. It will be highly helpful in working with the repetitive pattern making with less time by avoiding the process of drafting on to the paper again and again. The key features in use of CAD for pattern making are pattern design system and pattern generation system [17]. Pattern design system helps in speeding up the pattern making process and thereby increases the accuracy of the process. This supports in drafting the pattern blocks and constructing the same by placing them in place on simulation. Pattern generation system is used to create the pattern using the components of the pattern automatically by the system (Figure 7.3).

The consumers of the current market prefer to have the personalised outfit rather than the ready-to-wear kind of apparels [18]. This is mainly due to the advancements and shift of the apparel industries from mass production garment and made-to-measure technique to mass customisation concept [19, 20]. This sophistication of mass customisation increases the demand for the personalised style of garments from the customers. The increased demand for mass customisation leads to the need for the customised fit among consumers. In order to meet the increased demand, many researches have been carried out in meeting the fit problems with the help of computer-aided designing software.

Kaixuan Liu et al. proposed a 3D interactive garment pattern making technology (Figure 7.4) to address the above-said problem with the existing pattern making techniques. In this system, they first extract the three-dimensional garment outlines with the help of gravitational virtual environment. This garment outline is then adjusted according to the design requirements by smoothing out the folds and wrinkles in the garment. Then in the adjusted outline, the construction curves and lines are incorporated on to the surface based on the design details. From this design, 2D patterns are developed by unfolding the 3D surfaces of the garment outline. Then, the 2D is further stretched or shrank based on the elasticity properties of the fabric and the pattern can be used for the drafting of the new garment [21].

They also claimed that this research has the advantages of developing the pattern with an interaction between the 2D pattern design and 3D garment visualisation

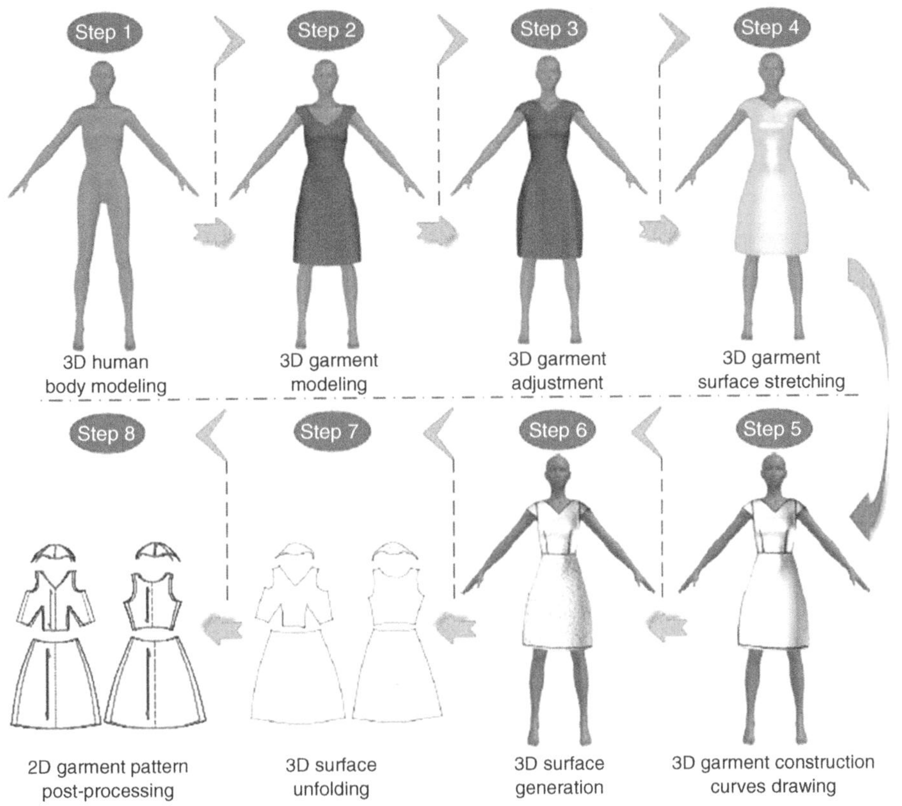


**FIGURE 7.3** Computerised pattern making.

**Source: Ref. [17].**

without the assistance of pattern making skills, can be used for any type of garment fit, reduces the trial and error by which cost and time consumption has been reduced and includes the fabric parameters in the drafting of pattern.

Mok et al. discussed about the intelligent pattern design module to develop a clothing pattern from the sketch design using software. Traditional flat pattern making technique is adopted for the development of pattern design. This includes the processes such as basic block development, fashion silhouette modification and fashion feature creation. Initially, basic block is developed based on the individual's body measurement including the wearing ease of measurement between the human body and garment for basic movement of the body. Secondly, the basic block is modified into secondary blocks according to the required silhouette by including the design ease and design features that leads to final appearance of the garment [22]. The pattern making module used for the intelligent pattern design module includes sequential workflow, flat pattern techniques, garment construction rules and seam information.

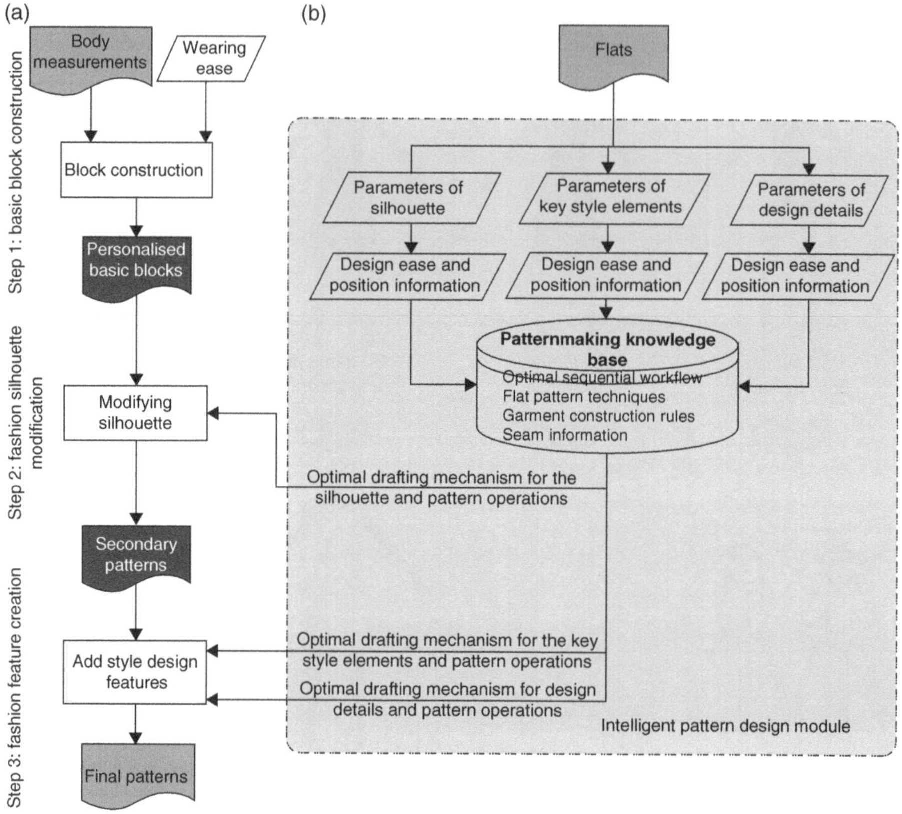


**FIGURE 7.4** 3D Interactive garment pattern making technology.

**Source Ref. [21].**

## 7.4 CONCLUSION

Pattern making is one of the crucial stages in the development of garment from design sketches. The advancements in the technology and various innovations in the development of software for pattern making lead to the sophistication in the process enabling customer requirements to be fulfilled with less time and reduced fabric wastage. This is because pattern making is the process that supports in approval of the sample for the order by providing expected fit of the buyer. Thus, adoption of either manual or computerised pattern making can lead to the development of pattern that fits the particular size without many alterations and produce satisfactory results with less iterations.

## REFERENCES


1. Obinnim E, and Pongo N A, The significance of flat pattern making in fashion designing: A case study of dressmakers in the Ho municipality of Ghana, *International Journal of Innovative Research in Science, Engineering and Technology*, 2015, 4(4), 1850–1857.
2. Fischer A, *Basics: Fashion Design Construction*, West Sussex: AVA Publishing SA, 2009.
3. Shoben M M and Ward J P, *Pattern Cutting and Making Up*, New Delhi: CBS Publishers, 1999.
4. MacDonald N M, *Principles of Flat-Pattern Design*, New York: Fairchild Books, 2010.
5. Arya N, Jamal Z and Yadav N, Advances in apparel designing techniques, *International Journal of Recent Scientific Research*, 2016, 7(7), 2506–12510.
6. Unal N and Utkun E, Impacts of female body shapes on patternmaking, *Industria Textile*, 2018, 69(6), 519–523.
7. Cai H, On the effect of human body shape on fashion pattern making, *Proceedings of the 2014 International Conference on Education Technology and Social Science (ICETSS)*, Advances in Social Science, Education and Humanities Research 2014, 251–254.
8. Tama D and Öndoğan Z, Fitting evaluation of pattern making systems according to female body shapes, *Fibres & Textiles in Eastern Europe*, 2014, 22(4), 107–111.
9. Ahmed M, Alrushaydan T, Gill S, Hayes S G, and Brubacher K, The suitability of body scanning measurement in pattern drafting methods, Proceedings of 3DBODY. TECH 10th Int. Conference and Exhibition on 3D Body Scanning and Processing Technologies, 2019, 22–23.
10. Hamilton W, Pattern Alteration Guide C-228, NM State University, 2015.
11. Lesko L J, *Bodice Fit compared: Conventional Alterations with and without Graphic Somatometry*, Iowa State University, United States, 1982.
12. Shen L and Huck J, Bodice pattern development using somatographic and physical data, *International Journal of Clothing Science and Technology*, 1993, 5, 6–16.
13. Pandarum R and Yu W, Garment sizing and fit, *Garment Manufacturing Technology*, 2015, 8(2), 187–204.
14. Bye E, Labat K L, and Delong M R, Analysis of body measurement systems for apparel, *Clothing and Textiles Research Journal*, 2006, 24(2), 66–79.
15. Yu W, Subjective assessment of clothing fit, in J. Fan, W. Yu and L. Hunter Yu W (Eds.), *Clothing Appearance and Fit: Science and Technology*, Woodhead Publishing Limited, 2004.
16. Suryani H, The effectiveness of clothing pattern making training with CAD-based system on fashion students, *Advances in Social Science, Education and Humanities Research (ASSEHR)*, 2018, 201, 311–316.
17. Datta D B and Seal P, Various approaches in pattern making for garment sector, *Journal of Fashion Technology & Textile Engineering*, 2018, 4(1), 29–34.
18. Gu B, Liu G, and Xu B, Individualizing women's suit patterns using body measurements from two-dimensional images, *Textile Research Journal*, 2016, 87(6), 669–681.
19. Apeagyei P R and Otieno R, Usability of pattern customising technology in the achievement and testing of fit for mass customisation, *Journal of Fashion Marketing and Management*, 2007, 11(3), 349–365.

20. Cho Y, Komatsu T, Inui S, Takatera M, Shimizu Y, and Park H, Individual pattern making using computerized draping method for clothing, *Textile Research Journal* 2006, 76(8), 646–654.
21. Liu K, Zeng X, Bruniaux P, Tao X, Yao X, Li V, Wang, 3D Interactive garment patternmaking technology. *Computer-Aided Design*, 2018, 104, 113–124.
22. Mok P Y, Xu J, and Wu Y Y, Fashion design using evolutionary algorithms and fuzzy set theory a case to realize skirt design customizations, in Tsan-Ming Choi (Ed.), *Information Systems for the Fashion and Apparel Industry*, Woodhead Publishing Series in Textiles, 163–197.

# 8 Garment Production Techniques

## *Pre-Sewing Processes*

*Surjit Ramamoorthy and*
*Rathinamoorthy Ramasamy*

## 8.1 INTRODUCTION TO PRE-SEWING PROCESSES

The pre-sewing processes play a very important and major role in getting a garment with the required quality parameters. All the pre-sewing processes are very much essential in ensuring that the right pieces are sewn and required garments are obtained at the lowest possible time. The efficiency of each of the activities in pre-sewing aids in ease of sewing and results in quick turnaround times. The sewing process originated even before spinning and weaving in the Old Stone Age. However, different materials were used in and for sewing during those times. Even before the sewing machines were invented, hand sewing was carried out as needles were used for it and it was 20,000 years ago. The sewing machines came into existence in the 18th and 19th centuries but the pre-processes existed for a long time before that. It is because the clothes have to be cut into different shapes as per the pattern before the sewing is done either by hand or machine. It was in the mid-19th century that mass manufacturing of garments began where it was essential for effective coordination and execution of all the pre-sewing processes to ensure that sewing was done on time and the products reach the customer promptly. The pre-processes had to be planned well in advance for order execution and it was essential to create new technologies in the pre-sewing processes for managing mass production. It is essential to know the various pre-sewing processes carried out before finding out how they are carried out along with their latest advancements.

### 8.1.1 Pre-Sewing Processes

The pre-sewing processes are the activities or steps carried out to get the material ready for sewing the required garments. The various pre-sewing processes carried out are shown in Figure 8.1.

The first step in the pre-sewing process is the receipt of the tech pack or the technical specification sheet from the buyer stating all the requirements. In some cases, samples are sent by the buyer along with the tech pack for better understanding and reproduction of the sample in bulk form. Once the tech pack is received, the apparel company personnel carry out two jobs in tandem. One is that the marker planning

DOI: 10.1201/9781003432739-8

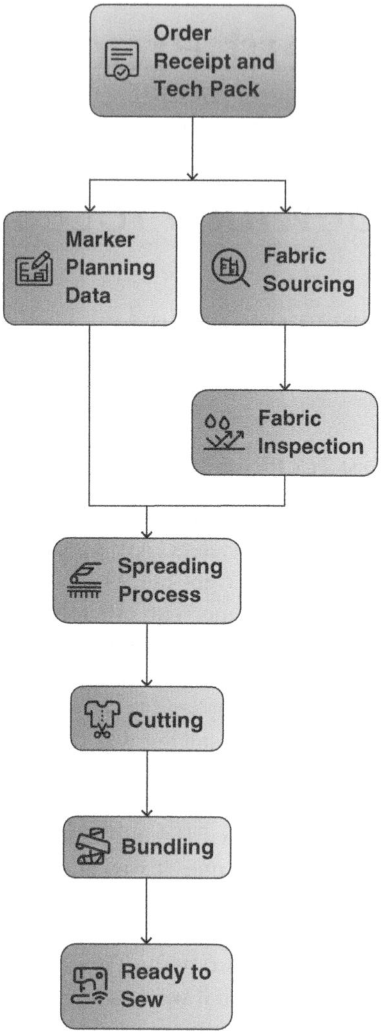


**FIGURE 8.1** Flow process chart of pre-sewing processes.

is done based on the different patterns required for the garment construction. It is a very intricate step as it involves immense assessment of the best usage of the cloth with minimum wastage and also requires avoiding any quality issue due to wrong placement of pattern pieces. This process is done manually in the case of small order quantity, and computer-based pattern-making tools are now available to reduce manual labour and increase efficiency. This process has to be efficient for maximum utilisation of the fabric with reduced wastage. However, the fabric quality should also not be compromised as in using wrong grain lines in the resultant garment. Hence, interactive marker-making is done using software where the software arranges the pattern pieces in the right order and calculates the efficiency of the marker and states it. Then the experienced pattern master or makers can alter based on the requirement to achieve maximum quality and quantity. There are elements like digitiser and plotter used for these operations, which will be discussed in the later sections. Once the pattern is ready, then based on the cutting machine used, they are either plotted on a pattern sheet or stored in the hard disk for feeding in the computerised cutting machines.

In the meantime, the fabric sourcing department looks at the order quantity and orders the required fabric for the order execution. Once the fabric is received, it is inspected for defects. The inspection is done manually or using automated machines. There are various advancements in these inspection machines, which will be discussed in the subsequent sections. The inspection is also done at the supplier's end and that data is available in real time for the apparel company if they use ERP software. The supplier inspects the fabric and that data can be seen by the manufacturer and if he/she is convinced, then the fabric inspection before spreading can be avoided. In today's context, when vendor-managed inventory is in practice and efficient supply partners are available, such inspections done at the supplier's end will reduce the workload and increase the efficiency of the overall supply chain. It will also reduce the time involved in multiple inspections. However, it is still in practice in many industries where as a part of the legacy system, the inspection is carried out before spreading the fabric. Once the inspection is done and the order is ready for execution, then the fabric spreading is done.

The spreading process involves laying the fabric on the table for cutting. The number of layers depends on the order quantity, fabric thickness, machine type, etc. The spreading process is done by hand and also using machines. In this process, the required number of layers is spread on the table and then the pattern sheet with the markings is kept on the top layer for cutting. However, placing the pattern sheet is based on the type of cutting machine used as computerised cutting machines may not require patterns as mentioned earlier. Nowadays, labellers are used. These labellers give labels based on the marker plan and these labels are automatically applied on top of the lay using adhesive. They indicate the component names and the number of pieces in them and reduce the usage of paper. After spreading, the cutting is carried out using different machines according to the requirement. There are stationary and portable cutting machines operated by hand and machine. Computerised machines are also available. In this cutting process, using a cutting machine, the garment pieces laid on the table are cut based on the marker plan. The time involved in cutting depends on the number of layers, the intricacy of the patterns, and the type of cutting machine used. Once the cutting is done, the cut garment parts or components are bundled and then kept ready to send to the sewing section. The components are segregated and then sent to the sewing line to the appropriate operator for production. The bundling process is more simplified when the labellers are used as they give complete information about the component. These pre-sewing processes also involve high technical requirements as marker planning, spreading, and cutting need to be done at high precision and all the processes should be carried out based on the tech pack instructions given by the buyer. If the tech pack instructions are not followed, then the sewn garment may not be accepted by the buyer. Hence all these processes must be carried out without any defects and errors for successful execution of an order. The above points indicate the importance of the pre-sewing processes and it is imperative to understand the entire process in detail. In the subsequent sections, each of the pre-spinning processes is explained.

## 8.2 TECHNICAL SPECIFICATION SHEET OR TECH PACK

In the apparel industry, the role of the tech pack is very critical as it contains the most important information about the styles that are going to be produced. The technical specification sheet is developed in two different situations as reported below,

i. Firstly, the customer (also known as a buyer) will develop a tech pack. As they decide to develop a style either from the forecasting process or by some inspiration, by focusing on the forthcoming season. Later this information will be passed to the manufacturers to reproduce the developed styles in the bulk quantity.
ii. In the second case, the buyer might have been inspired by some styles from fashion shows or some other places. In that case, they will send the design requirements or photographs of the styles to the manufacturer. The manufacturer develops the technical specification sheet. This is further used for the bulk production.

In the apparel industry, the merchandiser or merchandising department usually handles the buyer communication and initial processing of orders. When a merchandiser is in connect with some buyer, they will receive the technical pack as a process of order follow-up. It is their source document to process a particular order. The document provides a complete technical specification and materials required to produce a particular style of garment. It was developed by designers, technical designers, and product developers. This document will aid as a communication medium between the buyer, manufacturer and even in between the different departments of the manufacturing industry. This will eliminate unnecessary confusion at different manufacturing stages and help in error-free production of the garments.

### 8.2.1 Components of Tech Pack

The following are the components of technical specification packs. Each of the components is detailed in the subsequent sections.

1. Order details in the header and footer
2. Technical drawings/images of the apparel
3. Style description
4. Fabric specification
5. Colour specification
6. Embroidery/print design and specifications
7. Bill of materials/material details
8. Measurements and grading with specification sketch
9. Construction details

10. Reference images that explain how to make a garment
11. Testing methods
12. Packing instruction
13. Costing sheets.

#### 8.2.1.1 Order Details in the Header/Footer

Figure 8.2 represents the sample image of the first page of the typical tech pack. Specifically, all the pages of the tech pack will carry the order details like order quantity, buyer name, style number, and other details that will be provided in the header/footer of the page.

This information is given on all pages of the tech pack as a header or footer. This will help the merchandiser or production department personnel to know the order details of a certain order. It will also avoid unnecessary confusion with other existing orders and this will help the users even if individual sheets are separated from the file to identify the order details.

#### 8.2.1.2 Technical Drawings/Image of the Apparel

Other than header and footer, the first page of the tech pack will typically contain the technical sketch or photograph of the proposed style/garment. This image will provide a clear view and style details to the manufacturers/merchandisers to reproduce the product. As provided in Figure 8.2, both the front and back views of the proposed garment will be given for reference.

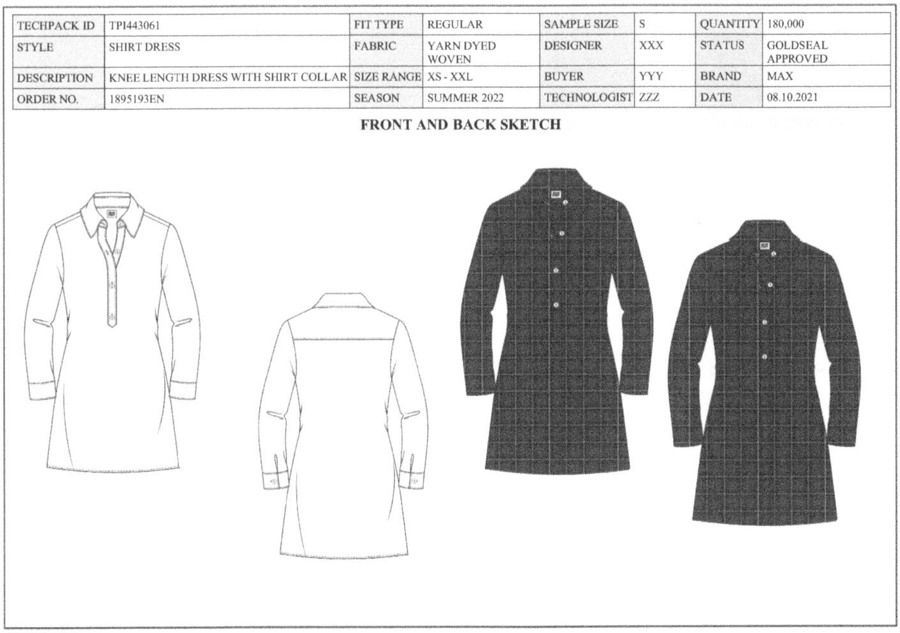

| | | | | | | | |
|---|---|---|---|---|---|---|---|
| TECHPACK ID | TP1443061 | FIT TYPE | REGULAR | SAMPLE SIZE | S | QUANTITY | 180,000 |
| STYLE | SHIRT DRESS | FABRIC | YARN DYED WOVEN | DESIGNER | XXX | STATUS | GOLDSEAL APPROVED |
| DESCRIPTION | KNEE LENGTH DRESS WITH SHIRT COLLAR | SIZE RANGE | XS - XXL | BUYER | YYY | BRAND | MAX |
| ORDER NO. | 1895193EN | SEASON | SUMMER 2022 | TECHNOLOGIST | ZZZ | DATE | 08.10.2021 |

FRONT AND BACK SKETCH

**FIGURE 8.2** The first page of the tech pack with header details.

### 8.2.1.3 Style Description

In the majority of the cases, the style description details along with the sample approval status will be provided on the first page of the tech pack. In some cases, the details related to given styles are presented on the second page as shown in Figure 8.3. The information that is provided on this page includes brand name, merchandise group, style number, theme, product, selling period, body type, fit, size range, sample size, verbal description, approval status, date of approval, fabrics, and trims.

### 8.2.1.4 Fabric and Colour Specification

The third page of the tech pack typically contains a complete specification detail of the fabrics used in the apparel. Where the fabric swatch or the close-up or microscopical image of the fabric will be provided along with their corresponding Pantone shades. This page also contains details related to the fabric construction parameters like main fabric colour, fabric composition, fabric structure, dyeing method, yarn type, yarn count, ends per inch (EPI), picks per inch (PPI), fabric weight (GSM), fabric width, applied design (if any), finishes provided at fabric stages, and care instruction that need to apply during the user phase will be provided as shown in Figure 8.4. If more fabrics were used in the garment, the tech pack will complete the details of all the fabrics in the garment as mentioned above.

| | | | | | | | |
|---|---|---|---|---|---|---|---|
| TECHPACK ID | TPI443061 | FIT TYPE | REGULAR | SAMPLE SIZE | S | QUANTITY | 180,000 |
| STYLE | SHIRT DRESS | FABRIC | YARN DYED WOVEN | DESIGNER | XXX | STATUS | GOLDSEAL APPROVED |
| DESCRIPTION | KNEE LENGTH DRESS WITH SHIRT COLLAR | SIZE RANGE | XS - XXL | BUYER | YYY | BRAND | MAX |
| ORDER NO. | 1895193EN | SEASON | SUMMER 2022 | TECHNOLOGIST | ZZZ | DATE | 08.10.2021 |

**DESCRIPTION**

| | |
|---|---|
| BRAND NAME | MAX |
| MERCHANDISE GROUP | FORMAL WEAR |
| STYLE NUMBER | SD C_31 |
| THEME | CHECK AND FIT |
| PRODUCT | SHIRT DRESS |
| SELLING PERIOD | SUMMER 2022 |
| BODY TYPE | WOMEN'S WEAR |
| FIT | REGULAR FIT |
| SIZE RANGE | XS - XXL |
| SAMPLE SIZE | S |
| VERBAL DESCRIPTION | KNEE LENGTH DRESS WITH SHIRT COLLAR AND YOKE IN THE BACK; BUTTON PLACKETS IN THE FRONT NECKLINE; CUFFS AND BUTTON PLACKET IN THE SLEEVE HEM |
| APPROVAL STATUS | GOLD SEAL SAMPLE APPROVED |
| DATE OF APPROVAL | 08.10.2021 |
| FABRICS | YARN DYED WOVEN FABRIC - PLAIN WEAVE |
| TRIMS | PLASTIC BUTTONS, BRAND LABELS |

**FIGURE 8.3** Style description of the given garment is provided on the second page of the tech pack.

| TECHPACK ID | TPI443061 | FIT TYPE | REGULAR | SAMPLE SIZE | S | QUANTITY | 180,000 |
|---|---|---|---|---|---|---|---|
| STYLE | SHIRT DRESS | FABRIC | YARN DYED WOVEN | DESIGNER | XXX | STATUS | GOLDSEAL APPROVED |
| DESCRIPTION | KNEE LENGTH DRESS WITH SHIRT COLLAR | SIZE RANGE | XS - XXL | BUYER | YYY | BRAND | MAX |
| ORDER NO. | 1895193EN | SEASON | SUMMER 2022 | TECHNOLOGIST | ZZZ | DATE | 08.10.2021 |

**FABRIC DETAILS**

| | |
|---|---|
| MAIN FABRIC | CHECKERED FABRIC |
| COLOUR | BLUE AND BLACK CHECKS |
| PLACEMENT | WHOLE GARMENT |
| FABRIC COMPOSITION | 100% COTTON |
| FABRIC STRUCTURE | 2/1 WARP FACED TWILL |
| DYEING METHOD | YARN DYED FABRICS |
| YARN TYPE | SINGLE PLY YARN |
| YARN COUNT | $40^{S}$ $Ne_{C}$ |
| EPI | 65 |
| PPI | 55 |
| FABRIC WEIGHT (GSM) | 170 |
| FABRIC WIDTH (IN) | 54 |
| APPLIED DESIGN | NA |
| FINISHES | CREASE RESISTANT FINISH; PRE-SHRINKING FINISH |
| CARE INSTRUCTION | MACHINE WASH; WARM CYCLE |
| SUPPLIER | AAA |
| SUPPLIER LOCATION | XYZ |

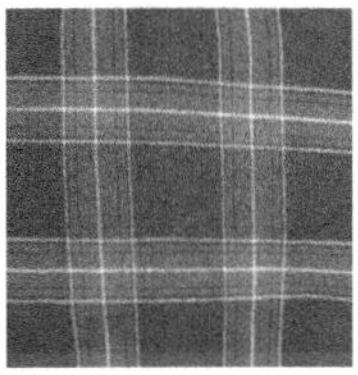

PANTONE NEUTRAL BLACK C
HEX: #2D2926
RGB: (45,41,38)

PANTONE 653 C
HEX: #326295
RGB: (50,98,149)

PANTONE 319 C
HEX: #2DCCD3
RGB: (45,204,211)

PANTONE 000 C
HEX: #ffffff
RGB: (255,255,255)

**FIGURE 8.4** Fabric description of the given garment is provided on the third page of the tech pack.

In the case of a very small number of colours used in the fabric or a smaller number of colourways, the related Pantone shades were provided on the fabric page itself. However, if a particular order has more colourways, meaning that the same style needs to be produced in different colours, then a separate page will be dedicated to mention the different colours and their corresponding Pantone. This is also true in the case of the garment with multiple colour combos. In that case, the details of the colours of each colourway will be provided on a separate page of the tech pack with the required details as shown in Figure 8.5.

### 8.2.1.5 Embroidery/Print Design and Specifications

The very next page in the tech pack will have complete details regarding the surface ornamentation. First of all, the tech pack will have a detailed illustration of any surface ornamentation like print or embroidery or patch with its size details and positions on the garment as provided in Figure 8.6. Additionally, as shown in Figure 8.6, the details related to the different print colours/thread colours in embroidery will be provided with corresponding Pantone/shade numbers. Irrespective of the surface ornamentation type, the page is very crucial for manufacturers as it largely affects the aesthetic look of the garment. Sometimes, based on the complexity of the garment styles, required instructions were also provided here on this page.

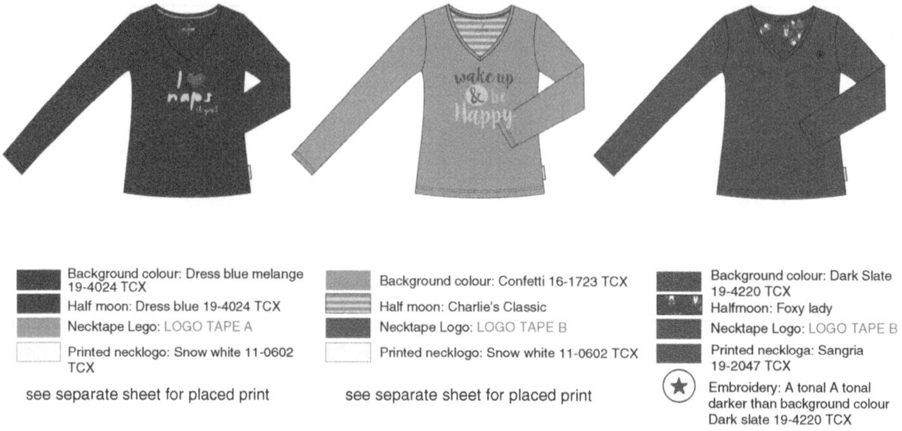


**FIGURE 8.5** Different colour ways of the same garment with Pantone numbers (usually provided on the fourth page [separately] of the tech pack).

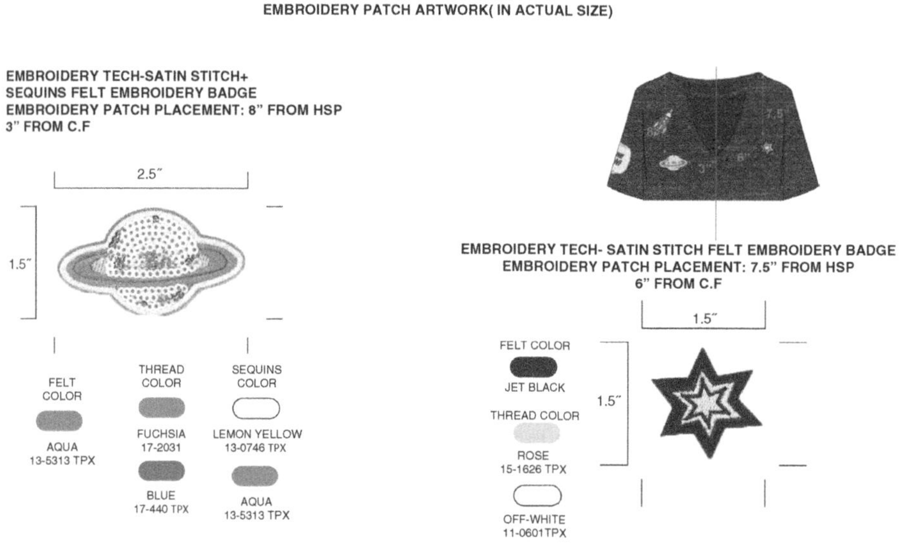


**FIGURE 8.6** Embroidery/print/patch details with its position and colour shades on the fifth page.

### 8.2.1.6 Bill of Materials/Material Details

The details related to the raw materials used in the garment, other than the fabrics like buttons, labels, etc. are detailed on the sixth page. Based on the importance of the style feature, the technical details of the trims and accessories are provided and those are more than enough to reproduce the same material for bulk

| | | | | | | | |
|---|---|---|---|---|---|---|---|
| TECHPACK ID | TPI443061 | FIT TYPE | REGULAR | SAMPLE SIZE | S | QUANTITY | 180,000 |
| STYLE | SHIRT DRESS | FABRIC | YARN DYED WOVEN | DESIGNER | XXX | STATUS | GOLDSEAL APPROVED |
| DESCRIPTION | KNEE LENGTH DRESS WITH SHIRT COLLAR | SIZE RANGE | XS - XXL | BUYER | YYY | BRAND | MAX |
| ORDER NO. | 1895193EN | SEASON | SUMMER 2022 | TECHNOLOGIST | ZZZ | DATE | 08.10.2021 |

**TRIM DETAILS - 1**

| | |
|---|---|
| TRIM 1 | BUTTON |
| TYPE | SEW THROUGH BUTTON |
| DESCRIPTION | FLAT BUTTONS WITH FOUR HOLES |
| SIZE | 14 L |
| MATERIAL | POLYAMIDE |
| COLOUR | BLACK |
| SUPPLIER | BBB |
| SUPPLIER LOCATION | ABCD |
| PLACMENT | PLACKETS IN NECKLINE AND SLEEVE |
| CONSUMPTION DETAIL | 8 BUTTONS PER GARMENT |

4- HOLE SEW THROUGH BUTTONS

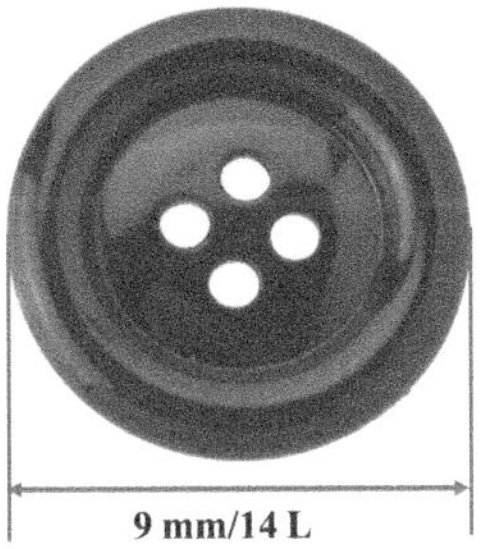


**FIGURE 8.7** The details of the trims and accessories that need to be reproduced are provided on the sixth page.

production. Figure 8.7 shows the details of a button that was provided by the customer/buyer. The basic details like raw material that a trim is made of, colour, method of manufacturing, position details, and also the consumption per garment will be specified. This will help the manufacturer to estimate the required quantities that need to be produced to meet the requirement of the particular order. The same will be provided for all the various trims and accessories that are required to complete the order.

### 8.2.1.7 Measurements and Specification Sketch

The next section of the tech pack will have detailed measurements of all the components of the garment with a technical flat sketch. The measurements will be graded for all the different sizes and provided with markings. The detailed illustration with the marking of every specific component of the garment will help the manufacturer to stick to the requirements and needs of the buyer/customers. An example of these measurements, how it is illustrated in the tech pack, is provided in Figure 8.8.

### 8.2.1.8 Construction Details

The next section in the tech pack is the construction details. After the measurements, the tech pack will have detailed instructions about the different seams and stitches if anything is required specifically. The instructions (if required) will have a complete list of operations and seams with illustrations along with stitches per inch based on the requirement. A sample of such an illustration is provided in Figure 8.9. In this

(a)

| | | | | | | | |
|---|---|---|---|---|---|---|---|
| TECHPACK ID | TP1443061 | FIT TYPE | REGULAR | SAMPLE SIZE | S | QUANTITY | 180,000 |
| STYLE | SHIRT DRESS | FABRIC | YARN DYED WOVEN | DESIGNER | XXX | STATUS | GOLDSEAL APPROVED |
| DESCRIPTION | KNEE LENGTH DRESS WITH SHIRT COLLAR | SIZE RANGE | XS - XXL | BUYER | YYY | BRAND | MAX |
| ORDER NO. | 1895193EN | SEASON | SUMMER 2022 | TECHNOLOGIST | ZZZ | DATE | 08.10.2021 |

**MEASUREMENT GUIDELINE SKETCH**

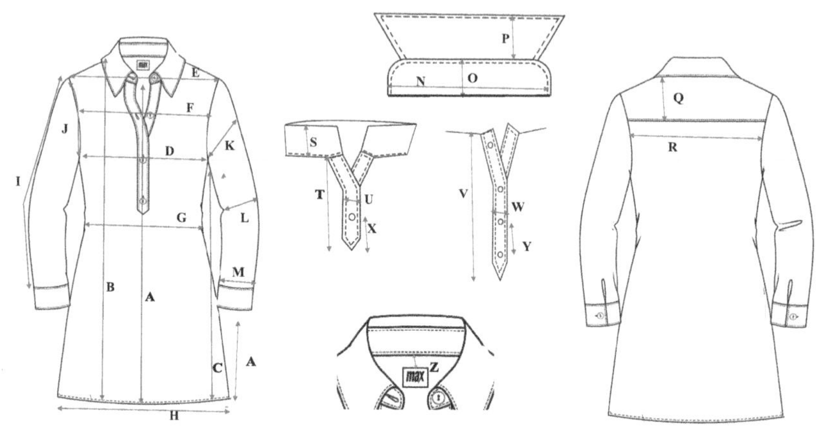


(b)

| | | | | | | | |
|---|---|---|---|---|---|---|---|
| TECHPACK ID | TP1443061 | FIT TYPE | REGULAR | SAMPLE SIZE | S | QUANTITY | 180,000 |
| STYLE | SHIRT DRESS | FABRIC | YARN DYED WOVEN | DESIGNER | XXX | STATUS | GOLDSEAL APPROVED |
| DESCRIPTION | KNEE LENGTH DRESS WITH SHIRT COLLAR | SIZE RANGE | XS - XXL | BUYER | YYY | BRAND | MAX |
| ORDER NO. | 1895193EN | SEASON | SUMMER 2022 | TECHNOLOGIST | ZZZ | DATE | 08.10.2021 |

**MEASUREMENT CHART**

| S.NO | CODE | MEASUREMENT NAME | TOLERANCE | XS | S | M | L | XL | XXL |
|---|---|---|---|---|---|---|---|---|---|
| 1 | A | CENTER FRONT LENGTH | ± 0 | 81 | 81 | 82.5 | 82.5 | 84 | 84 |
| 2 | B | FULL LENGTH | ± 0 | 86 | 86 | 88 | 88 | 89 | 89 |
| 3 | C | SIDE SEAM LENGTH | ± 0 | 64 | 64 | 66 | 66 | 68 | 68 |
| 4 | D | BUST WIDTH | ± 0 | 51.5 | 54 | 56.5 | 59 | 61.5 | 64 |
| 5 | E | ACROSS SHOULDER | ± 0 | 35 | 36 | 37 | 38 | 39 | 40 |
| 6 | F | ACROSS FRONT | ± 0.5 | 38 | 39 | 40 | 41 | 42 | 43 |
| 7 | G | WAIST WIDTH | ± 0.5 | 37.5 | 40 | 42.5 | 45 | 47.5 | 50 |
| 8 | H | HEM WIDTH | ± 0.5 | 53 | 54 | 55 | 56 | 57 | 58 |
| 9 | I | FULL SLEEVE LENGTH | ± 0.25 | 55.5 | 56 | 56.5 | 57 | 57.5 | 58 |
| 10 | J | ARM HOLE CURVE | ± 0 | 20.5 | 22 | 23.5 | 25 | 26.5 | 28 |
| 11 | K | BICEP WIDTH | ± 0 | 17.5 | 18 | 18 | 18.5 | 18.5 | 18.5 |
| 12 | L | ELBOW WIDTH | ± 0 | 14 | 14 | 14.5 | 14.5 | 15 | 15 |
| 13 | M | CUFF WIDTH | ± 0 | 13 | 13 | 13 | 13.5 | 13.5 | 13.5 |
| 14 | N | COLLAR CLOSED AROUND | ± 0 | 43 | 43 | 44 | 44 | 45 | 45 |
| ALL THE MEASUREMENTS ARE IN CENTIMETERS | | | | | | | | | |

**FIGURE 8.8** (a, b) The technical specification and measurement details of the proposed garment in the seventh page.

| TECHPACK ID | TPI443061 | FIT TYPE | REGULAR | SAMPLE SIZE | S | QUANTITY | 180,000 |
|---|---|---|---|---|---|---|---|
| STYLE | SHIRT DRESS | FABRIC | YARN DYED WOVEN | DESIGNER | XXX | STATUS | GOLDSEAL APPROVED |
| DESCRIPTION | KNEE LENGTH DRESS WITH SHIRT COLLAR | SIZE RANGE | XS - XXL | BUYER | YYY | BRAND | MAX |
| ORDER NO. | 1895193EN | SEASON | SUMMER 2022 | TECHNOLOGIST | ZZZ | DATE | 08.10.2021 |

**OPERATION DETAILS – STITCHES AND SEAMS**

| SLEEVE PREPARATION | | | | | | |
|---|---|---|---|---|---|---|
| S.NO | OPERATION | STITCH TYPE | SEAM TYPE | ILLUSTRATION | SPI | MACHINE USED |
| 1 | PLACKET ATTACHMENT | CLASS 301 - SNLS | BOUND SEAM | | 16 | SNLS MACHINE |
| 2 | CUFF ATTACHMENT | CLASS 301 - SNLS | BINDING (CLEAN FINISH) | | 16 | SNLS MACHINE |
| 3 | INNER SEAM | CLASS 301 - SNLS / CLASS 515 - 4 THREAD OVERLOCK | SUPER IMPOSED SEAM | | 12 | SNLS/ OVERLOCK MACHINE |

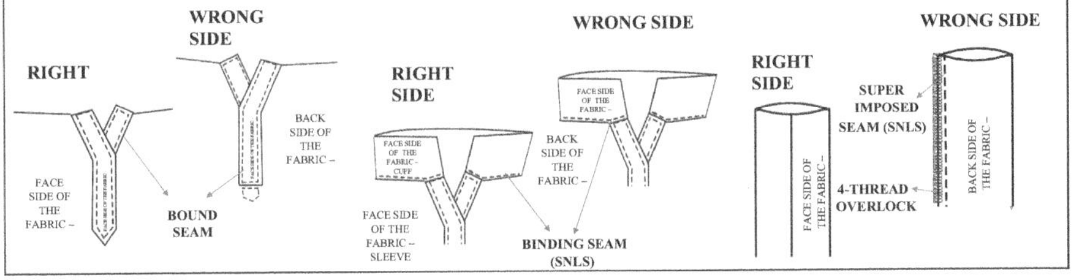


**FIGURE 8.9** List of seams and stitch types required for every operation.

section, information about the stitch or seam class and the types of machines that need to be used also will be provided.

### 8.2.1.9 Reference Images That Explain How to Make a Garment

After the construction procedures, the very next page will list the reference images of a few operations or sections of the garment with specifications. This will help the manufacturers to understand the customised requirements of the buyers. As provided in Figure 8.10, uncommon details related to the given garments will be provided in this section.

### 8.2.1.10 Testing Methods and Quality Evaluation Details

In this section of the tech pack, complete testing methods related to the different materials used in the apparel will be listed. The main function of this section is to educate the manufacturers regarding the quality standard that the buyer is expecting. This will aid the manufacturers in verifying the quality of raw materials like fabric, trims, etc. during the sourcing or manufacturing process itself. A deviation in the quality specification may lead to the rejection of the order and will result in loss. Hence, this section will be carefully evaluated by the merchandiser and monitored by the quality department. Figure 8.11 represents a sample tech pack sheet that requested a few basic tests for the proposed garment.

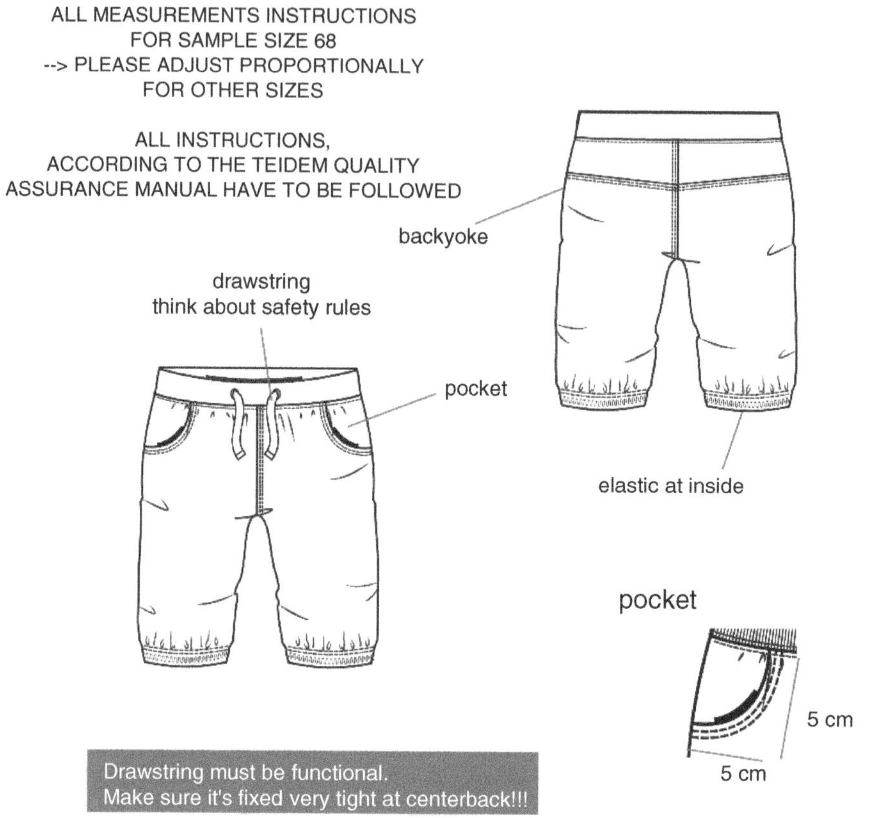


**FIGURE 8.10** Reference images provided by the buyer that explain the quality specification of the garment.

This section also will have the details related to the inspection methods requested by the customer during manufacturing and also mention the acceptance quality lever specified by the buyer to accept the order.

### 8.2.1.11 Packing Material and Instruction

The last section of the tech pack is the packing details. This section will have the details of packing accessories that need to be used in the final packing of the product. Additionally, it will also enumerate the quality details of the materials that need to be used in it. For example, a carton box will be provided with its board GSM/thickness including several plies to be used to manufacture it. The dimensions of the carton box along with other specifications will be provided. An example of such a list of accessories is provided in Figure 8.12 for better understanding.

This section also contains the details related to the packing, commonly known as packing instructions. During the packing process, based on the buyer's requirement, the garments can be packed either in an assorted manner or solid manner. Further, how the primary and secondary packing needs to be done and how many accessories

| TECHPACK ID | TPI443061 | FIT TYPE | REGULAR | SAMPLE SIZE | S | QUANTITY | 180,000 |
|---|---|---|---|---|---|---|---|
| STYLE | SHIRT DRESS | FABRIC | YARN DYED WOVEN | DESIGNER | XXX | STATUS | GOLDSEAL APPROVED |
| DESCRIPTION | KNEE LENGTH DRESS WITH SHIRT COLLAR | SIZE RANGE | XS - XXL | BUYER | YYY | BRAND | MAX |
| ORDER NO. | 1895193EN | SEASON | SUMMER 2022 | TECHNOLOGIST | ZZZ | DATE | 08.10.2021 |

**FABRIC - INSPECTION AND TESTING DETAILS**

**FABRIC INSPECTION** : 4 POINT SYSTEM

**ACCEPTANCE** : < 40 points/sq.yards

| S.NO | TEST CATEGORY | TEST | STANDARD | ACCEPTANCE VALUE | TOLERANCE |
|---|---|---|---|---|---|
| 1 | IDENTIFICATION TESTS | FIBER COMPOSITION | ISO/TR 11827:2012 | 100 % COTTON | ± 0% |
| 2 | | FABRIC WEIGHT | ASTM D3776-96 | 170 g/m² | ± 5% |
| 3 | STRENGTH TESTS | TENSILE STRENGTH | ASTM D5035 | 25 lbs/in | ± 5% |
| 4 | | TEARING STRENGTH | ASTM D1424 | 2200 kg | ± 3% |
| 5 | | ABRASION RESISTANCE | ASTM D4966 | Weight loss - 0% | + 5% |
| 6 | COMFORT TESTS | MOISTURE ABSORBENCY | ISO 20158:2018 | 5 s | ± 1% |
| 7 | | AIR PERMEABILITY | ASTM D737-04 | 98.5 $cm^3 cm^{-2} sec^{-1}$ | ± 3% |
| 8 | DURABILITY TESTS | DIMENSIONAL STABILITY | ISO 6330 | Dimension Change – 0% | ± 3% |
| 9 | | COLOUR FASTNESS TO RUBBING | ISO 105 X12 | 5 | -0.5 |
| 10 | | COLOUR FASTNESS TO WASHING | ISO 105 C06 | 5 | -0.5 |
| 11 | | COLOUR FASTNESS TO LIGHT | ISO 105 B02 | 5 | -0.5 |
| 12 | | COLOUR FASTNESS TO STEAMING | ISO 105 E11 | 5 | -0.5 |
| 13 | | COLOUR FASTNESS TO PERSPIRATION | ISO 105 E04 | 5 | -0.5 |
| 14 | | SEAM STRENGTH | ASTM D1683 | 100 N | ± 8% |

**FIGURE 8.11** Testing methods adopted/requested by the buyer to evaluate the quality of the product.

| TECHPACK ID | TPI443061 | FIT TYPE | REGULAR | SAMPLE SIZE | S | QUANTITY | 180,000 |
|---|---|---|---|---|---|---|---|
| STYLE | SHIRT DRESS | FABRIC | YARN DYED WOVEN | DESIGNER | XXX | STATUS | GOLDSEAL APPROVED |
| DESCRIPTION | KNEE LENGTH DRESS WITH SHIRT COLLAR | SIZE RANGE | XS - XXL | BUYER | YYY | BRAND | MAX |
| ORDER NO. | 1895193EN | SEASON | SUMMER 2022 | TECHNOLOGIST | ZZZ | DATE | 08.10.2021 |

**PACKING MATERIALS**

| S.NO | PACKING ITEM | REFERENCE IMAGE | MATERIAL | SIZE | QUANTITY | PLACEMENT | SUPPLIER | SUPPLIER LOCATION |
|---|---|---|---|---|---|---|---|---|
| 1 | POLY BAGS |  | POLYETHYLENE | 25 cm X 18 cm | 1 PER GARMENT | - | PPPP | SSSS |
| 2 | EAN STICKER | | 100% PAPER | 5 cm X 3 cm | 1 PER GARMENT | PLACED ON THE RIGHT CORNER OF POLYBAGS | PPPP | SSSS |
| 3 | HANG TAGS |  | 100% PAPER | 8 cm X 10 cm | 1 PER GARMENT | ATTACHED TO BUTTON OF THE GARMENT WITH STRING | PPPP | SSSS |
| 4 | STRINGS |  | NYLON | 10 cm | 1 PER GARMENT | USED TO ATTACH HANG TAGS WITH GARMENT | PPPP | SSSS |
| 5 | CARTON BOXES |  | 100% PAPER (3 LAYERS) | 100 cm x 50 cm x 50 cm | 1 PER 30 GARMENTS | - | PPPP | SSSS |

**PACKING INSTRUCTIONS**

**FIGURE 8.12** The packing accessories list provided in the tech pack.

| | | | | | | | |
|---|---|---|---|---|---|---|---|
| TECHPACK ID | TPI443061 | FIT TYPE | REGULAR | SAMPLE SIZE | S | QUANTITY | 180,000 |
| STYLE | SHIRT DRESS | FABRIC | YARN DYED WOVEN | DESIGNER | XXX | STATUS | GOLDSEAL APPROVED |
| DESCRIPTION | KNEE LENGTH DRESS WITH SHIRT COLLAR | SIZE RANGE | XS - XXL | BUYER | YYY | BRAND | MAX |
| ORDER NO. | 1895193EN | SEASON | SUMMER 2022 | TECHNOLOGIST | ZZZ | DATE | 08.10.2021 |

**PACKING INSTRUCTIONS**

| S.NO | DETAILS |
|---|---|
| 1 | EACH GARMENT (AFTER SEWING CARE LABEL AND SIZE CUM BRAND LABEL) SHOULD BE ATTACHED WITH HAND TAGS |
| 2 | HAND TAGS ARE ATTACHED IN THE BUTTON IN THE FRONT PLACKET WITH NYLON STRING TAGS |
| 3 | EACH GARMENT IS FOLDED WITHOUT USAGE OF SUPPORTING ITEMS LIKE BOARDS AND PINS |
| 4 | EACH GARMENT IS PACKED IN A SINGLE POLYBAG |
| 5 | EACH POLY BAG IS STICKED WITH EAN STICKER AT THE BOTTOM RIGHT CORNER |
| 6 | 30 POLYBAGS WITH GARMENTS SHOULD BE PACKED IN A CARTON BOX |
| 7 | SIZE WISE ASSORTMENT - EACH BOX SHOULD HAVE 5 PIECES OF EACH SIZE; NO COLOURWISE ASSORTMENT |

Page 23 of 25

**FIGURE 8.13** Packing instructions in the last section of the tech pack.

like brand tags, and price tags, will be specified. Even the complete packing procedure will be provided by the buyer. A detailed example of actual packing instructions is provided in Figure 8.13 for better understanding.

### 8.2.1.12 Costing Sheet

The costing sheet is one of the optional information in the cost sheet. In this sheet, all the raw materials and process costs will be estimated and the price that was quoted to the buyer during the acceptance of the order will be provided including all the taxes and the manufacturing country. However, due to the confidentiality of the costing process, this section will be kept along with the merchandisers and will not be passed to other contributors involved in the manufacturing process.

A complete tech pack will have all the essential details to reproduce a particular style without any inconvenience. Buyers will prefer to have a similar garment that they have requested. Hence, no changes will be accepted in the approved styles further. The main objective of the tech pack is to advise the individual manufacturing department with step-by-step instructions. Hence, this document will be always considered an important reference document to complete the order successfully.

## 8.3 MARKER PLANNING IN THE APPAREL INDUSTRY

The marker planning process in the apparel industry is one of the preproduction activities, performed to estimate the most efficient layout of the pattern pieces for a specific style, fabric, and distribution of sizes (Pamuk and Yildiz 2016). As fabric is the most important raw material that decides the cost of the garment, efforts will be taken at this stage to reduce the wastage of the fabric to the possible extent. The percentage of waste generated in the cutting process is mainly based on the style feature and size of the garment, fabric width, and assortment plan. A reduction in fabric wastage will reflect directly on the product cost and result in a higher profit. Hence it is necessary to take possible care in the marker planning process as it determines the wastage in the cutting process. Marker can be defined as the arrangement of the garment patterns in the most efficient way by considering fabric grain, the visual orientation of the finished surface of fabrics, colour and design matching between adjacent fabrics, pile direction, etc. (Tanaka and Wachi 1973). Marker planning is the plan that is used for cutting the fabric plies. It is the process of arranging all the patterns of components of one or more garments on the area of fabric lay (Wong et al. 2005). In the apparel industry, fabric contributes around 50%–60% of the total manufacturing cost (Tsao et al. 2022). Hence, marker planning is crucial as an effective marker plan can reduce fabric wastage during the cutting process.

## 8.3.1 Marker Efficiency

Marker efficiency is the percentage of the ratio between the total area of all patterns on the marker and the total area of the marker (Rahman et al. 2017).

$$\text{Marker Efficiency} = \frac{\text{Area of all patterns on the marker}}{\text{Area of marker}} \times 100$$

### 8.3.1.1 Factors Affecting Marker Efficiency

Marker plan efficiency can be greatly influenced by different factors such as fabric width, fabric spreading modes (single ply/face to face), and bundling option (all bundle same direction/alternate bundle alternate direction) (Pamuk and Yildiz 2016). The individual parameters are detailed in the following section.

#### *8.3.1.1.1 Fabric Width*

The efficiency of the marker planning process largely depends upon the fabric width. As the marker length can be decided by the planning team, the width is preset during the fabric manufacturing itself. If an order was processed with different fabrics produced with different widths, ultimately, the lower fabric width will be considered for the panning process. This will lead to a lot of fabric wastage. Hence, preplanning the fabric widths in advance will resolve this issue very well. Unfortunately, this becomes a crucial factor in deciding the fabric wastage if the manufacturer purchases fabric from different vendors.

#### *8.3.1.1.2 Fabric Type and Design*

The type of fabrics used in the marker planning is one of the main factors that decide the effectiveness of the marker plan. Fabrics with stripes or checks require the matching of patterns between different components. This can reduce the marker efficiency. Whereas in the case of plain fabrics, the efficiency will be higher (Vilumsone-Nemes 2018). In general, higher marker efficiencies were obtained in the case of symmetrical fabrics, whereas for the asymmetrical fabrics, lower marker efficiency is achieved comparatively.

##### *8.3.1.1.3 Fabric Spreading Modes*

The spreading of fabric will be performed in different modes, and that will be detailed in the forthcoming section of this chapter. Based on the design or pattern or nature of the fabric, the fabric will be spread in either one direction or two directions. Similarly, the way it was spread, based on its face side direction, the efficiency of the marker plan is largely affected.

##### *8.3.1.1.4 Size Ratio*

The number of sizes in the order and the size proportions will have a major influence on the marker planning process. When the sizes are distributed in a balanced manner from small to large size, there is a higher possibility to get increased efficiency.

##### *8.3.1.1.5 Garment Style and Design*

The garment style and the design influence the marker efficiency. For instance, researchers reported T-shirts with set-in sleeves had more marker efficiency than the Raglan sleeve T-shirt and the Kimono sleeve T-shirt (Islam et al. 2019). Similarly garments with a large number of panels with different shapes, especially smaller panels will have lower efficiency compared to a garment with larger and fewer panels.

## 8.3.2 Manual and Computerised Marker Planning

### 8.3.2.1 Manual Marker Planning

In the manual marker planning process, there is no fixed rule. The effectiveness of the planning depends upon the individual who plans the pattern arrangement. In this process, once the patterns of the particular style are drafted for all the sizes, based on the lay length and width, the patterns will be arranged on the top surface of the spread. Based on the total area and pattern area, the efficiency of the marker plan will be estimated. If further increment is required, the arrangement will be modified for better combinations. In this process, it is important to ensure the following points as this process is performed manually.

i. The correctness of the pattern sizes and number of components per garment/ correct number of patterns.
ii. Type of marker – open and closed layout.
iii. Orientation of the patterns concerning the individual component requirements.
iv. Quality specifications like stripe matching, pocket alignment, etc. need to be checked carefully.

### 8.3.2.2 Computerised Marker Planning

In the computerised marker planning process, though the patterns are made manually, the details related to the components like the pattern size, pattern pieces, grade rule, spread length, width, and other required information are fed into the software. The drafted patterns were digitised or converted into digital format to feed into the software. This process will ensure the accuracy of the measurements for each component of all sizes. Once the required components are fed into the computer, based on the software types, either by providing the efficiency level or by providing the lay length and width one can create a marker planning.

Software also will have the provisions to control and change the size combinations to be planned on the lay. So machine operator can combine the required sizes to get the maximum efficiency with the given length and width of the fabric. The developed markers will be printed out to the required length and width using plotters. The plotters are capable of plotting all types of patterns, individually or graded nests of patterns with complete annotations as required by the firm. The plotters will have multiple head jets that can print variable line density and width, text identification information, and bar codes. However, the costs associated with plotters limit the number of plotters a firm can have. Hence it is one of the bottlenecks in the preproduction stage, when the company needs multiple copies of the marker plans or when multiple orders are run. Though the higher initial investments and requirement of skilled operators are noted as disadvantages, the following benefits are noted with the computerised marker planning process.

- Computerised marker planning is suitable for large-scale industries that produce frequent style changes.
- Irrespective of styles used, a higher marker efficiency can be achieved and hence less wastage reported.
- The plans can be stored digitally and can be reproduced at any time.
- Difficulties associated with the grading process are eliminated; based on the input parameter, the system grades the patterns automatically.
- Higher production and requirement of lower manpower, especially the skilled operators.

## 8.4 FABRIC INSPECTION TECHNIQUES

Fabric inspection plays an important role in quality control of the textile and fashion industry as they have a greater impact on the costs and grading of final products. Nearly 85% of the second-quality products in the garment industry are caused by fabric defects that lead to a loss in revenue for manufacturers. Therefore, fabric inspection is given more priority in the textile industry for the production of quality products. Usually, fabric inspection is carried out by textile professionals and they evaluate the products subjectively. As the production rates are higher nowadays, the manufacturers must be able to identify the defects and their source of occurrence to reduce the number of second-quality fabrics. Therefore, to identify the defects and to produce high-quality fabrics, a wide range of online automated fabric inspection systems is taken into concern. They have several advantages, which include improved safety, reduced labour costs, elimination of human error, and the creation of timely statistical product data (Goyal 2018).

### 8.4.1 Conventional Fabric Inspection Techniques

The widely used fabric inspection systems are the Dallas system, Graniteville '78' system, 4-point system, and 10-point system.

#### 8.4.1.1 Dallas System

This system was developed in the year 1970 specifically for knitted fabrics and it was also approved by the Dallas Manufacturers Association. In this system, when the garment has a defect, it will be termed as 'second'.

#### 8.4.1.2 Graniteville '78' System

This system was developed in 1975, especially for the field of grading. In this system, the defects are classified into major and minor types. When the defect is very obvious and it leads the fabric to second quality, it is termed a major defect. A minor defect does not lead the fabric to second quality and it also depends on the location of the defect. Table 8.1 shows the length of the defect and the assigned penalty points using this system (Rana 2012).

#### 8.4.1.3 4-Point System

The most common and widely used fabric quality inspection system for all types of fabrics in the textile and apparel industry is the 4-point fabric inspection system. This system is also called as the American Apparel Manufacturer's point grading system for the determination of fabric quality.

It assigns 1, 2, 3, and 4 penalty points according to the size and significance of the defect in the fabric (Table 8.2).

#### 8.4.1.4 10-Point System

The 10-point fabric inspection system is a point-per-fault system and it gives a measurable guide to quality grading per roll (Islam 2015). In this system, if the total penalty point of the piece does not exceed the total yardage, it is graded as 'first'. When the penalty point exceeds the total yardage, it is graded as 'second'. As the points per length are different for warp and weft defects, this system is a bit complicated and difficult to use in practice. The penalty points assigned to warp and weft defects are shown in Table 8.3.

### 8.4.2 Manual Fabric Inspection Machine

In fabric inspection, rolls of fabric are opened and run through a machine with a light under the fabric, and a fabric checker visually inspects the fabric. The fabric checker

**TABLE 8.1**
**Penalty point allocation for the various lengths of defects in the Graniteville '78' system**

| Length of defect | Allotted penalty points |
|---|---|
| Up to 9 inches | 1 |
| 9–18 inches | 2 |
| 18–27 inches | 3 |
| 27–36 inches | 4 |

**TABLE 8.2**
**Penalty point allocation for the various lengths of defects in the 4-point system**

| Length of defect | Allotted penalty points |
|---|---|
| Up to 3 inches | 1 |
| 3–6 inches | 2 |
| 6–9 inches | 3 |
| Over 9 inches | 4 |
| Woven fabric hole | |
| 1 inch or less | 2 |
| More than 1 inch | 4 |
| Knitted fabric | |
| Any hole | 4 |

**TABLE 8.3**
**Penalty point allocation for the various lengths of defects in the 10-point system**

| Length of defect | Allotted penalty points |
|---|---|
| **Warp defect** | |
| Up to 1 inch | 1 |
| 1–5 inches | 3 |
| 5–10 inches | 5 |
| 10–36 inches | 10 |
| **Weft defect** | |
| Up to 1 inch | 1 |
| 1–5 inches | 3 |
| 5 inches to half the width of the fabric | 5 |
| Full width | 10 |

marks the defects with chalk or stickers while inspecting the fabric. Figure 8.14 represents the conventional manual fabric inspection system used in the apparel industry.

The inspection report is made at the end of checking the fabric roll. Then the rolls are graded under different categories.

- Passed fabric rolls are ready for use and sent to the cutting department.
- Failed fabric rolls are not ready for use and correction should be done.
- Rejected fabric rolls are not acceptable and are sent back to the suppliers.
- Decision-pending approval should be done by the authorised person whether to pass/fail or reject the fabric roll.

FIGURE 8.14 Conventional fabric inspection table.

### 8.4.3 Automation Technology for Fabric Inspection System

Human inspection is generally used in the textile industry due to its cost-effectiveness. The major disadvantage of human inspection is that the work is repetitive and it leads to an increase in inspection errors. On the other hand, automated inspection can solve this issue by reducing errors and reducing the demand for highly skilled fabric inspectors. The development of an Automated Visual Inspection system for weaving defect detection based on image processing and recognition algorithms helps in effective fabric inspection. The purpose of the developed system is the inspection of fabric in the textile industry. In this system, the location, size, and image of the defect present in the fabric are recorded. After the process of inspection, the fabric will be graded in terms of the severity of defects and the detailed report will be printed. This system not only uses the standard image processing functions, but it also uses neural network techniques for the classification of weaving defects. This system also provides accuracy for the fabric inspection process (Chan et al. 1998). Several companies in the market have now developed fabric inspection systems that can detect and mark the fabric defect in the fabric. With the advancements in technology, automated inspection machines can inspect the fabric more accurately, quickly, and consistently. Due to the highly efficient production, automated inspection machines minimise the amount of wastage during the process. The anti-slip mechanism and edge guide help in maintaining the centreline of the fabric to measure fabric rolls with better accuracy and efficiency. The automation technology comes with both online and offline integration. During online inspection, the system allows the fabric to assimilate into the production process itself. Further to add, the offline integration also helps in production/inspection speed (Btex 2022).

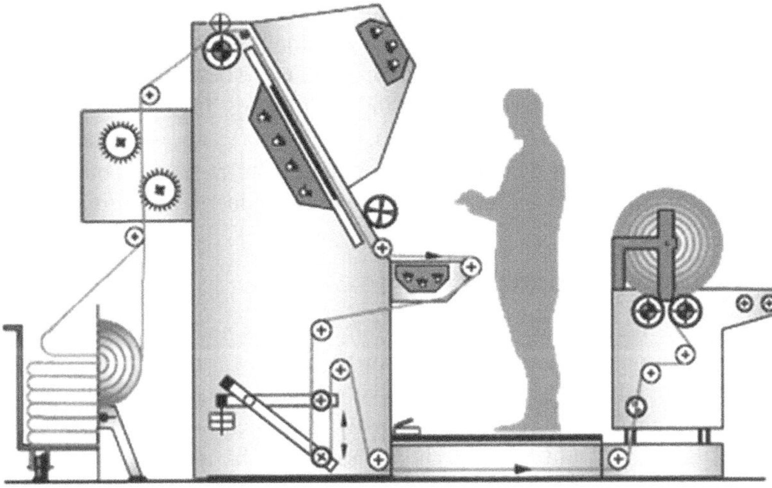

**FIGURE 8.15** Process flow of fabric in an automated fabric inspection machine.

The approximate speed of the automated inspection machine is in the range of 80–100 m/minute. The machines can accommodate any width of fabric without any restriction and are suitable for both knit and woven fabrics. The systems were developed to create invisible ink marks or labels on the point of defect to identify the defects in the later stages. Figure 8.15 represents fabric flow in the automatic fabric inspection machine. The grey line indicates fabric and the dark grey boxes show the position of sensing cameras to understand the defects. The fabric from the back side of the machine passes through the sensing cameras and is wound on the delivery package. Though the machines are provided with a platform for the operator, it is not necessary to allocate individual operators to every machine. A single operator can attend to multiple machines for machine stoppage, power failure, breakdown, etc.

## 8.5 SPREADING

It is the process carried out once the order is ready for execution. The marker plan and fabric inspection should be completed before the start of the spreading process. Spreading is the process of laying the fabric for cutting. The fabric is unwound from a cloth roll or folded form as a single layer (double layer in the case of tubular fabric) and laid on the table to enable the cutting process. The fabrics are laid with each layer superimposed on the other, and according to the marker plan, the number of layers for a particular order is decided. The number of layers is based on the order quantity and fabric characteristics, which include design direction, fabric type, thickness, and the capacity of spreading and cutting machine. When the fabric is spread, there are few requirements to be taken care of even spreading and the process involves time based on the order quantity. There are different modes of spreading used according to the final design requirement. The spreading is done either by hand or by machines that are manually operated and computer operated. There are different types of spreading tables used to enable easier cutting after laying. The cutting is mostly done on the same table where the fabric is spread. Once a fabric is spread, then the labeller is used to label the various parts of the fabric before cutting so that bundling can be done easily after cutting based on the labels thereby reducing the human effort.

### 8.5.1 Requirements of Spreading

Since the spreading process involves the fabric that is needed to get a garment, it has to be carried out without any defects or problems. In this regard, there are few requirements of this process. Firstly, the alignment of the fabric plies should be proper concerning the marker plan. Secondly, the plies or the layers should have correct tension and should be flat. There should not be any ridges in the spread. Thirdly, the ply direction should be correct as per the requirement, which means that the nap direction should be taken care of. The nap direction and its importance will be discussed in the modes of spreading. Fifthly, the fabric should be free from faults or defects and these faults are defects should be removed in the inspection process. Sixthly, the fusion of plies (especially in the case of synthetic fibres) should be avoided due to heat generated by the cutting knives during cutting. This can be done by using anti-fusion paper, reduction of lay height, and by using lubricant on the blade. Lastly, static electricity if any should be eliminated (in the case of synthetic fibre fabrics) as friction causes static electricity generation. Increasing the humidity can avoid static electricity generation (Tyler 2009; Rathinamoorthy and Surjit 2015).

### 8.5.2 Modes of Spreading

There are different modes of spreading based on the type of garment and fabric. The modes of spreading are classified based on the nap direction and fabric facing. The nap direction is the orientation of the fabric surface and the fabric facing is how the fabrics laid will be placed concerning each other. The fabric facing refers to the layers facing each other or not facing each other. The spreading based on fabric facing is a result of the fabric and garment symmetry. The symmetry of the fabric and garment plays a major role in spreading. It is necessary to understand what symmetry refers to before knowing the modes of spreading. In the case of fabric, it can be symmetrical or asymmetrical. A fabric is said to be symmetric when there is no change in the appearance of the fabric when it is turned 180°. If it is a solid-coloured fabric, mostly it is symmetrical. If the front and back of the fabric are the same, then it is symmetrical. If it is different, then it is called asymmetrical. In some garments like shirts and trousers, the outside part of the garment is viewed and it gains importance and no one needs to care about the inside view of the fabric. So in those cases, fabric can be symmetrical or asymmetrical. But in some cases like pile structures, a difference in the face and back structure is predominantly noted wherein the back cannot be used as the face while stitching the garment. So in those cases, spreading should be done in such a way that the pile structure is facing one direction always.

**FIGURE 8.16** (a) Symmetrical garment and (b) asymmetrical garment.

This indicates that the spreading has to be changed based on the fabric symmetry. Similar to fabric symmetry, garment symmetry also plays a role in spreading. An example of symmetrical and asymmetrical garments is given in Figure 8.16. In the case of Figure 8.16a, the garment is symmetrical, wherein the left and the right sides of the garment are the same (i.e., each side is the mirror image of the other side). When the spreading is done, it can be done in such a way that the fabrics can lie face to face; thereby when cut, every pair of plies will give the left and right patterns of each garment reducing the cost of cutting by half. A half set of patterns can give the complete garment and such markers are called as closed markers. In the case of Figure 8.16b, the left and right sides of the garment are different from each other and it is known as an asymmetrical garment. Unlike a symmetrical garment, cutting a right front will not yield a left front in the asymmetrical garment as they are different and so an open marker has to be used in such cases. An open marker is one where all components of the garment are obtained separately and no pairs of right and left are obtained together.

The nap direction also plays a role in deciding the spreading mode. Nap direction refers to the direction in which the fibres in the surface naturally lay. This is mainly viewed in the case of pile fabrics and printed fabrics. In the case of asymmetric fabrics, there is a down direction and an up direction. So if a change in the orientation of direction causes a change in the garment's appearance, then the fabric should be spread in one direction. Hence, the modes of spreading have to be changed based on the nap direction being up and down.

There are six modes of spreading based on the fabric type and garment type. They are

a. Face one way, nap one way mode (FOW, NOW)
b. Face one way, nap up down (FOW, NUD)
c. Face to face, nap up down (FF, NUD)
d. Face to face, nap one way (FF, NOW)
e. Closed face to face, nap one way (CFF, NOW)
f. Closed face to face, nap up down (CFF, NUD)

There are significant differences between these modes of spreading and it is highlighted in Table 8.4. The characteristics of the spreading modes are given in Table 8.4. In the case of quality of spreading, FOW, NOW has the highest quality as the fabric is laid from left to right and then the fabric end is cut and then the spreader moves back to its original position to lay the next layer. In the case of FOW, NUD mode, the spreader moves from left to right, then the fabric end is cut, the roll is rotated, and then the fabric is laid from the right to left. Although the nap direction is altered, the fabric facing is one way and it gives the second-best quality of spreading. In the case of FF, NUD and CFF, NUD modes, moderate quality is achieved as the fabric is kept face to face in both cases and nap direction is both up and down, which means that nap direction is not playing a major role in the garment. CFF modes are used in the case of knitted fabrics where tubular fabrics are obtained. In the case of FF, NOW and CFF, NOW, the quality is lowest as the fabric is kept face to face and it is not possible to identify the defects. When it comes to the spreading speed, FOW modes have the lowest speed as the spreading is done in such a way that the face is always in one way thereby causing either deadheading or roll rotation. In the case of FF, NOW mode too, the speed will be less as there is deadheading and roll rotation, but since the cutting of a left or right ply yields both left and right side pattern (closed marker), they are considered to be comparatively faster to FOW modes. The FF modes are the fastest as they do not have deadheading or roll rotation, and also in the case of CFF, NOW mode, two layers of fabrics are cut for one time spread as each time spreader moves from left to right, two layers are laid as it is a tubular fabric.

In the case of FOW modes, the fabric is always viewed face up thereby leading to easy fault identification, whereas face up and down are viewed in the case of FF modes thereby leading to difficulty in identifying the faults as the face is not viewed on the top of every layer.

The nap direction is one way in all one-way modes thereby causing no nap issues. Since the nap direction is one way, pile structures and asymmetric fabrics can be spread using these modes. If nap direction does not play any role in the final garment, then NUD modes can be used for spreading. For asymmetric fabrics and garments with a one-way nap, open markers are required and hence NOW modes will be handy. However, in the case of FF, NUD modes, both open and closed markers can be obtained. If the fabrics are asymmetric, then only NOW modes of spreading can be used. The end is cut before the next layer spreading in all modes except FF, NUD and CFF, NUD modes. The roll rotation is done in the case of FOW, NUD and FF, NOW modes. The deadheading is there in all NOW modes. Deadheading refers

**TABLE 8.4**
**Classification of spreading modes**

| Characteristics | FOW, NOW | FOW, NUD | FF, NUD | FF, NOW | CFF, NOW | CFF, NUD |
|---|---|---|---|---|---|---|
| **Quality** | Highest | 2nd highest | Moderate | Lowest/moderate | Lowest/moderate | Moderate |
| **Speed** | Slowest | 2nd slowest | Fastest | 3rd slowest | Fast | Fastest |
| **Viewing** | Face up | Face up | Up/down | Up/down | Up/down | Up/down |
| **Fault identifying** | Easy | Easy | Difficult | Difficult | Difficult | Difficult |
| **Nap direction** | One way | Two way | Two way | One way | One way | Two way |
| **Nap issues** | No | Yes | Yes | No | No | Yes |
| **Marker type** | Open | Open | Open/closed | Open/closed | Open/closed | Open/closed |
| **Fabric symmetry (S)/asymmetry (A)** | S and A | S | S | S and A | S and A | S |
| **Cut at end** | Yes | Yes | No | Yes | Yes | No |
| **Roll rotate** | No | Yes | No | Yes | No | No |
| **Deadheading** | Yes | No | No | Yes | Yes | No |
| **Appearance** | 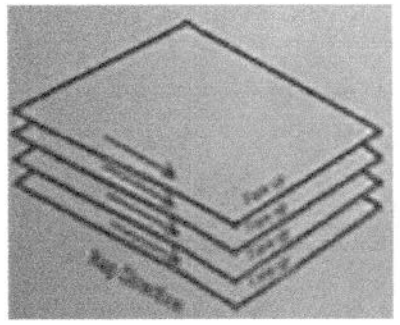 | 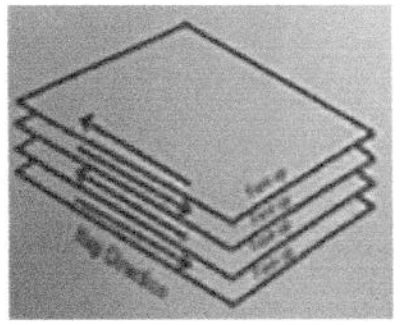 | 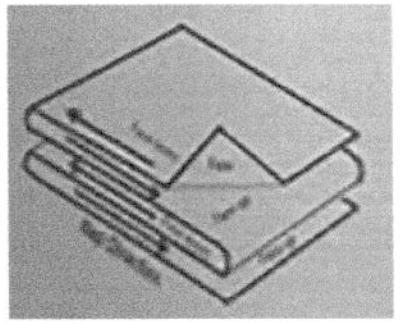 | 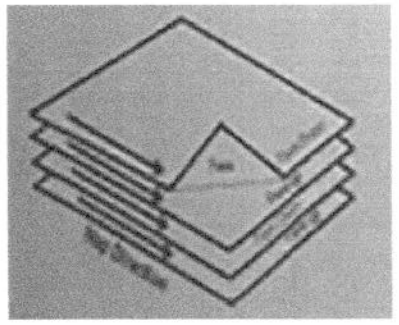 | 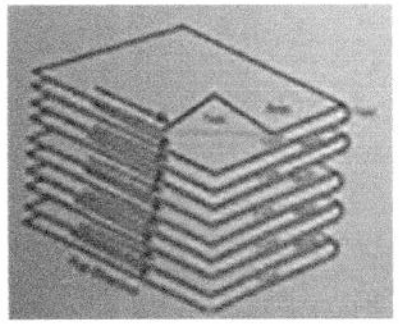 |  |

to the spreader spreading fabrics only in one direction i.e. from left to right and then returning to the left to lay the next layer. During return, it does not lay the fabric as such garments require a nap in one way. This leads to reduced production. The appearance of different modes of spreading is also shown in Table 8.4.

### 8.5.3 Spreading Methods

It is required to know the various methods of spreading before knowing the process of spreading. The spreading methods refer to how the fabric is spread on the table. The means used for the spreading are what it refers to and the means determine the time taken for spreading. The spreading is done in a manual or computerised way. In the case of the manual method, spreading is done by hand or with the help of a spreading truck with an operator. In the case of spreading by hand, two personnel stand on either side of the table and they spread the fabric roll on the table as per the length and width required. Then a scissor is used and the end is cut and again the next layer is laid. This is used for small quantities of garment manufacturing and samples as productivity is very low. When a spreading truck is used with an operator, then a truck loaded with a fabric roll is placed on one side of the table. Then it is moved by hand to the other side and the fabric is opened out from the roll and spread on the table along the movement according to the length and width required.

Since the above methods consume more time and less productivity, automatic spreading machines are the need of the hour for bulk production with increased efficiency and utilisation. The maintaining of ply tension is also done automatically when a spreading machine is used. An example of a computerised spreading machine is shown in Figure 8.17. It is the spreader machine that is computerised where the fabrics can be fed as rolls or in folded form.

**FIGURE 8.17** Computerised automatic spreading machine.

These spreading machines consist of a cradle, which is the feed system where rolls are kept for the spreading purpose. Then, there are various sensors available like fabric tensioning, positioning, and width indicating sensors that automatically govern the tension, position, and width of the fabric based on the details fed on the computer. There are also damage control mark sensors to identify damages in the fabric. This sensor is used to stop the machine when a damaged or fabric fault is located. Then there is a table on which the cradle is placed and fabric is laid. At the other end of the table, an end catcher and fabric cutter are attached to hold and cut the fabric based on the spreading mode. These machines have a control panel as shown in Figure 8.17 and it has all the details related to lay length, number of plies, starting point of the spread, spreading mode, speed, deadheading speed, fabric tension, and fabric cutting speed. These machines are equipped with safety systems to avoid hazards and accidents. An operator can stand on the side platform and move along the machine when it spreads fabric on the table. The machine setup moves on the table based on the spreading length set in the control panel. The spreading tables are of different types. There are smooth surface tables, tables with air flotation, and vacuum tables. These tables are made of metal with a bi-laminated top. The table heights are adjustable and are around 90 cm. There are special tables available with 10 × 10 grid holes with pins positioned for aligning lays correctly in the case of stripes and checks.

### 8.5.4 Spreading Process

The spreading process involves the spreading of lays onto the table using any one of the methods mentioned above. Based on the marker plan, the number of lays required along with the length and width of the marker is known before spreading. The types of lays are flat and stepped lays, with the flat lay being one layer or multiple layers depending on the requirement and stepped lays are multiple lays with different lengths that are used for getting more marker efficiency based on the garment to be manufactured. Once these details are known, then they are fed onto the computerised spreading machine or the details are provided to the operator if the manual method of spreading is used. In the case of automatic spreaders, the cradle is loaded with the fabric roll. Then the roll is unwound and passed over a tension roller and then it is positioned in such a way to fall on the table. Now, the spreader that includes a cradle with roll and tension roller is moved from left to right and then the fabric is laid on the table. Now once the required length is reached, then the end catcher positioned at the other end will catch the fabric and an end cutter, which is normally a round knife, will automatically move across the table to cut the fabric. In the case of spreading modes where cutting is not required, the fabric is folded and the next layer is laid without cutting. If there is a roll rotation required, then it is done and again the spreader moves to the other side laying the fabric. In the case of FOW, NOW mode, the material is laid from one end to another end, then the fabric is cut and the spreader moves back to left without laying the fabric. Then again the spreader moves forward laying the next layer. In the case of FOW, NUD mode, the spreader moves from right to left, then the fabric end is cut and the roll is rotated by 180° and the fabric is laid from right to left. In the case of FF, NUD mode, the fabric is laid from left to right, then it is folded and again laid from right to left. A similar way is followed for CFF, NUD mode where it is a tubular fabric with two layers laid for every movement of the spreader in one direction. In the case of FF, NOW mode, where asymmetric fabrics can also be spread, the spreader moves from left to right and lays the fabric, then the end is cut, the roll is rotated, and the fabric is not laid as it is deadheading. After return, the fabric is again spread from left to right. Here a closed marker can be used. A similar laying method is followed for CFF, NOW where roll rotation alone is not required as it is a tubular fabric. In these different modes, the fabric is spread and then it is used for cutting. On the same tables, cutting can be done or the materials can be moved through conveyors to the cutting table as per the requirement. Before the cutting operation, the labels are stuck on the top layer of the fabric using the labelling machine in the case of computerised cutting thereby making the bundling process easier.

### 8.5.5 Labelling Machine

The labelling machines are used for identifying the cut fabric bundles as it avoids marker printing, and costs associated with marker paper and ink, and also their impact on the environment is negated. Figure 8.18 shows a labelling machine used in the apparel industry where the machine runs across the table on which the fabric is laid and sticks the labels on top of the lay for identification after the fabric is cut.

In conventional label printers, labels are printed separately and are manually stuck on the garment components whereas, in modern machines, the labels are printed

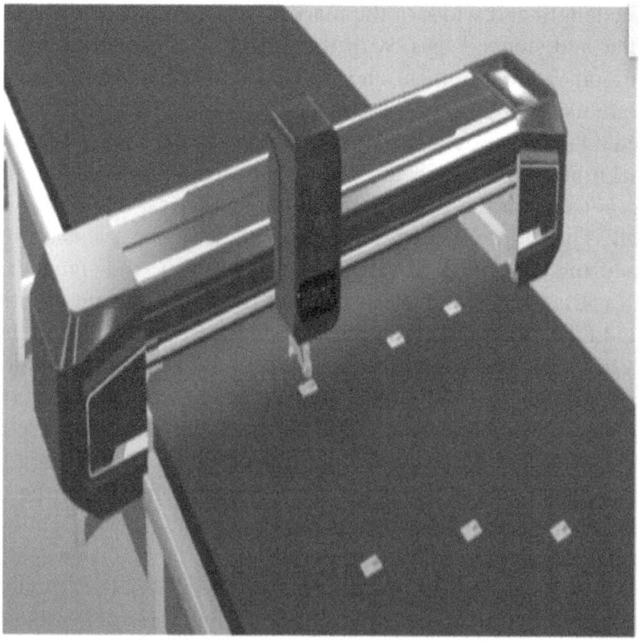

**FIGURE 8.18** Labelling machine.

using a thermal printer and the adhesive labels are applied on top of the lay. The labels are of various sizes based on the data to be printed on them. These machines are compatible with any CAD machines and can also run independently. Software is provided to correctly affix the labels in the right areas and there is also provision for the operator to move them based on his requirement. Once the labelling is done, the material is ready for the cutting process.

## 8.6 CUTTING

The cutting process is one of the most important pre-sewing processes. Cutting is the process where the spread fabrics are cut as per the marker plan for further usage in the construction of the required garment as per specifications. In many cases, cutting operations play a major role in the final quality of the garment. There is always a need for proper cutting of the garment components for further usage without any issues. The cutting process plays a significant role in the final garment manufacturing as wrong cutting may lead to incorrect garment components thereby making the final sewing difficult. There are various requirements in the cutting process. The cutting should be accurate and precise. When samples are manually cut using cutting machines, the accuracy depends on the knife sharpness and operator skill; whereas in the case of computerised cutting machines and die-cutting machines, the software controls the cutting accuracy. Another requirement of cutting is consistency in cutting. The garment parts cut should be of the same size and shape in all the lays and there should not be any difference in the measurement as that will lead to the production of defective garments. The fused edges should be avoided during cutting as more heat generation of cutting blades may lead to the fusing of layers in the case of synthetic fibre fabrics. The fused edges can be avoided by reducing lay height, cutting speed, lubricating cutting knife, and using anti-fusion paper in the lay. Also, the lay support is an important requirement wherein the tables should be capable of supporting the lay well for proper cutting to take place (Tyler 2009).

### 8.6.1 Cutting Machines

The cutting process can be carried out using different machines according to the requirement of cutting. It is imperative to understand the various machines used for cutting before understanding the process followed in cutting. The various machines used in cutting are classified based on the type of cutting and movement of the machine. The types of cutting are manual and machine operated. The manual cutting is done by using scissors and in the case of machine cutting, various machines are used. Machine cutting is classified as stationary and portable cutting. The portable cutters are straight knife-cutting machines, round knife-cutting machines, and rotary shears. The stationary cutters are band knife-cutting machines, die-cutting machines, and computerised cutting machines. Along with these machines, there are also other cutting machines in the stationary category like ultrasonic cutting machines, laser cutting machines, and water jet cutting machines, which are used to a limited extent compared to the other three stationary cutting machines. There are also allied small cutters employed for strip cutting, rag cutting, and end cutting called strip cutters, rag cutters, and end cutters, respectively. Apart from these cutters, there are position markers that are used for marking the position for placement of darts, pockets, etc. They are notchers, drills, and thread markers. Among these machines, the most commonly and majorly used machines are straight knife-cutting machines, band knife-cutting machines, round knife-cutting machines, computerised cutting machines, and die-cutting machines.

The straight knife-cutting machine consists of a straight knife attached to a frame fitted with a motor. In these types of machines, the machine along with the knife is moved inside the fabric lay to cut the required components. The oscillating knife is moved inside the lay manually according to the pattern shape and cutting is done. It involves skilled operators to cut garment parts effectively and correctly. It can be used for small-scale production and where the number of lays is less. However, it can handle the lay height of around 25 cm with ease and operates at around 3,000 rpm. The round knife-cutting machine is used for cutting smaller parts and small lays as it can handle 20–30 layers. These machines also have a small motor fitted to it and can be moved through the fabric for cutting as round knives of 6 to 20 cm diameter with different shapes like octagon and hexagon are used based on the type of cutting required. These machines operate at a speed of 1,200 rpm. In the case of stationary cutters, a band knife is a major cutter that is used as it is the only cutter where the material and the knife are moved during the cutting operation. In all other cutters, either the material is moved or the cutter is moved and in the band knife, the endless knife runs continuously and the material to be cut is moved by the operator according to the pattern shape to complete the cutting process. Band knife machine is extensively used for cutting smaller garment parts and intricate garment parts as it is easy to move small parts with more precision for cutting in this machine. It is possible to cut fabrics at right angles too with this knife. It has high efficiency and productivity. The die-cutting machines have very high precision and accuracy as dies are created first according to the shape to be cut and then the die is pressed on the fabric to perform the cutting process. It can be used when huge quantities of products are produced as making dies for small quantity manufacturing is costly and also the productivity is less.

The most versatile cutting machine is the computerised cutting machine that can run at a speed of 6,000 rpm and execute cutting with the highest precision and accuracy because of computerised control. This machine is shown in Figure 8.19. These machines have nylon bristles on the table for holding the lay tightly and the lay is laid on top of it. These machines don't need marker sheets to be placed on top as the details are fed to the software and based on that the cutting knife is placed in the starting point and the cutting takes place as per the marker plan that is available in the system. It has a router and plunger to perform the cutting process effectively along with a round knife. The entire unit of cutting moves horizontally across and front and back with the help of two rails fixed at right angles to each other. Higher speeds are achieved in this machine and the marker efficiency can also be increased as there is no wastage due to improper cutting or space for dies is also eliminated. These machines are used for effective cutting processes and the process of cutting is explained in the next sub-section.

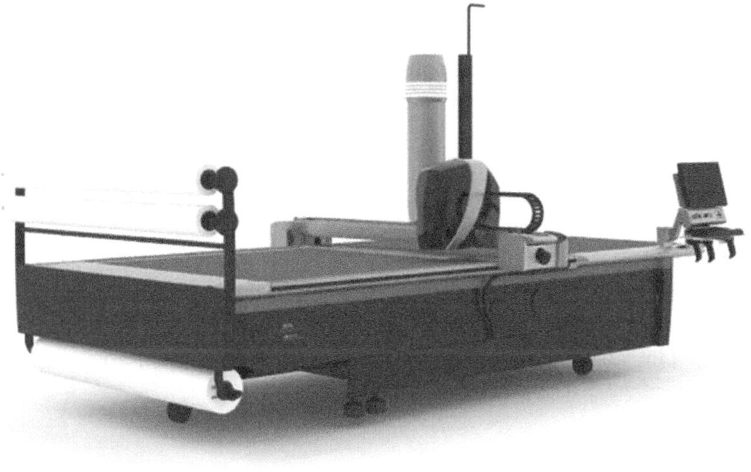

**FIGURE 8.19** Computerised cutting machine.

### 8.6.2 Cutting Process

The cutting process involves cutting the fabric lays that are spread on the table. The spreading machine spreads the fabric as per the marker plan and then the marker sheet with the component drawings is placed on the top. Then the cutting is done according to the shape drawn on the sheet. Then the cut components are bundled together and taken for the sewing process. The cutting process is based on the type of machine used, the type of fabrics cut, and the intricacy of the pattern to be cut. If in case, there are deep curves to be cut, then a straight knife-cutting machine is used to first cut the larger garment parts and then the smaller intricate parts are cut using a band knife-cutting machine. If the lay size is smaller, a round knife-cutting machine can be used to complete the cutting operation. In the case of mass-produced small parts, die-cutting machine can be used for the cutting process. Apart from the cutting operation, notches and drills can also be created in the lay using notchers and drillers. However, computerised cutting machines will avoid the usage of separate notchers and drillers as these machines have an inbuilt mechanism to create notches and drill holes. When a computerised cutting machine is used, then the process becomes simple as the machine detects the starting point from where the cutting has to take place and then based on the data available in the software, it does the cutting operation. Instead of a pattern sheet, a nylon cover is laid on top of the fabric when the computerised cutting machine is used for cutting. Even delicate intricate parts can be cut with ease when computerised machines are used. The process of cutting has become simple due to the advent of computerised machines. However, in the case of straight knife cutting, band knife cutting, and round knife cutting, the operator's skill plays a major role as they have to move the material correctly for cutting. Also, enough precautions should be taken to avoid damage to the personnel and machine.

The quality of cutting is also important as wrong cutting can lead to entire order cancellation and huge financial losses. It is required that utmost care is taken during the cutting operations for efficient pore-sewing process completion. The use of CAD/CAM helps in achieving better pre-sewing material for further sewing and the usage of these CAD/CAM tools will be discussed in the next section.

## 8.7 CAD/CAM INTERVENTION IN CUTTING ROOM OPERATIONS

The CAD/CAM usage in cutting rooms is extensive and their usage is not limited to cutting and they have become an integral part of all the pre-sewing processes. The CAD/CAM usage is there in fabric inspection, marker planning, spreading, labelling, and cutting. The various usages of CAD/CAM in these areas are detailed below:

### 8.7.1 Fabric Inspection

In the fabric inspection domain, the usage of computers and artificial intelligence (AI) has increased. A new fabric inspection system developed by M/s. Oshima Co. Ltd., named EagleAI, is used in fabric inspection (Oshima 2022). The AI-based inspection systems use very strong lights and high-definition cameras to capture the defects as they compare the fabric that is inspected with the fabric database to identify whether there are any defects. They are well equipped to handle bulk materials inspection in quick time thereby removing the need of workers for inspection and also the data is immediately available online for the manager to check and order for correction if needed. The data is obtained in real time using their software that can be viewed from anywhere. Such systems lead to higher profits, decreased manpower, less risk of human error, high accuracy in inspection, and fewer product recalls. The software of these machines has data related to all the faults that come in the fabric like contamination due to colour, flux, oil, holes, knots, warp and weft defects, hooked yarns, etc. It is reported that these machines have an accuracy rate of 94% for identifying the defects in the inspected fabrics. The accuracy is 70% even in the case of a new fabric inspection wherein that fabric is not available in its database. These systems have an online information management platform and are integrated with the output from the inspection system for real-time information availability.

### 8.7.2 Marker Planning

The usage of CAD in marker making is very huge and all the medium- to large-scale apparel industries use such software for getting more marker efficiency. It helps in marker making, pattern development, pattern grading, and digitisation of manual patterns. There are various CAD software available. Some of the major software available are Modaris from Lectra, Gerber's Accumark pattern design software, Cosma software, Optitex pattern design software, TukaCAD, PAD system, REACH CAD, and V stitcher. Manual pattern making and marker planning is a tedious task and complex to execute. The usage of CAD software has minimised human intervention and has aided garment making at a very high level. The garment components are fed into the system along with the garment style. The details of order quantity, fabric length and width, and other relevant details are fed into the software. Now, the software will come out with the probable marker plan that will ensure the quality and quantity of the final garment produced. It will calculate the marker efficiency to the best possible extent. It has options for the pattern master to modify the pattern placements and accordingly, the marker plan will be done and the efficiency will be calculated. Now such developed markers can be printed using a plotter or stored digitally. All these software providers have connectivity with the cutting machines where they operate based on the marker plan stored in the software. Along with this, this software can also be used to digitise the older garment patterns that were developed before the use of computer-aided marker making. Using digitisers, the old paper patterns can be converted into digital patterns and can be used in future production.

### 8.7.3 Spreading

The spreading machines used in spreading processes are automated and CAD/CAM software is used for ensuring even spreading without manual intervention. The new computerised spreading machines are equipped with encoders where end catchers are not required and the length of each layer of spread can be set in the software. Automated and interconnected production is possible as the software in the machine integrates with external systems by collecting data related to the marker plan and transferring data related to spreading to the cutting machines. Internet connectivity is enabled in the machines to do remote diagnostics and provide remote service for the machine. The different modes of spreading can be selected in the software and according to the selected mode, the spreading will be carried out. Along with spreading, labelling machines are also used with CAD/CAM software where the details are fed to the software and the labels are printed automatically. They are pasted onto the spread fabric using CAM software available in the labelling machine.

### 8.7.4 Cutting

Among all the pre-sewing processes, extensive usage of CAM software is evident in the cutting machines. Computerised cutting machines are available for precision and accurate cutting at high speeds. The CAM is used for large markers efficiently. In the computerised cutting machines, all the controls are automatically executed using the software. The CAD files of the marker plan are fed to the machine software and based on the marker plan, the machine plans the cutting process. In many machines, the machine is integrated with marker-making software and a spreading machine. The new cutting machines are equipped with more than 150 sensors that aid in effective cutting management as they are linked to the software. The cutting machines can be remotely monitored and troubleshooting can be done by the manufacturer online thereby increasing the utilisation of the machine for the cutting process. All the functions of the machine are digitally controlled thereby aiding in increased efficiency of the cutting process. These machines have a control centre with an ergonomic graphical interface that helps in control of cutting speed, blade rotation speed, vacuum, etc. The new machines have an optimiser that optimises the distribution points in the patterns and also identifies the common lines and symmetries. It automatically decides the optimum blade entry point and critical areas where speeds have to be reduced. The preview shows the pieces cut, the ones to be cut, and the position of the cutting blade dynamically thereby making the cutting process smooth and efficient.

## 8.8 CONCLUSION

In this chapter, the pre-sewing processes are detailed elaborately. The processes involved, their scope, machines used, requirements, and various modes and methods are explained. The process of tech pack development, marker planning, and fabric inspection is described with examples. A complete tech pack is presented for a better understanding of the readers. The various modes of spreading, their classification, machines used, and the spreading process along with labelling and its need are discussed in this chapter. The usage of a labelling machine is provided. The process of cutting with its requirements and the machines used are highlighted. The usage of CAD/CAM in the pre-sewing processes is mentioned with their scope. This chapter gives a clear view of how the pre-sewing processes work in the apparel industry and will help the readers in gaining a better understanding of the processes followed.

## REFERENCES

Btex, 2022. Automated fabric inspection in textile industry, Available from https://www.btexindia.com/automated-fabric-inspection-in-textile-industry/. Accessed on 30th December 2022.

Chan, C.-H., Liu, H., Kwan, T., & Pang, G., 1998. Automation technology for fabric inspection system, Available from https://doi.org/10.1007/s00170-003-1819-3

Goyal, A., 2018. *4-Automation in Fabric Inspection*, pages 75–107. Automation in Garment Manufacturing, the Textile Institute Book Series. Woodhead Publishers, Cambridge, England.

Islam, M., 2015. 10 point system for fabric inspection in garment industry, Available from https://textilelearner.blogspot.in/2015/07/10-points-system-for-fabric-inspection.html. Accessed on 30th December 2022.

Islam, M., Saha, P.K., Islam, M.N., Saha, S.K., & Biswas, S.K., 2019. Impact of sleeve types on marker efficiency and fabric consumption, *Global Journal of Researches in Engineering: JGeneral Engineering*, 19(4), 11–16, Version 1.0 Year 2019.

Oshima, 2022. AI inspection machine, Available from https://www.titas.tw/Upload/UserData/TITAS%202022%20Exhibitor%20Introduction%20-%20OSHIMA.pdf. Accessed on 31st December 2022.

Pamuk, O., & Yildiz, E.Z., 2016. A study about parameters affecting the marker plan efficiency, *TEKSTİL ve KONFEKSİYON*, 26(4), 430–435.

Rahman, M.F., Rashid, M.R., Haque, E., & Hasan, Md. Z., 2017. Effect of garment size ratio and marker width variation on marker efficiency for both manual and computerized marker, *European Journal of Advances in Engineering and Technology*, 4(10), 765–772.

Rana, N., 2012. Fabric inspection systems for apparel industry, *Indian Textile Journal*, 122(11), 57.

Rathinamoorthy, R., & Surjit, R., 2015. *Apparel Machinery and Equipments*. CRC Press.

Tanaka, M., & Wachi, T., 1973. Computerized marker making, *Journal of the Textile Machinery Society of Japan*, 19(3), 74–81.

Tsao, T.-C., & Delicia, M., 2022. Thuy-Linh Vu, Marker planning problem in the apparel industry: Hybrid PSO-based heuristics, *Applied Soft Computing*, 123, 108928. https://doi.org/10.1016/j.asoc.2022.108928

Tyler, D.J. ed., 2009. *Carr and Latham's Technology of Clothing Manufacture*. John Wiley & Sons.

Vilumsone-Nemes, I., 2018. Industrial cutting of textile materials – Lay planning and marker making in textile cutting operations, 13–27. https://doi.org/10.1016/B978-0-08-102122-4.00003-2

Wong, W., Leung, S., & Au, K., 2005. Real-time GA-based rescheduling approach for the pre-sewing stage of an apparel manufacturing process, *AMT*, 25, 180–188.

# 9 Garment Production Machines
## *Sewing Operations*

*Anu H. Gupta*

### 9.1 INTRODUCTION

Garment production is a vital phase in the apparel manufacturing process in which two-dimensional fabric cut pieces are put together to make a three-dimensional product or garment. Sewing, or stitching, is the process of uniting plies of cloth to create a product. It is also referred as assembling garment components, as well as attaching embellishments such as fasteners, laces, and piping to the product. Stitching may be carried out manually or using a machine. Embroidery, quilting, surface ornamentation, and other sewing-related crafts are some of the examples of the many uses for both hand and machine sewing. The garment-sewing process takes place in a room or area typically referred to as the Production Room or Production Floor in a garment factory (Figure 9.1). This area often has a variety of sewing machines to execute the numerous tasks necessary to stitch a garment.

Plies of fabric are joined together to create a three-dimensional form with a row of stitches referred to as seam (Figure 9.2a). The importance of seams in the construction of clothes, accessories, footwear, home textiles, athleisure, and many other lifestyle products cannot be overstated. Technology may play a significant role in the swift and accurate completion of this task. Specialised sewing machines are one of the examples of the technology. With the use of such machines, for instance, a side seam in sewing of shirts (Figure 9.2b) may be created efficiently and in less time if the operator utilises a feed-off-the-arm machine as compared to a single-needle lock stitch machine.

(a) (b)

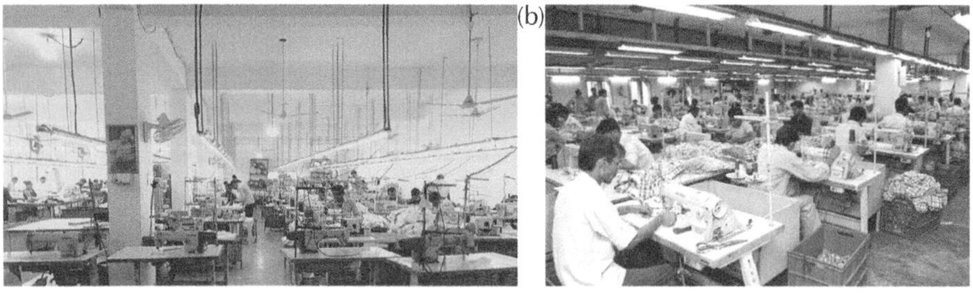

**FIGURE 9.1** Garment sewing by assembling cut pieces of garments on production floor.



DOI: 10.1201/9781003432739-9

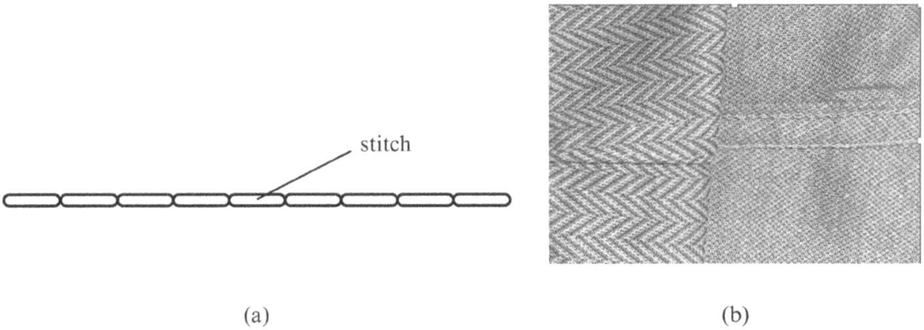


**FIGURE 9.2** Seam: (a) Diagram of row of stitches in a seam; (b) right and wrong sides of a garment showing row of stitches.

The assembly of components into a product or garment comprises three vital components: stitches, seams, and technology. A number of factors can influence the strength, durability, and visual appeal of a finished garment or product, including the type of seam used, the stitch, and the technology or sewing machine used in its production. The following sections examine stitches and stitch categorisation, seams, types of machines, and recent technological advancements in sewing room.

## 9.2 SEAM AND STITCH CLASSIFICATION

### 9.2.1 Seam Classification

A seam is the result of a row of stitches connecting two or more pieces of material. Seams can be classified on the basis of their type (superimposed, lapped, bound, flat, decorative, etc.) and their placement or location in the product or garment (side seam, shoulder seam, centre front/back seam, inseam). Every seam has a seam allowance, which refers to the additional fabric extending from the raw edge of the fabric to the stitch line.

The width of seam allowance is mostly determined by the fibre content of the fabric, the type of seam being stitched, and also the location of the seam or the shape of the seam (a circular or curved seam has a smaller seam allowance as compared to a straight seam). To prevent the edges of the seam allowance from ravelling and to give the garment a tidy or neat appearance from inside of the garment or wrong side, the allowance is finished in many ways.

As the seam should have the same physical properties as the material being sewn, choosing the type of seam used in the production of goods or clothing is crucial. The type of fabric used in the garment determines the choice of seam. A French seam can be used on sheer fabrics but not on heavy-weight or thick fabrics. For heavy-weight fabrics like denim, a lapped or flat-felled seam is recommended. Depending on the garment's intended use, seam strength and flexibility must be taken into account while selecting a seam. A wedding gown will have different seam finishes than a biker's jacket, and infant wear or lingerie or inner wear has different seam requirements than an outer garment. Seams at the crotch and opening of a pocket require special attention because if these are weak or are not reinforced, the product will tear. The strength, extensibility, durability, and comfort of seams all play a significant role in seam performance. The stitches per inch (SPI), or stitch density; strength, elasticity, and tension of the thread used; cover factor; thread count; and density of the fabric all affect seam's strength. The strength of a seam can be measured in the two directions – one parallel to the seam and other perpendicular to the seam. The choice of stitch type and sewing thread for a seam is also crucial, as the seam should complement the garment's aesthetics.

The suitability of a seam is dependent upon its intended use, with the particular type of seam being determined by the kind of the product being stitched. The seam in a contemporary manufacturing setting needs to ensure both economical production and the required appearance and performance standards. A garment's seams should be comfortable to wear and simple to produce or stitched by the operator, cost-effective; and at the same time availability of the machine or equipment for making the seam is also important (e.g., for fell seam, feed-off-the-arm machine is required). Hence it is important to know about the different seam types, stitch types, uses, finishes, and variations so that the right seam can be chosen for a garment that is both strong and is aesthetically appealing.

Seams are classified based on the type and number of fabric components used. Eight classes of seams are defined by ISO 4916:1991. Seams can be classified (Chuter 1995; c.f. Karthik, Ganesan and Gopalakrishnan 2017) as follows:

Class 1: Superimposed Seam (SS)
Class 2: Lapped Seam (LS)
Class 3: Bound Seam (BS)
Class 4: Flat Seam (FS)
Class 5: Decorative/Ornamental Seam (OS)
Class 6: Edge Finishing/Neatening
Class 7: Applied Seam/attaching of separate items
Class 8: Enclosed Seam/single-ply construction

#### 9.2.1.1 Class 1: Superimposed Seam

SSs are made up of one, two, or more layers of material that are joined at a certain distance from the edge (seam allowance) with one or more rows of stitches. SSs are widely used in the making of apparel, accessories, or lifestyle products like shirts, skirts, tops, trousers, bags, cushions, and shoes. There are several types of SSs, including plain seams, French seams, and double machine seams. In case of a plain seam, the raw edges of seam allowance can be finished with overlocking (Figure 9.3a).

An SS can be sewn with various types of stitches – lock stitch, chain stitch, over-edge stitch, or safety stitch. These seams can be made with stitch types 301 or 401 in the case of a plain seam. For knitted garments such as t-shirts, SS can also be prepared with stitch class 500 (over-edge stitch) and combination stitches such as class 516. The stitch classes are discussed in the next section of this chapter.

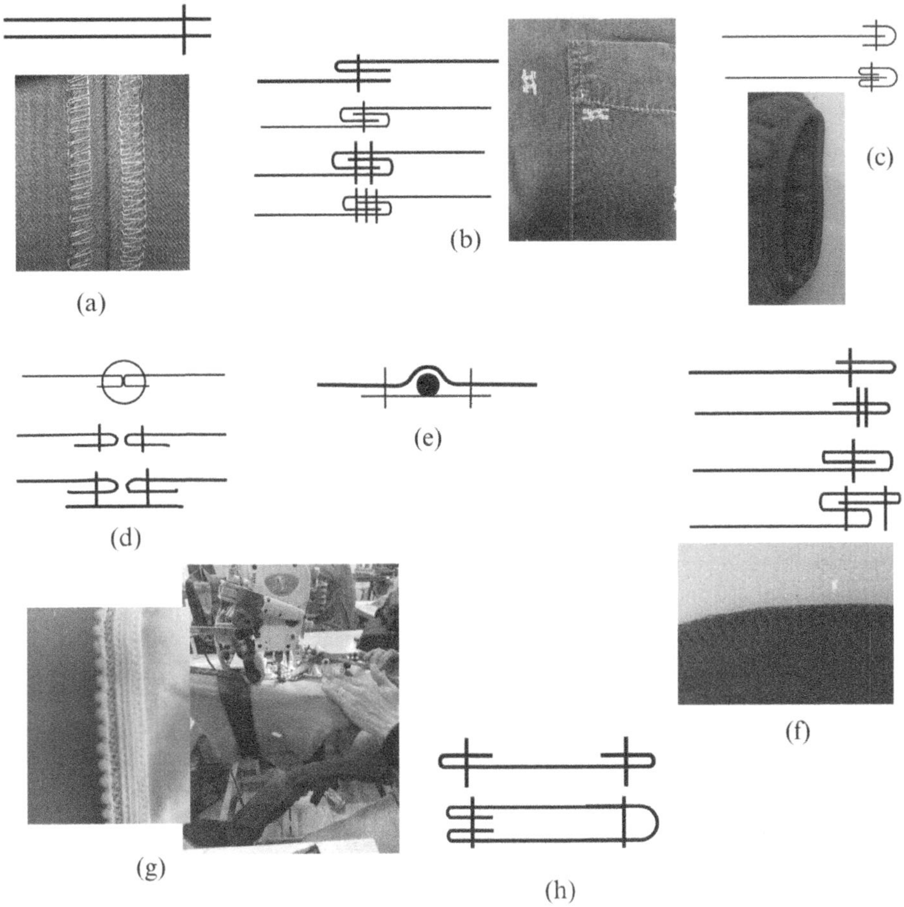

**FIGURE 9.3** Seam types: (a) Diagram of superimposed seam and superimposed seam (plain) with both seam allowances flattened and finished with overlocking; (b) diagram of different types of lapped seam and lapped seam at an edge of a patch pocket; (c) diagram of different types of bound seam and application of binding at sleeve opening; (d) diagram of different types of flat seam; (e) diagram of an ornamental seam; (f) diagram of some edge finishings and overlock edge neatening; (g) elastic attachment to an edge of foundation garment; (h) diagram of single-ply construction.

### 9.2.1.2 Class 2: Lapped Seam

LS is formed by lapping two layers of fabric and then seaming with either one top seam or more number of top seams. This seam is commonly used in the attachment of yoke, gusset, joining of lace, patch pocket, the side seam of garments like jeans and men's shirts, etc. (Figure 9.3b). It can be stitched with a single-needle lock stitch (stitch class 100) or chain stitch machine (stitch class 300 or 400), but not with an overlock machine (Shaeffer 2014).

#### 9.2.1.3 Class 3: Bound Seam

A BS is created by applying binding (strip) to a raw edge of fabric or fabric plies. A binding strip is folded along the outer edges of the base fabric or layers of fabrics, followed by stitching along the edge of folded strip either with one or more rows of stitches (Figure 9.3c). The BS finish can be achieved using the 301 or 401 stitch classes. This seam is commonly used to finish the neckline of a t-shirt. Sleeve hemlines, necklines, seams of unlined garment like jackets and coats, and finishing raw edges of fabrics like chenille, velvet, and satin all use BSs (Figure 9.3c). This seam is also used for inner raw edge finishing in designer and haute couture garments.

#### 9.2.1.4 Class 4: Flat Seam

As the name indicates, these seams are flat in nature. The FS, or butt seam, involves bringing two fabric edges together, whether flat or folded, and then over sewing with stitches to create a joint with no extra thickness, which is especially useful in foundation and fine knitted garments (Figure 9.3d). This seam is prepared with a zigzag lock stitch, chain stitch, or cover stitch. A decorative element can be created with a covering stitch using a decorative cover thread, but at the same time, the strength of the cover thread is important to provide strength to the seam, and the looper thread to have elasticity and strength.

#### 9.2.1.5 Class 5: Decorative/Ornamental Seam

An OS is made up of a row of stitches that follow an ornamental or decorative design. This seam can be worked on a straight or curved seam with one or more plies of fabric. Some examples of decorative seams are the stitches on a patch pocket of jeans, seams with cord inserts, and pintucks made on garments like tops and blouses (Figure 9.3e).

#### 9.2.1.6 Class 6: Edge Finishing/Neatening

This seam class includes the finishing or neatening of fabric edges with stitches. It does not include binding. This seam category also includes folding hems and edges. The most common method for neatening or finishing up an edge is through overedging or serging (Figure 9.3f). Neatening can also be accomplished in a number of ways: one method is to neatly fold the edge of the fabric at the hemline followed by blind stitching; other method is to fold the edge of a skirt or trousers at the hemline with a folder; another method is to fold the edge like an overlap placket on a men's shirt.

#### 9.2.1.7 Class 7: Applied Seams/Attaching of Separate Items

This seam class requires the application of an additional item or a component along the edge of a garment or fabric ply. The additional component has a distinct edge on both sides, e.g., elastic braid or lace. Such additions are attached to the foundation garments using edge stitch and attachments on the sewing machine (Figure 9.3g).

#### 9.2.1.8 Class 8: Enclosed Seam/Single-Ply Construction

This type of seam class utilises a single ply or strip of cloth, which is turned on both edges (Figure 9.3h). Applications for enclosed seams include belt loops or belts that can be attached with a machine-operated folder.

### 9.2.2 Stitch Classification

The fundamental building block for creating seams is a stitch. A stitch can be formed either within a material, through a material, or on the surface of a material. Stitching is performed by hand or machine, and the process of stitch formation entails first piercing or penetrating a fabric or surface by needle, then carrying thread (which is in the eye of the needle) through it, and then completing the stitch by any of the processes namely intra-looping, inter-looping, or interlacing threads (threads that come from lower mechanism of sewing machine with loopers or bobbin). Following is a description of the three distinct steps required to create a stitch:

- **Intra Looping**: This method involves passing one loop of thread into another loop formed from the same thread supply to create a seam, such as a single-thread chain stitch (101) (Figure 9.4a).
- **Inter Looping**: Inter looping involves passing one loop of thread through a second loop made with a different supply of thread, like in the double-lock chain stitch (401) (Figure 9.4b).
- **Interlacing**: Interlacing entails crossing one thread over or around another thread or a loop of another thread supply for instance, the lock stitch (301) (Figure 9.4c).

The qualities and attributes of a created stitch are determined by the sewing machine and kind of fabric. Fabric, stitch, seam type, needle, thread, and machine settings must be compatible in order to manufacture or sew a high-quality product. Therefore, it is essential to pay close attention to the stitch properties: stitch size and stitch consistency.

Stitch size includes stitch length, width, and depth. SPI determines stitch length; the longer the stitch, the fewer stitches are there per inch. Decision for stitch length depends on the fabric weight. A stitch length of 2.5 mm is suggested for light- to medium-weight fabrics, 3 mm for medium- to heavy-weight fabrics, and 3.5–4 mm for thicker and heavier weight fabrics, such as denim (Figure 9.5). Stitch width refers to the space between the most outermost stitch lines. The space between the top and lower halves of a stitch is referred to as its depth.

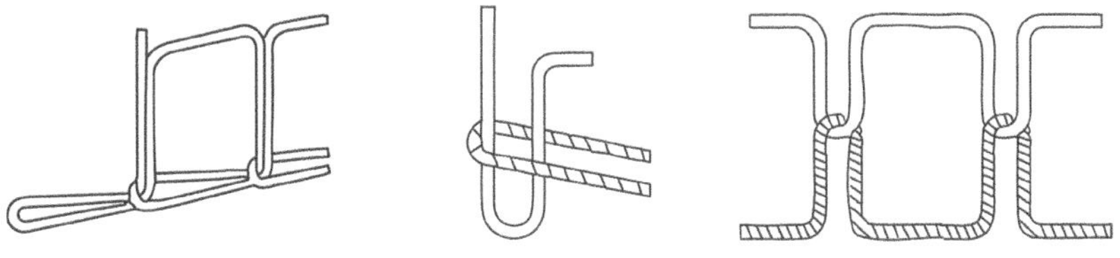

(a) (b) (c)

**FIGURE 9.4** Creation of stitch: (a) intra looping of a yarn; (b) inter looping of yarns; (c) interlacing of yarns.

**FIGURE 9.5** Varied stitch size in different weight fabrics.

Stitch consistency refers to the uniformity or regularity of each stitch generated in a seam or row of stitches. Stitch consistency contributes to the appearance or look of the seam and the garment and is also important for the structural integrity of the garment. The integrity of a seam may be affected by irregular stitching. In high-quality and luxurious clothing, attention is especially paid to the uniformity of the stitching.

The needles, loopers, and threads that form a row of stitches determine the stitch arrangement for each kind of sewing machine. Each of these configurations or arrangements is known as a 'stitch type,' and it is classified according to its main features (Hayes, McLoughlin and Fairclough 2012). According to BS 3870/ASTM D-6193/ISO 4915:1991, stitches may be classified into six classes, with a number assigned to each class (Tyler 2008). To ensure consistency, the International Organization for Standardization uses this numbering for each stitch class, which is as follows:

Class 100: Chain Stitches
Class 200: Hand Stitches
Class 300: Lock Stitches
Class 400: Lock Chain Stitches
Class 500: Over Edging Stitches
Class 600: Covering Stitches

#### 9.2.2.1 Class 100: Chain Stitches

Class 100 or chain stitches are distinguished by intra looping and are made with one or more needle threads. After the thread has passed through the material, it is intra looped with a subsequent loop or loops (Figure 9.6a). Chain stitches are thicker and more elastic than lock stitches that are discussed further. Basting, tacking, labelling, and button sewing are all examples of this stitch class. This kind of stitch is unsecure

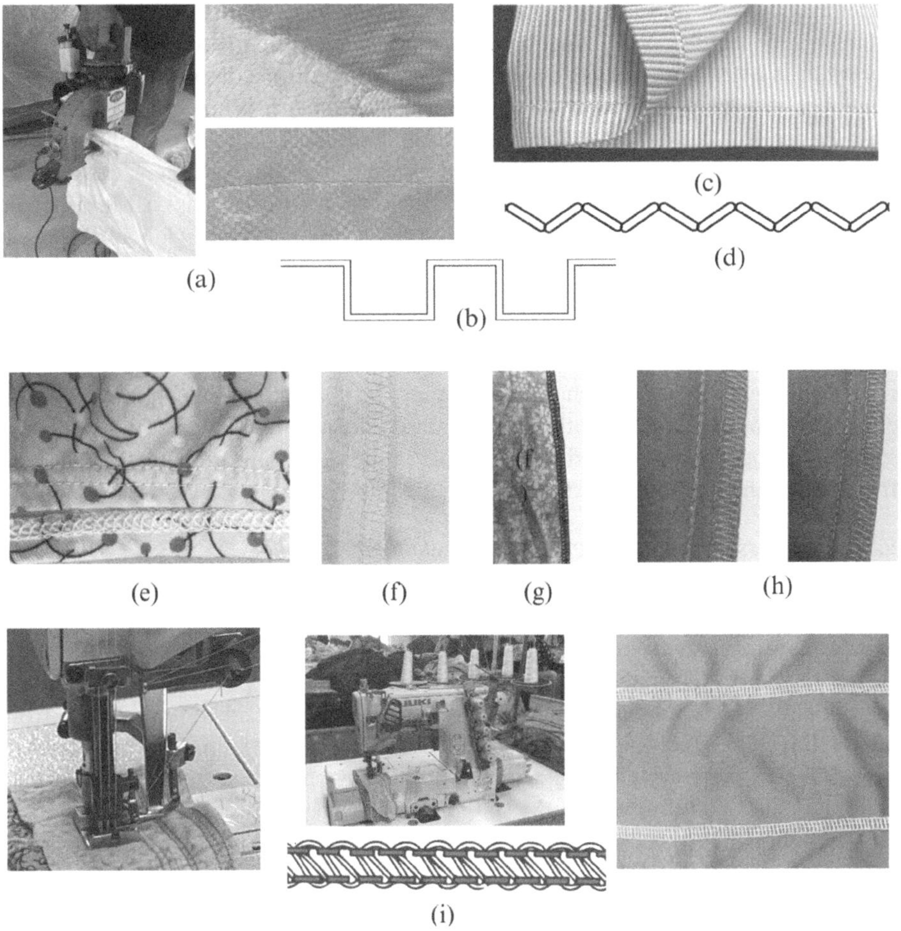


**FIGURE 9.6** Stitch class: (a) Bag stitching using stitch class 100; right and wrong sides of 100 stitch class used in bag sewing; (b) diagrammatic representation of running stitch; (c) straight lock stitches for finishing hemline of a garment; (d) diagram of zigzag stitch; (e) front and back views of multi-thread chain stitch; (f) over-edge stitch finishing raw edge of seam allowance; (g) mock-safety stitch; (h) safety stitch – front and back views; (i) cover stitch formation.

because each loop is dependent on the one before it. If the end of the thread is not passed through the last loop or the fabric, or if a stitch is broken, it is easy to unravel. The stitch classes 101, 102, 103, 104, 105, 107, and 108 are additional subcategories of the stitch class 100. The two that are used the most are 101 and 103.

#### *9.2.2.1.1 Class 101: Single-Thread Chain Stitch*

The 101 stitch, which consists of a single thread, is one of the simplest kinds of stitches. On the right or face side of the fabric or material, it appears as straight stitches or lock stitches, but on the wrong or reverse side, it appears as loops or chain. The backside's looped structure offers the seam flexibility but also makes it simple to unravel or remove, making it useful for 'basting' operations, which temporarily retain fabric parts before the final stitch is secured. Because this stitch is merely temporary, it is often not used for seaming. However, multi-needle machines typically employ it as a temporary stitch to seal industrial-use bags. When used with numerous needles, this pattern may form a beautiful chain-like stitch on the right or facing side of the cloth, which is commonly found on women's blouses. More variety can be achieved with the use of fewer or more needles (up to 65 in some cases). Multi-needle elastication, which employs elastic thread in the needles, may provide an ornamental stretchable effect on cuffs, yokes, and the waist. To prevent the threads from sliding back, the end of the seam must be fastened or secured, or the cloth edges must be joined with another securing the ends of the chain stitches. Stitch class 101 also includes securing button attachment and button holing, where zigzag stitching is done without bobbin. The chain stitch cannot be chained off, which implies that it cannot be sewed or created without fabric under the needle. The stitching cannot be suddenly stopped or completed in the middle of the cloth. This kind of stitch must go from one cloth edge to the next.

##### *9.2.2.1.2 Class 103: Blind Hemming or Felling*

Blind hemming or felling, which is also called stitch class 103, is another subcategory of the 100 stitch class. The stitch is executed using a single thread and a curved needle that traverses from left to right, passing through and exiting from the same side of the fabric. The needle thread is caught by a blind looper. It is used for stitching of belt loops, felling, and hemming. The seam must be tacked to prevent it from unravelling. The stitch is insecure, as in 101, but it can be improved in numerous ways. One of the way is to use a textured or hairy surface sewing thread instead of smooth-surfaced thread as it doesn't slip easily when broken because it tends to create friction. This keeps the stitches from coming apart. Another way is the stitch cramming mechanism, which shortens the length of the stitch at the beginning and end of the seam.

#### 9.2.2.2 Class 200: Hand Stitches

The category for this class consists of hand stitches. A needle is inserted into the material from one side to the other using a single or double thread to create these stitches (Figure 9.6b). This kind of stitch is exclusive to high-priced apparel since it is handmade and involves substantial labour. Such stitches are used for surface enhancement on designer clothing, edges of hemline of sleeves and garments, lapels or edges of jackets, coats, home textiles such as bed spreads and pillowcases, and several other luxury items.

#### 9.2.2.3 Class 300: Lock Stitches

Class 300 stitches are interlaced stitches created using a lock stitch machine by interlacing two or more sets of sewing threads. The needle thread is the upper thread that emerges from the spool, passes through thread guides, and finally through the eye of the needle, while the bobbin thread is the lower thread that originates from the bobbin. Because the needle thread and bobbin thread are interlaced, a break in a single stitch will not let the seam unravel entirely. This interlacing makes the seam secure and hard to unravel. Furthermore, the seam is well-integrated or bedded into the fabric, which increases its abrasion resistance. For a high-quality seam or a perfect lock stitch, it is crucial that both threads be evenly tensioned and balanced. The kind of thread used on a lock stitch machine might affect the strength and resilience of the stitch or seam. Lock stitch sews consistently at an angle of 90 degrees while turning the cloth at the needle point for topstitching, especially in pockets, collars, and cuffs. Due to these characteristics, the lock stitch is widely used to seam almost all garment components, such as side seams, shoulder seams, sleeve attachments, collars, cuffs, pockets, and neck and sleeve facings.

Lock stitch can be executed with a single or multiple needles. In contrast to stitches using a looper, this type of stitch exhausts the thread package, especially the bottom thread, necessitating regular bobbin refilling and replacement. This results in more manufacturing time being consumed or wasted. Therefore, lock stitch is the most common stitch in low-volume manufacture. Lock stitch gives adequate flexibility – up to 30% – for comfort-stretch clothes; however, this stretch capacity is limited, especially when adapting modern high-stretch fabrics, which is another limitation or drawback of this stitch type (McLoughlin 2018).

While sewing garments using twin- or multiple-needle machines, the loss of stitches on one row owing to the exhaustion of one bobbin would be a big concern. Moreover, it is not feasible to have multiple bobbins as they occupy a lot of space. Therefore, the most common needle count for lock stitch machines is two. Optical bobbin run-out detectors are used to address the issue of bobbins running out of thread. As the thread starts to run out, these detectors detect a reflection from a flat side in the centre of a circular bobbin. In buttonholing, where restitching is possible in case the hole has not been cut, machines often stop without cutting the hole if the bobbin runs out of thread or the needle thread breaks.

Subclasses 301, 304, and 306 are commonly used stitches in stitch class 300.

Straight lock stitch (301) is the most popular and commonly used stitch in the apparel industry. It is used for almost all straight, angled, and curved seam joining, garment assembly, and small-volume garment production or manufacturing. It uses a single needle and bobbin thread. When performing a balanced lock stitch, the face and reverse sides of the fabric appear similar (Figure 9.6c). Straight lock stitch offers a stronger seam and is used to attach practically all types of fabric plies in garments. The strength of the stitch depends on the selection of thread and the amount of stretch required.

Zigzag lock stitch (Class 304) (Figure 9.6d) has several applications including attachment of trimmings, fagotting, and creating beautiful patterns using zigzag topstitch. Lingerie may be adorned with laces and elastic using a large row of zigzag stitching. Two fabric pieces are positioned next to each other and secured together using a zigzag stitch on the right side of the fabric, with a portion of the zigzag thread on one fabric piece and a portion of the zigzag stitch on the second fabric piece. The zigzag stitch is also used as a decorative stitch on yokes and many patterns on tops, blouses, etc.

The 306 stitch type provides a lock stitch that is used as a blind stitch. This is more secure than the standard 103 stitch type. Hemming or attaching lining to shell cloth is a common application. Other subclasses of stitch class 300, such as 308 and 315, are likewise zigzag, but have a higher number of stitches before changing direction, resulting in broader stitches, more coverage, and stretch, which are used for sewing, topstitching, and cover stitching in knits or woven fabrics.

#### 9.2.2.4 Class 400: Multi-Thread Chain Stitches

Multi-thread chain stitches consist of two or more groups of threads and are distinguished by the inter looping of two sets of threads, i.e., needle thread and looper thread, on the reverse or underside of the cloth (Figure 9.6e). Loops of one set of threads are carried through the fabric (using sewing needles) and fastened by interlacing and inter looping with loops of another set of threads (looper threads). Multi-thread chain stitches are used for stitching on all kinds of clothing. It is not really a lock stitch, but is commonly called a 'double lock stitch' since each loop of the needle thread is tied to two loops of the same single under thread. Because of the reduced static thread tension and the inter-looping threads, it is less prone to pucker. It has good strength and has better extension and recovery properties. The looper's ability to provide a continuous length of thread makes it useful for long seams as well. If the thread chain was not stitched across at the end when creating the garment, one centimetre should be left. Although the stitches may be compressed, they do not back tack well. The thread tensions used in these machines are often lower than those used in lock stitch. This allows the machine to function at faster speeds while simultaneously minimising the likelihood of the thread breaking. This stitch class's most common subclasses are 401, 402, 404, 406, and 407.

The simplest form of this class is Stitch 401. On top, it looks to be a lock stitch, but the looper thread provides a double chain effect on the bottom. To balance the stitch, the needle thread is drawn through the chain, which is normally situated under the surface of the fabric. Because the chain stitch's loop form extends when stretched, it is used for flexible seams such as attaching sleeves and joining elastic fabrics. Stitch type 401 and an over-edge stitch are often used in mass production for cost-effective seam joining and neatening.

The stitch type 402 or cording stitch is usually used to stitch permanent creases. It uses two needle threads to stitch two rows of stitches throughout the front or face side of the fabric. The two needle threads on the back of the fabric are joined by a looper thread. This creates a ridge or crease between the needle threads on the front side of the fabric. The application of stitch class 402 in sportswear serves the purpose of preserving creases, particularly on the posterior surface of gloves.

Types 406 (two-needle version) and 407 (three-needle version) bottom cover stitches comprise a looper thread that covers the underside of the fabric between the two or three needles. These stitches are employed to conceal or cover seams or unfinished edges on the inside of clothing to keep them flat, to attach embellishments such as braids and lace to clothing, and to attach elastic edging to briefs where the fabric edge has to be neatened.

#### 9.2.2.5 Class 500: Over-Edge Chain Stitches

Class 500 over-edge chain stitches are made with one or more groups of threads, and at least one of the sewing threads goes over or around the edge of the fabric being sewn. They often trim or clean the edge(s) of cloth to make them seem neater. Before stitching, a crisp edge is ensured with machine's trimming knife, which also makes it possible to remove extra material. The process of utilising these stitches to finish the edges is known as 'overedging' (Figure 9.6f). This row of stitches has elasticity and resists unravelling. The strength of this seam is derived from the needle thread, whereas the flexibility of the stitch is derived from the looper threads. The loopers' sewing threads may be chosen based on their softness and appearance. Common names for these stitches include over-edge, overcast, overlock, serge, and merrow. Over-edge chain stitches are used for both bolt end seaming and serging.

The number of sewing threads used in an overlock stitch can range from one to five, and each has a specific application (Karthik, Ganesan and Gopalakrishnan 2017), which is explained as follows:

- Butt-seaming is done with overlock stitches with one thread.
- Merrowing, an overlock stitch with two or three threads, is used to neaten seams and edges on knit and woven garments.
- Mock-safety stitches, four-thread formations, provide additional strength while maintaining flexibility.
- Safety stitches, which use two needle threads in five-thread formations to create stronger seams, are utilised in the production of apparel.

The stitch types 502 and 503 use two threads, a needle thread and a looper thread. The 502 stitch class is used for seaming outside edge of a bag, whereas the 503 stitch class is used for serging a single-ply fabric. The two-thread construction prevents pressing from leaving an edge impression, hence it is considered ideal for trousers.

Class 504 and 505 over-edge stitches consist of three threads: one needle thread and two looper threads. Type 504 is a secure but highly extensible stitch that works well as a seam for knitwear such as T-shirts and sweaters; however, its narrow bight, which typically runs from 3 to 5 mm, does not give adequate strength for fabrics that fray or slip. This bight may be adjusted within these parameters based on the nature of the knitted fabric joining in order to get a suitable appearance and an acceptable amount of bulk. The operator must leave a thread chain when joining materials and seam ends, particularly at the edge of the garment to be secured using lock stitch tack, bar tack, or small zigzag stitch. Decorative effects can be achieved in a variety of ways using this stitch, including: high stitch density and a narrow bight to create a neatened decorative effect; different-coloured thread that spreads out to form a good cover over; and stretching the edge while sewing to create a soft frill on suitable fabrics.

Four threads (two needle threads and two looper threads) are used to create the mock-safety stitches 512 and 514 (Figure 9.6g). These may be used to sew together knits with elasticity and great recovery, as well as light-weight to medium-weight wovens, especially lingerie and shirts.

Safety stitches are stitch types 515, 516, and 519, which are a combination of an over-edge stitch and a 401 chain stitch. A second row of tight over-edge stitches supports the chain stitch nearest to the seam (Figure 9.6h). The safety stitch finds its application in the manufacturing of shirts, blouses or tops, jackets, jeans, etc.

### 9.2.2.6 Class 600: Covering Chain Stitches

Class 600 stitch type consists of two or more groups of threads, with two groups typically covering both sides of the fabric. Three sets of sewing thread from needle, looper, and spreader create these stitches (Figure 9.6i). The needle threads generate loops that are then threaded through the loops formed by the spreader threads, thereby enclosing them on the fabric surface. After that, these are passed through the cloth and looped using looper thread on the underside of the cloth (Tyler 2008). These strong and elastic stitches are used to cover raw edges and prevent ravelling. Cover-seam chain stitches are utilised for binding operations on many garments, like knitted undergarments, sport shirts, athletic wear, and new born apparel, and the attachment of flat-knit or rib-knit collars (Colovic 2015), and also for joining of braids and elastics to the foundation garments with decorative top cover stitch and the bottom cover stitch over the raw edge of the garment fabric.

The only stitch type that doesn't follow this rule is type 601. In this stitch, one of the threads in the first group does the work of the third group, so only two groups of thread are used.

Stitch class 602 consists of four threads: two needle threads, one looper thread, and one top-covering thread. Class 605 is achieved with three needle threads, one looper thread, and one top-covering thread. The cover stitches 602 and 605 are extremely durable and elastic and are used for butt seams.

Another complex stitch in this category is stitch type 606, or flat lock. Its application includes a butt seam or FS. This type of stitch is used in knitted garments, especially undergarments. The flat-seaming stitch type 607 consists of nine threads (four needle threads, four looper threads, and one top-covering thread). The stitch class makes a seam with very little bulk that can be worn against the skin. It is used to join elastic, braid, or binding to the edges of clothes like briefs and to make the flat, butted seams that are popular for knitted underwear (Clayton 2008).

## 9.3 SEWING MACHINE TYPES

Sewing is the craft of fastening or attaching objects using stitches made with a needle and thread. Crafting using this method dates back to the Palaeolithic period, making it one of the earliest textile arts. The practice of joining plies of fabric and converting a two-dimensional surface into a three-dimensional form was primarily done by hand. It was a tedious and time-consuming process, but the revolution came with the invention of the sewing machine in 1790, and labour became easier and quicker. People involved in sewing practices can now sew with good finishing and speed. With the changing times, sewing moved from homes to professional tailors and then to mass production. The technology involved in upgrading sewing machines has played a great role in revolutionising the sewing and garment industry. Innovative technologies are being introduced in home sewing machines as well as industrial sewing machines. Engineers, scientists, researchers, and technologists are all working together to produce high-quality garments in less time. The incorporation of pneumatics and hydraulics has also made it possible to make things much faster.

A large variety of sewing machines are used in the industry, with varied bed types and different feed mechanisms, mechanical, computerised, or automated depending on the material to be sewn, the final product to be produced, or the kind of sewing operation. Hundreds of machines have been developed for varied operations in the production of variety of products ranging from apparel to accessories to foot wear, etc. As a garment or product manufacturer, it is important to install and use the correct sewing and other machines in the industry to achieve quality and fast production.

Sewing machines are classified by three main characteristics that distinguish the various kinds of machines for various seaming operations. These include the machine bed, the feed mechanism, and the stitch type. Machines can also be classified on the basis of their control. It can be manually operated, semiautomatic, or automatic. Some of the details are discussed in the next sections.

## 9.3.1 Sewing Machine Classification Based on Machine Beds

The bed of a sewing machine is the surface of machine on which material or fabric is placed and fed into the needle for stitching, and hence it helps or facilitates easier and smoother movement of material between the needle and needle plate. This part of the machine also has a feed dog, bobbin, or looper placed below it. The bed can vary in shape, size, and height depending upon the fabric and the type of product to be stitched; for example, the stitching of a leather shoe requires a bed that is different from that used for sewing jeans. Sewing machine beds can be classified as depicted in Figure 9.7.

### 9.3.1.1 Flat Bed

As the name indicates, a flat bed is flat and is on the same level as the sewing table or top (Figure 9.8a). This bed allows large and two-dimensional parts of a garment to be stitched flat on the sewing machine, e.g. attachment of sleeve with the bodice of a garment, preparation of a collar, preparation of a patch pocket and its attachment on a men's shirt, and the application of binding on a bib. This is the most widely used bed for stitching, e.g. lock stitch machine.

### 9.3.1.2 Raised Bed

The machine bed is often raised or mounted above the table top by up to 4 inches (Figure 9.8b) where there is a requirement for the garment to fall or drape away from the needle for better or easier handling by the operator. This elevated pedestal top makes it easier to thread loopers on machines with stitch classes 400, 500, and 600. In addition, the elevated bed facilitates the stitching of tubular items with a

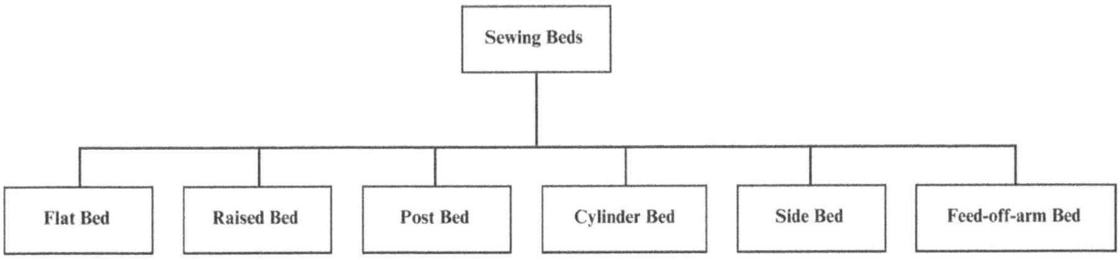


FIGURE 9.7 Classification of sewing machines according to types of machine beds.

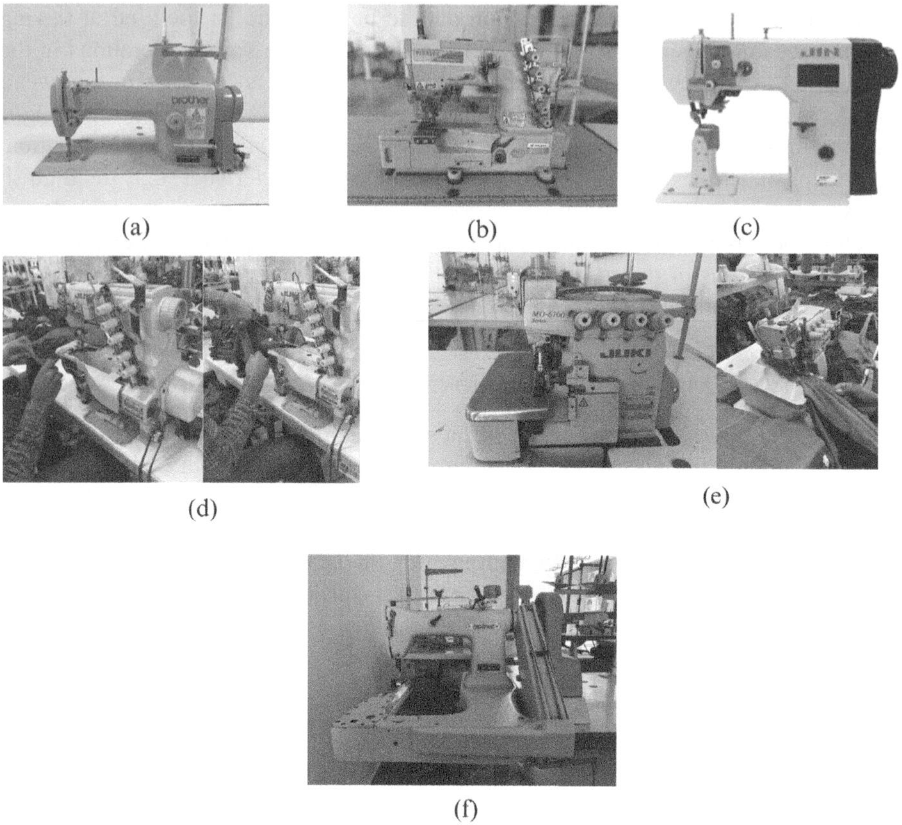

FIGURE 9.8 Machine beds: (a) Flat bed – single-needle lock stitch machine; (b) raised bed – flatlock machine; (c) post bed sewing machine; (d) cylinder-bed sewing machine (left), sewing of tubular component of a garment on cylinder-bed sewing machine (right); (e) side bed – overlock machine; (f) feed-off-the-arm machine.

large circumference, such as the hemming of T-shirts. Many attachments for specific functions, such as smocking, are attached to such beds. This bed facilitates pre-sewn component assembly and allows for operations such as buttonholing and button attachment.

#### 9.3.1.3 Post Bed

This machine has a tall, elevated vertical post or column over the flat machine bed (Figure 9.8c). Its vertical column has a needle plate, foot, bobbin, feed dog, or looper for sewing small, irregularly shaped, three-dimensional products like shoes, bags, glove assemblies, emblems, automobile upholstery, etc.

#### 9.3.1.4 Cylinder Bed

Cylinder-bed machines are used to assemble products with small, tubular, curved, or cylindrical elements (Figure 9.8d) that need to be rotated around a cylindrical shape for stitching. This machine's horizontal, arm-shaped bed and increased working height enable tubular components like sleeves, cuffs, trouser legs, armholes in small garments, round neck openings, etc. to be stitched with ease and quality.

#### 9.3.1.5 Side Bed

The bed of the machine for holding and sewing the fabric is on one side of the machine, like in overlock machines (Figure 9.8e). This type of bed is ideal for edge sewing and serging. Jana (2015) mentions this side bed as a mono block and a variation of raised-bed machine with no space on the right side of the needle.

#### 9.3.1.6 Feed-Off-the-Arm Bed

This bed has a cylindrical shape that extends out from the machine perpendicular to the direction of the cylinder-bed machine or parallel to the direction of sewing (Figure 9.8f). The component to be sewn is wrapped around the cylindrical bed and is fed away from the operator, off the end of the bed while sewing. This bed enables the joining of two opposite edges of a flat-shaped fabric to form a garment part that becomes a cylinder, tube, or hollow cylindrical-shaped item. This bed simplifies the creation of FS and fell seam. The operator can easily prepare the side seams of men's shirts and jeans with feed-off-the-arm machine.

Jana (2015) talks about a feed-up-the-arm bed, similar to a feed-off-the-arm bed. In this case, the open end of the cylinder faces the operator, while the attachment to the machine frame faces away from the operator. The axis of the cylinder bed is parallel to the direction of stitching. Sewing is accomplished in the opposite direction of what is done in the feed-off arm. The tubular stitched objects accumulate in the cylinder-shaped machine arm and do not fall off automatically, as they do in the feed-off-the-arm bed. When the sewing is accomplished, the operator has to lift the presser foot to take the items that have been sewn off the cylinder.

### 9.3.2 Classification of Sewing Machines on the Basis of Operator Control

On the basis of operator control, sewing machines can be classified as shown in Figure 9.9.

#### 9.3.2.1 Manually Operated Sewing Machines

This category encompasses all machines that are controlled manually by an operator (Figure 9.10). Typical components of such machines include a stand, a table, an electric motor, a single-stitch head, a bed (one of the various shapes), and the ability for the user to manage the sewing speed, stitch density, and presser foot position. A diverse range of attachments for these machines are available and extensively used, including work aids designed to spare the operator from some manual tasks or handling operations associated with sewing processes, e.g. replacing the presser foot with a ruffler foot aids in preparing ruffles with even dimension and maintaining quality.

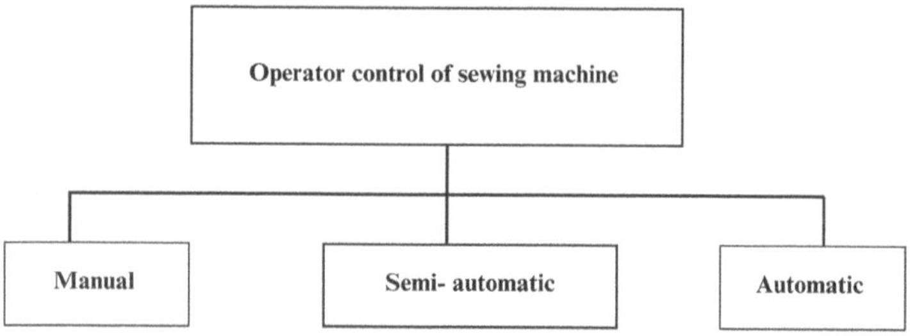


**FIGURE 9.9** Sewing machine classification according to the operator control.

**FIGURE 9.10** Manual operation of treadle sewing machine.

### 9.3.2.2 Semi-Automatic Sewing Machines

This category includes all sewing machines in which some part of an operation is automated, such as a buttonhole machine in which the operator sets or feeds the fabric under the needle in a specific position (Figure 9.11) and then the machine handles the rest of the process of working on and finishing the buttonhole. The machine is usually cam-controlled and produces only one sewing configuration. Button sewers, bar-tack machines, and label sewers are a few examples. The machine determines the shape of the sewing line, ensuring consistency, but the operator must still perform all handling both before and after sewing.

This category includes all sewing machines in which some part of an operation is automated, such as a buttonhole machine in which the operator sets or feeds the fabric under the needle in a specific position.

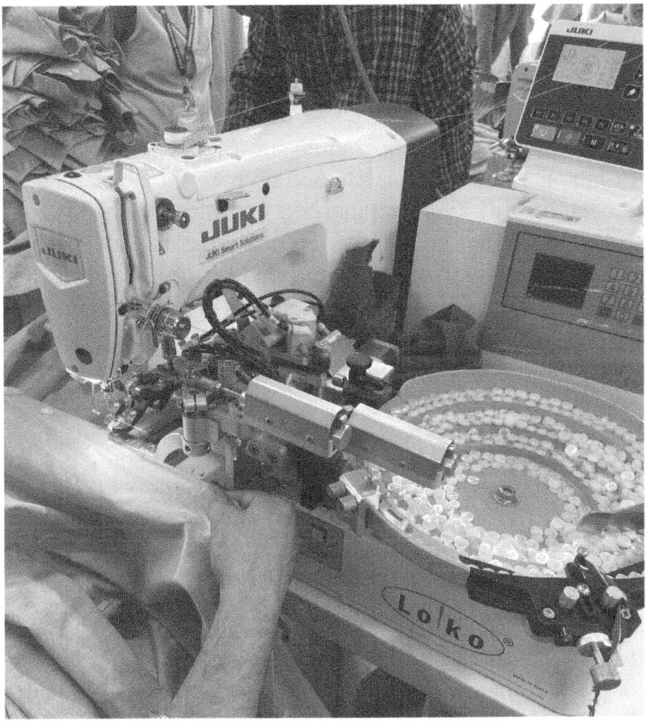


FIGURE 9.11 Button-sewing machine.

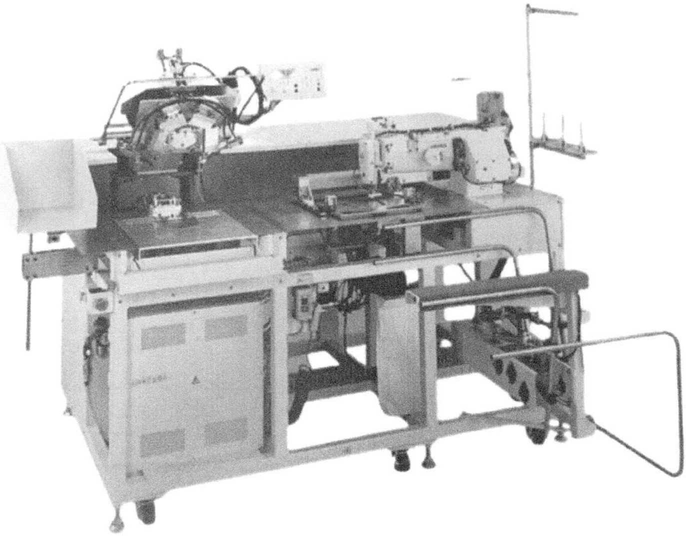

FIGURE 9.12 Automatic pocket setter.

#### 9.3.2.3 Fully Automatic Sewing Machine

These machines just need to be loaded with clothing components; the rest of the operations, such as picking up the pieces, sewing, and final disposal or disposal of finished pieces, are handled by the machine. These extremely expensive machines are used by the industry when specialised operations are needed (Figure 9.12). Additionally, the amount of labour needed to operate and prepare the goods is decreased in this category and machine utilisation can increase considerably.

Automatic Transfer Lines are another cutting edge technology that comprises of several machines or workstations interconnected by work handling devices in order to transfer components automatically. The cut parts or garment components enter one end of the automated flow line, where they are systematically processed at several stations, and the final product exits the other end. McLoughlin (2018) explains about the challenges of automatic transfer lines that robotic systems for manufacturing have been developed, but they have not been successful because it is difficult for robots to handle limp materials, such as apparel fabrics, and they cannot match the dexterity of human fingers, particularly when handling special fabrics.

## 9.4 SEWING MACHINE FEED MECHANISMS

Machines can also be classified on the basis of their feed mechanisms. Feed mechanism refers to the way the fabric is fed into the machine and how it moves while stitching. It is the feed mechanism that contributes to the good quality and performance of a seam in a product. The feed mechanism's purpose is to control the feeding speed of the fabrics during sewing, moving material from one stitch position to the next with optimum pressure on the material. Hence this mechanism controls the proper stitching, seam appearance, SPI, and movement of fabric forward or backward. A sewing machine's feed mechanism consists of a throat plate or needle plate, or feed plate, a feed dog, and a pressure or presser foot (Figure 9.13).

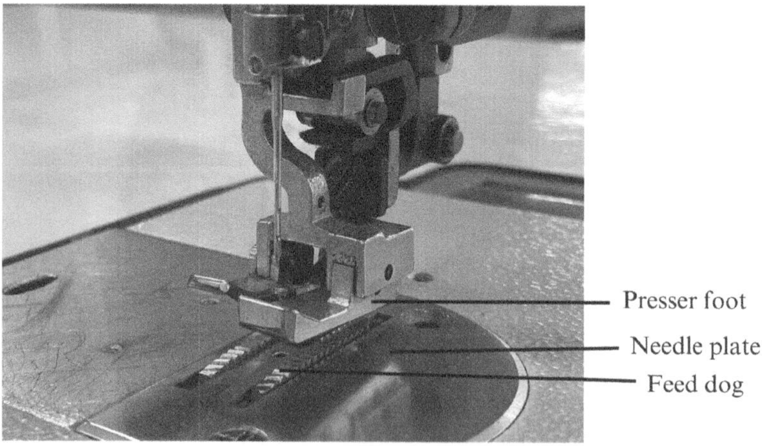


**FIGURE 9.13** Feed mechanism in a single-needle lock stitch machine showing feed dog, throat or needle plate, and presser foot.

Throat plate or needle plate is a smooth metal plate under the presser foot that has a hole (its size should not exceed 30% of the needle size) for needle penetration through the fabric and a slot for the feed dog. The right size of hole is important for the right kind of stitching. An increase in the hole size can lead to a problem called flagging, where some part of the fabric moves into the needle hole while stitching because of the bigger needle hole. This plate also facilitates the movement of fabric.

A feed dog is a piece of metal that is attached under the needle plate and has teeth or serrations on it. When the sewing machine is turned on, it moves up and forward through a slot in the throat plate. This moves the fabric being stitched forward by one stitch length. The main goal of this part is to move the fabric or material that has been stitched or sewn to a certain length. It also prevents fabric from slipping while being stitched. The stitch length, which is controlled by a stitch length regulator, determines how fast the feed dog moves.

The pressure/presser foot is a metallic, smooth piece that provides downward pressure and presses the material being stitched, and it works in conjunction with the throat plate and feed dog to achieve controlled movement of the material and correct stitch formation. The fabric is pressed between the pressure foot and needle plate and worked with a feed dog. Different types of feed dogs, presser feet, and throat plates are used based on how the sewing process and the material being sewn.

**Different Types of Feed Mechanisms Are**

- Manual feed
- Bottom feed or drop feed
- Needle feed or compound feed
- Differential feed
- Unison feed
- Variable top and bottom feed
- Differential bottom feed and variable top feed
- Puller feed
- Wheel feed

### 9.4.1 Manual Feed

It is also referred to as 'free motion' or 'darning feed.' In this mechanism, the operator controls the entire fabric feed while manipulating the material under the needle without any feed dog. Due to the absence of a feed dog, the operator has the freedom to adjust the stitch length. Usually, the feed dog must be removed from the sewing machine in order to get a manual feed. This feeding mechanism has many applications like darning, embroidery, freehand quilting, and jigging collars for men's shirts, etc.

### 9.4.2 Bottom Feed or Drop Feed

It is the most popular feed mechanism in which the feed dog lays just under the machine surface or bed and aids in the movement of the fabric with each stitch, which is pressed down by the presser foot against the needle plate (Figure 9.14a).

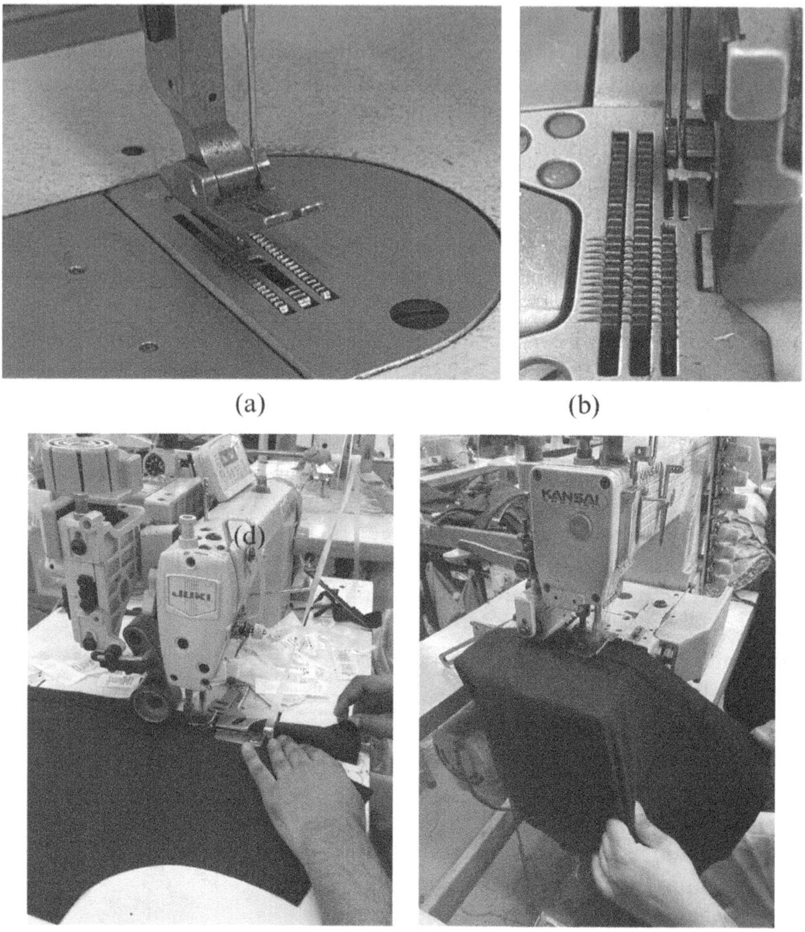


(c)

**FIGURE 9.14** Some feed mechanisms: (a) bottom feed mechanism; (b) differential feed in an overlock machine; (c) puller feed.

Bottom feed is also known as four-motion feed or four-motion drop feed. The feeding process consists of four movements or motions: rising above the plate to make contact with the material and advance it to the specified distance, descending under the plate to release the touch with the fabric or ‘drop’ beneath the material’s surface, and then returning beneath the plate to the required distance to continue the feeding process.

The serrated feed dog moves and extends slightly above the throat plate’s rectangular slot. These teeth contact the fabric from the underside, and when they move in accordance with the stitch length, they advance the material into the machine. As it continues to travel down and then up in a cyclical pattern, it continues to move the material or fabric, therefore creating rows of stitches. When fabric is being lifted and fed, the feed dog and needle are synchronised in such a way that the needle usually remains above the fabric. There may be variations in the number, location, and shape of the feed dogs, but the feed dogs' teeth are always slanted in the fabric-feeding direction. By exerting pressure on the fabric, the pressure foot keeps the fabric from sliding up and down with the needle during its up-and-down action. In the drop-feed mechanism, the fabric is fed under the bottom surface of the pressure foot, which stays stationary and has a smooth surface to permit fabric movement. The pitch of a feed dog, the distance from one peak to the next peak in a feed dog, is extremely important and varies depending on the type of fabric to be stitched: for very-light-weight fabric, the pitch is 1–1.25 mm, whereas for light- to medium-weight fabric, it is more and kept between 1.3 and 1.6 mm. Not only is pitch important, but also is the surface of the teeth: feed dogs with rubber-coated teeth are used for textiles that may be permanently marred by the sharp teeth of the feed dog.

The bottom feed system may facilitate both straight and curved stitching, although ply shifting and roping are common sewing machine problems with this mechanism. Because there is just a bottom feed, the bottom feed mechanism sometimes has uneven material feeding or inter-ply slippage, especially with slippery fabrics. While sewing two plies of cloth, it is possible for the top ply to slide because it is only pushed with a presser foot, whose surface is flat (producing little friction) and not serrated like a feed dog that contacts the bottom ply. When surface frictional levels are very high, inter-ply slippage is low. This phenomenon is also known as inter-ply shift and differential feeding pucker. Similarly, ply shifting during hem preparation may also sometimes result in roping.

### 9.4.3 Needle Feed/Compound Feed

Needle feed refers to the mechanism in which the needle itself moves forward and backward. There are two feeds in this mechanism: the needle feed and the bottom feed. The feed dog and needle feed simultaneously feed the fabric while the needle is positioned inside the cloth. In this system, the needle penetrates the fabric, goes backward to create one stitch while the feed dog advances the cloth, and then rises and proceeds forward to create the subsequent stitch. Thus, coordinated movement of the needle bar and bottom feed dog decreases slippage, reduces uneven material feeding, and allows the operator to stitch many layers of fabric. The feeding is intense and in case of tightness of the thread, the stitch may shrink. This method is ideal for sewing thick or thin fabrics and is especially useful for quilting through many layers, such as cloth and wadding.

### 9.4.4 Differential Feed

This feeding system consists of two feed dogs, one in front of the needle and the other behind it, both arranged in a straight line beneath the throat plate. The main feed dog is the one that is placed the furthest away from the operator, while the differential feed dog is the one that is placed closest to the operator. The feeding speed of each feed dog can be fixed or varied. When the main feed dog is set to move faster than the differential feed dog, the fabric stretches. However, when the differential feed dog is set to move faster than the main feed dog, the cloth gathers. The differential feed dog mechanism is useful for stitching stretch fabric by maintaining the correct dimension of fabric. This mechanism does this by driving the fabric into compression, which counter balances the stretch and ensures the correct shape and size of the seam. This kind of feed mechanism is found in overlock (Figure 9.14b) and cover stitch machines.

### 9.4.5 Unison Feed

Unison feed is a further combination of feeding mechanisms. This feed mechanism is designed such that the feed dog, needle, and a special presser foot, which is a two-part presser foot, move in unison or feed and sew the fabric while the needle is still inside the cloth. This has a unique presser foot with two components. One portion resembles a conventional presser foot, while the other is a little piece placed between the two feet of the presser foot and equipped with a hole for the needle to pass through. This little component feeds the fabric alongside the needle as it goes up and down. This system is also known as the 'walking foot system' because the presser foot has two independently driven parts that progress and maintain the plies' alignment. This kind of feed mechanism is ideal for many layers of fabric, such as tarpaulin, polyvinyl chloride, and leather, as well as bulky seams in heavy-duty material.

### 9.4.6 Variable Top and Bottom Feed

This feed mechanism has two feed dogs: one under the throat plate and one feed foot at the top. The feeding foot's toothed or serrated surface advances the top and bottom plies without puckering or seam slippage. This mechanism feeds when the needle is not in contact with the cloth. As the variable top feed is paired with a drop or bottom feed, it is possible to speed up the movement of the top ply to achieve shift-free stitching. Bottom and top feed can be adjusted depending on the fabric to be stitched or to create tension or slack in the plied fabric. This mechanism allows the sewing machine to adapt to a range of applications, which increases productivity and encourages the production of higher-quality items. This mechanism is often used for stitching high-friction fabrics, such as simulated leather and composites, where the use of a static presser foot is unsuitable.

### 9.4.7 Differential Bottom Feed and Variable Top Feed

When a differential bottom feed is coupled with a variable top feed, the quantity of material supplied from the top may be adjusted while the material fed from the bottom can be stretched and gathered. This mechanism makes it possible to sew without seam slippage or ply shifting and to provide ease or gathering in either the top or bottom ply. Consequently, by using overlock machines with differential bottom feed and variable top feed, one may produce the appearance of top-ply gathering attached to a flat bottom ply or vice versa, for instance, by constructing a garment with a waistline seam that is gathered and sewn with a smaller bodice.

### 9.4.8 X-Feed

Numerous Vetron/Typical sewing machines include an X-feed that is distinguished by a split bottom feed. It operates in tandem with needle feed. The feed dog is divided into two separate feed dogs. The needle and the first split or central section feed dog transport the material during the first half stitch length. While the needle is leaving the fabric and travelling ahead, the feed dog is descending and moving in the direction of stitching. The outer portion of the split feed dog transports the second half of the stitch length and completes the stitch. Due to the alternating motion of both feed dogs, material transport is continuous, and consistent seams with reduced thread tension are produced. Jana and Khan (2014) experimented on the sew ability of light-weight fabrics using X-feed mechanism and drop-feed mechanism. It was found that X-feed mechanism results in a flatter, pucker-free seam appearance and is better in handling slippage between top and bottom plies in comparison to drop feed.

### 9.4.9 Puller Feed

An auxiliary feed roller located at the rear or back of the presser foot pulls all the layers of material together and applies downward pressure to all the layers of material to drag them through the machine to achieve even sewing (Figure 9.14c). The additional roller or rollers behind the presser foot pull at the same rate as the feed dog; hence, through this mechanism, uneven material feeding is reduced and working properties are improved. Both rollers may be driven, or the top roller may be driven while the lower one idols. The top roller is generally driven by a machine, while the lower one moves due to the control and pressure of the top roller. Puller feed is primarily used in the seaming of heavy fabrics such as canvas tents or to keep tension in lighter weight materials. This kind of mechanism is useful in the multi-needle stitching of parts such as waistbands.

### 9.4.10 Cup Feed

This feed mechanism consists of two horizontally positioned feed wheels, allowing for an improved view of both fabric plies. These wheels are driven simultaneously in order to move the fabric from right to left through machines with horizontal needles. This mechanism is used for seaming skins, knitted goods, wetsuits, moccasins, etc.

### 9.4.11 Wheel Feed

Instead of using a presser foot, this mechanism uses a wheel to move the cloth and sew seams, particularly in leather and plastic. To improve friction for feeding leather or plastic, the wheel's surface is frequently knurled or coated.

Jana (2015) mentions auxiliary and modified feed types. The purpose of auxiliary feed types is to enhance or highlight the features of a main feed. One such example is the use of puller feed in conjunction with drop feed. Clamp feed systems, which retain the material and facilitate multidirectional stitching, are another example. In modified feed types, the form of the presser foot and/or feed dog is often altered to wheels or belts to improve material movement. For example, shoe uppers are sewn with a wheel feed, where the wheel takes the place of the feed dog, a needle feed, and a roller presser foot, where the roller takes the place of a normal presser foot.

## 9.5 LATEST TECHNOLOGIES IN SEWING ROOM

The advent of the sewing machine in the 18th century accelerated the process of clothing production. Since then, the technology has been continuously improved and evolved to the extent that products may now be manufactured using high-tech machinery at unprecedented speeds. The apparel business relies heavily on sophisticated technology for cutting, stitching, and finishing. The sewing process in the apparel sector is undergoing an extension of innovation and the advancements are transforming the garment manufacturers from simple, low-tech machinery to high-tech, digitally controlled machines, resulting in faster, higher-quality manufacturing. Recent advancements in garment-sewing technology have introduced remarkable breakthroughs to the apparel sector. There are certain categories of machines that are popularly or widely found in almost all apparel manufacturing units. Some of these are discussed further.

### 9.5.1 Single-Needle Lock Stitch Machine

This is the most common sewing machine used in a garment factory's sewing room or production floor. It is referred as SNLS machine. It has a single needle and produces a lock stitch. All plain seams in most of the applications use a single-needle lock stitch. These types of sewing machines are often used to stitch cloth, leather, and other materials and use two threads: an upper thread through the needle and a lower thread through the bobbin. The needle penetrates the cloth from the top in order to pass the needle thread through the fabric from the top to the bottom in order to create a lock stitch. This machine is also used with several attachments, such as a binder foot for applying binding to a neckline and a ruffler foot for producing ruffles, and tape attachment (Figure 9.15a). The wide applicability of this machine makes it popular in every apparel factory. In the most recent machines, one may find characteristics such as direct drive, as well as a number of ergonomic elements such as LED lighting (Figure 9.15b) for ease of vision and clarity.

### 9.5.2 Double-Needle Lock Stitch Machine

This is another lock stitch machine used for topstitching. It has two needles and two bobbin cases, allowing it to create two rows of stitches (Figure 9.15c), e.g. on a patch pocket attachment on a shirt and leg seam. These machines are suitable for sewing many operations in shirts, uniforms, jeans, jackets, and other similar garments, handbags, suitcases, gloves, footwear, etc.

### 9.5.3 Overlock Machine

These are over-edge neatening machines that trim and cover the raw edges of a fabric or material to provide a clean and neat look when seam allowances or seam edges are visible. While overlocking, cutters positioned or set at a slight distance from

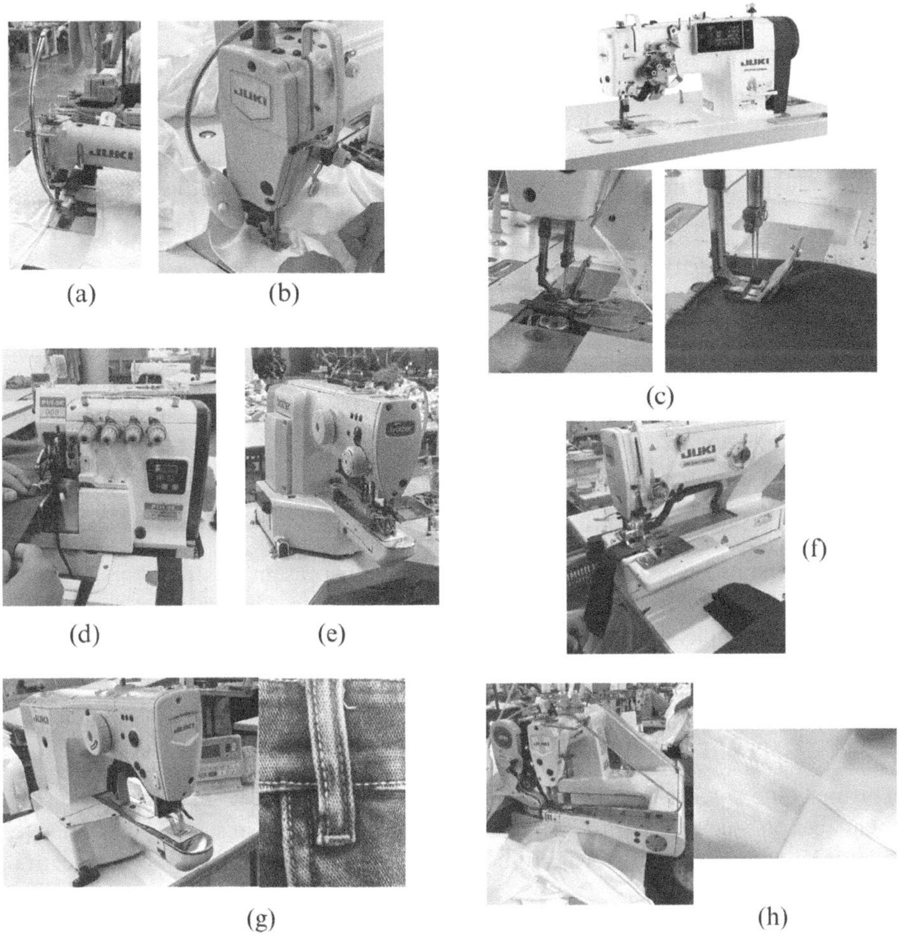


**FIGURE 9.15** (a) SNLS machine with tape attachment; (b) attachment of cuff with SNLS machine using compensating presser foot and LED light; (c) double-needle lock stitch machine; (d) overlock machine; (e) button-sewing machine; (f) buttonhole sewing machine; (g) bar-tacking machine and bar tack on belt loop; (h) feed-off-the-arm sewing machine and lap felled seam formed with the machine.

the needle remove the surplus or excess fabric and neaten the edge (Figure 9.15d). Overlocking or overedging is a standard method of seam finishing for many different types of clothing, including those for children, women, and men, as well as for home furnishings, textiles, and other lifestyle products. These machines may have between one and five spools of thread for varied applications and have larger looper thread supplies. These machines, sometimes known as sergers, are used to sew a variety of knitted fabrics and garments, including T-shirts, leggings, etc. The devices' specifics have already been covered in prior sections. Whether it is a huge garment manufacturing operation or a small one producing customised designer items, an overlock machine is essential.

### 9.5.4 Button-Sewing Machine

A button-sewing machine attaches buttons to garment plackets (Figure 9.15e). Buttons with holes or shanks may both be sewn on with this machine. The machine can also have hopper feed that feeds buttons directly to the clamp at the needle's tip. The needle clamp secures and places the button at the stitching location. The button-sewing machine makes a row of parallel stitches between the holes.

### 9.5.5 Buttonhole Sewing Machines

These machines create buttonholes (Figure 9.15f) in a wide range of garments, notably shirts, pants, tops, jackets, etc. These machines can lock stitch or chain stitch. These buttonholes may vary in appearance. A buttonhole machine may either cut the buttonhole first and then complete the raw edges with stitches, or it can finish the raw edges of the buttonhole first and then slash them.

### 9.5.6 Bar-Tacking Machine

This machine produces very small, close stitches that resemble small bars. This bar is specifically sewn at a spot in a garment that requires reinforcement, such as the ends of a buttonhole, the ends of the mouthlines of pockets, and the ends of belt loops (Figure 9.15g).

### 9.5.7 Feed-Off-the-Arm Sewing Machine

This machine stitches a lap felled seam in a tubular portion or small width of a garment section, such as a shirt (Figure 9.15h), trousers, or jeans, such that after the seam, the garment panels form a tube-like structure, such as the inner leg seam of jeans.

### 9.5.8 Blind Stitch Machine

A blind stitch is used to join two layers of fabric such that the stitch thread is not visible. This stitch catches just a few fabric strands and is concealed behind the cloth's folded edge. A blind stitch machine has a bent needle, so it may partially pierce the fabric while gathering up a small number of strands. It is used for hemming formal skirts and trousers, as well as attaching pockets, linings, facings, and trims to garments. This machine is useful for a variety of fabric types and weights.

### 9.5.9 Embroidery Machine

Embroidery machines are used to adorn the surface of fabrics with a variety of sophisticated embroidery patterns, which are then turned into luxurious fashion or lifestyle goods. These machines add value to the product. Numerous types of embroidery machines are used in the industry. The embroidery can be achieved using free-motion machine embroidery or computerised machine embroidery.

Free-motion machine embroidery is sometimes referred to as 'free hand stitching' or 'free machine embroidery.' It is produced using a single-needle, zigzag sewing machine without the automated characteristics of a specialised embroidery machine. By lowering the machine's feed dog, the operator may use the sewing needle to create a design on stretched fabric encased in a frame. This machine has a single needle that must be threaded repeatedly to change the colour of embroidery thread while making a multicoloured embroidered pattern, which is also a time-consuming task. Due to the fact that this is a manual design process, each design is one-of-a-kind, and there is a chance of variance among similar designs.

In order to facilitate mass manufacturing and the production of high-quality goods, recent embroidery machines are computer-controlled. Such machines can have a single or multiple heads with multiple needles (Figure 9.16). Using specialised software, an embroidery pattern is developed and then uploaded on a computerised machine. These machines have a frame that holds the framed section of fabric under tension and automatically moves it to generate the stored, pre-programmed pattern or design.

### 9.5.10 Automation in Sewing Machines

The increasing rate of production necessitates swift processes, whether it is sewing, embroidery, or finishing. Apparel manufacturers are always looking for ways to cut production costs, and efforts in the sewing or cutting department are critical because sewing accounts for between 35% and 40% of total apparel production costs (Gries and Lutz 2018). Along with sewing, the handling of cut or semi-sewn components or fabrics between and on the work stations is also important and should be taken care of. The handling activities are extensive and entail numerous tasks that can be streamlined by combining or eliminating many of them. Automation on the manufacturing floor can aid in decreasing this time while enhancing efficiency, output, and product quality. Nayak and Padhye (2015) also emphasised that in order to be successful in the competitive global market, apparel industry has to lay importance

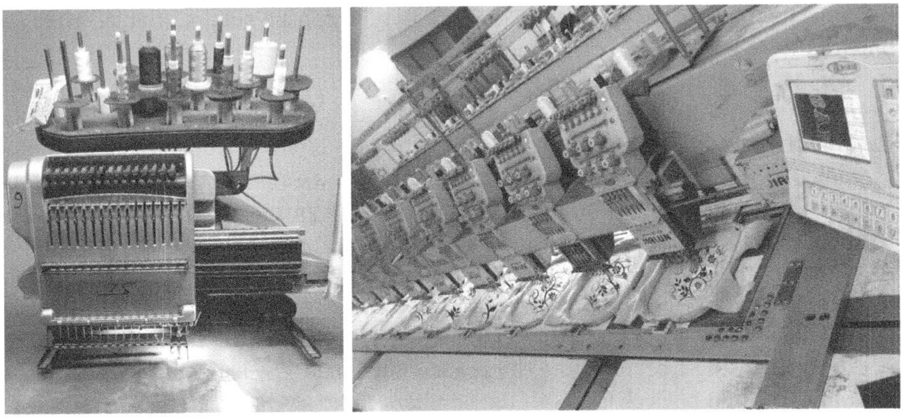

**FIGURE 9.16** Single-head (left) and multi-head (right) computerised embroidery machine.

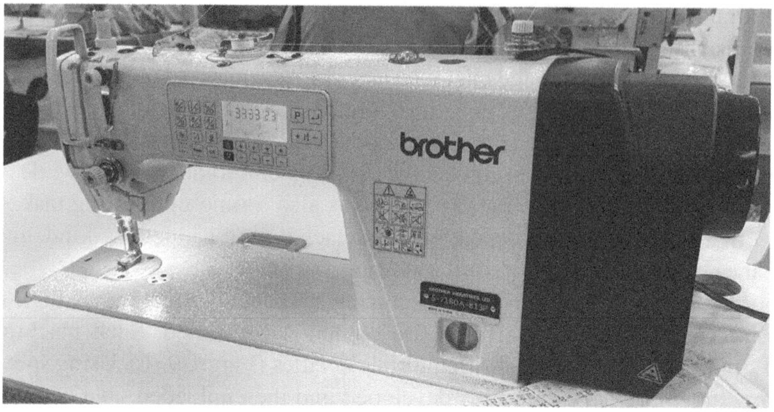


**FIGURE 9.17** Sewing machine with control panel.

on automation, worker training, modernisation for productivity and quality enhancement, and a decrease in waste and production delays.

As previously stated, electronic-controlled panel (Figure 9.17) or computer-controlled technology is used in many specialised machines for precise and high-quality products in addition to embroidery. The electronic control of a sewing machine enables a range of programmable functions, including the ability to alter the stitch pattern, the detection of bobbin run-out, and the automated functioning of needle position (NP) and an under-bed trimmer (UBT). Additional advantages include the reduction or minimisation of a number of non-sewing actions that contribute unnecessary time and effort (Jana 2015).

Programming of back-tack stitches in a computerised or electronic-controlled panel mounted on the head of a sewing machine helps make a back tack automatically at the beginning or end of a seam. This also reduces the sewing time and improves the stitch quality. The control panel allows the operator to change the number of stitches in a back tack. Furthermore, such technology can assist the operator in achieving various types of back tack, which can be single, double, triple, or quadruple.

Step programming (SP) is an additional feature that includes programming each step of a process or operation. One of its application is in sewing of patch pockets; the number of stitches on the three sides of the pocket are calculated and then programmed. After programming, when the operator sews a pocket, the machine stops automatically after one seam, the operator rotates or pivots the pocket and presses the pedal for the second row of stitches, the machine stops again after a set number of stitches at the corner, the operator rotates or pivots the pocket once more, and the operator completes the stitching with a back tack. In addition to SP, the ply sensor (PS) represents a substantial advancement. The PS detects the difference between plies in the sewing line, enabling the sewing machine to be programmed to stitch a certain number of plies or layers. If the number of plies is different, the machine will stop. Therefore, the operator may adjust the layers as necessary rather than redoing the whole process. It is feasible to sew a patch pocket to perfection using SP and a

PS, with the programming set to stitch three plies of material. Computerised control panels attached to many sewing machines have USB port configurations where many sewing patterns or stitching instructions can be uploaded using USB drives. This technology also enables software updating using USB storage devices.

Bobbin thread sensor, bobbin-thread-remaining detector, or bobbin run-out detector in a sewing machine detects automatically before the bobbin thread runs out. This function notifies the operator through a buzzer or light when a thread is ready to finish. This feature in the sewing machine enhances the product's quality and the operator's productivity. In addition to this feature, the bobbin changer is another important technological advancement. The completion of bobbin thread requires extra time for bobbin changing and bobbin winding. Moreover, the finishing of yarn reduces the quality of stitching, especially in the case of embroidery. Automatic bobbin changing systems save time when creating large designs on textiles. Sewing machines with this technology can hold multiple bobbins on a plate at once and exchange the bobbin automatically when needed.

The NP setting may automatically move the needle up (out of the material being stitched) or down (inside the material being stitched) when the machine stops. This eliminates the requirement for the operator to manually position the needle by turning the flywheel before each operation.

Sewing machines equipped with a UBT automatically clip the thread when the stitch is completed. To cut the sewing thread, the operator just has to touch a button or press the paddle back. Operators no longer need to carry a thread cutter or trimmer to cut the sewing thread after finishing a stitch. As a result, machines with UBT reduce sewing or production time, improve stitch quality, and reduce the amount of labour required for thread cutting, but such a feature raises the machine's cost.

Decorative sewing, including the precise stitching of patches, insignia, etc. onto the fabric of uniforms, garments, home furnishings, and lifestyle products, employs a vision sewing system consisting of a camera alignment system installed on the sewing machine. This machine can be a lock stitch, chain stitch, or pattern sewing machine that is electrically programmable. The patch or label is added to the base cloth by the operator. The machine's embedded digital camera takes a photograph of the setup. The vision sewing system then analyses the picture to determine the item's perimeter. This technology then develops an automated stitch pattern and creates a sew pattern around the object's perimeter. This technology allows users to customise the stitch type and size. This kind of method facilitates the application of intricate pieces or patches to the base with accuracy, consistency in quality, high productivity, and efficiency.

Aside from sewing, automated workstations execute a number of complex operations using electric, electronic, and pneumatic control, as well as modern conveyor and clamp technology (Tyler 2008). Preparation of welt pockets, patch pocket setting on jeans and other garments, and run stitching collars or flaps etc. are some examples. Uploading of the garment pieces, unloading of the sewn product, and some of the handling activities are taken care of by the operator, but the rest of the tasks are undertaken by the machine. Such automation increases the utilisation of sewing machines. Throughout the sewing cycle, the operator may do a range of handling tasks depending on the kind of machine. Due to the addition of handling time into the sewing time, a basic sewing machine's machine utilisation goes from 20% to 80% (Tyler 2008). In these machines, the size and shape of the stitched components can be changed, but the basic stitching operation remains the same.

There are many new technologies available that provide automation, seam performance, and high-quality products, which is the need of the hour. Pedal-less sewing, robotic technology, laser welding sewing technology, special grippers for handling, automated workstations, etc. are some of the examples of innovative solutions for increased efficiency in sewing. The latest sewing technology is continuously and constantly supporting the manufacturers to grow and create best of the products in less time.

## REFERENCES

Chuter, A. J. 1995. *Introduction to Clothing Production Management*. Oxford: Blackwell Scientific Publications.

Clayton, M. 2008. *Ultimate Sewing Bible—A Complete Reference with Step-by-Step Techniques*. London: Collins & Brown.

Colovic, G. 2015. Sewing, stitches and seams. In *Garment Manufacturing Technology*, eds. R. Nayak, and R. Padhye, 247–273. Cambridge: Woodhead Publishing.

Gries, T. and V. Lutz. 2018. Application of robotics in garment manufacturing. In *Automation in Garment Manufacturing*, eds. R. Nayak, and R. Padhye, 179–197. Cambridge: Woodhead Publishing.

Hayes, S., J. McLoughlin, and D. Fairclough. 2012. *Cooklin's Garment Technology for Fashion Designers*. New Jersey: John Wiley and Sons, Ltd.

Jana, P. 2015. Sewing equipment and work aids. In *Garment Manufacturing Technology*, eds. R. Nayak, and R. Padhye, 275–314. Cambridge: Woodhead Publishing.

Jana, P. and A.N. Khan. 2014. The sewability of lightweight fabrics using X-feed. *International Journal of Fashion Design and Technology Education* 7(2):133–142. https://doi.org/10.1080/17543266.2014.906661

Karthik, T., P. Ganesan, and D. Gopalakrishnan. 2017. *Apparel Manufacturing Technology*. Boca Raton: CRC Press.

McLoughlin, J. 2018. Stitches and seams. In *Textile and Clothing Design Technology*, eds. T. Cassidy, and P. Goswami, 420–440. Bota Raton: CRC Press.

Nayak, R. and R. Padhye. 2015. Introduction: The apparel industry. In *Garment Manufacturing Technology*, eds. R. Nayak, and R. Padhye, 1–15. New Jersey: Woodhead Publishing.

Shaeffer, C. 2014. *Sewing for the Apparel Industry*. London: Pearson Education Ltd.

Tyler, D. J. 2008. *Carr and Latham's Technology of Clothing Manufacture*. New Jersey: Blackwell Publishing Ltd.

# 10 Pressing Equipment
## *Conventional and Modern*

*Shipra Sharma*

## 10.1 INTRODUCTION

While alternative garment assembly processes such as fusing, adhesives, molding, and welding have been developed, sewing remains the primary method due to its superior seam appearance, flexibility, and speed. Conventional sewing techniques can have several drawbacks, including seam grinning at joints, accumulation of thicker material at the joining point, and constraints on production speed. Consequently, novel alternative processes are being devised and improved upon. These processes are proving to have a significant commercial impact on particular applications and are rapidly gaining wider market acceptance due to their potential to add value to the design and production of garments (Petrie, 2015). However, their successful use demands careful pre-production development and production process control.

## 10.2 ALTERNATIVE METHODS OF JOINING MATERIAL

The selection of an appropriate method for the fabric joining process is a crucial decision that impacts the entire production process and the final performance of the garment. The selection relies on various factors related to performance, production, and customer needs, including characteristics of materials used, manufacturing equipment, material cost, production speed for cost-effective joints, joint strength, the durability of the joint, visual appeal, and comfort while wearing (Petrie, 2015). Alternative methods of joining materials are successful in specific contexts and have a substantial commercial impact. Their advantages include increased strength, durability, and water resistance, and they are especially advantageous for materials that are difficult to sew. However, these approaches may necessitate specialized equipment and expertise, while rigorous testing and quality assurance procedures are crucial to ensure that they are executed correctly and consistently for each garment.

Nevertheless, their use requires careful pre-production development and process regulation during manufacture (Tyler, 2008). Fusing, welding, adhesives, and molding are viable substitutes for traditional material-joining methods. Among these, fusing is a process of joining fabrics with a fusion of interlining to the fabric by applying heat and pressure. Welding is a process that involves heating two or more materials to their melting points and joining them together. The resulting joint can withstand the

DOI: 10.1201/9781003432739-10

weight and provide strength and has garnered popularity in sportswear and lingerie. The adhesive application involves using thermoplastic materials to bond two surfaces together, creating a joint that can bear weight and provide strength. Molding involves shaping material by pouring it into a predefined mold, where it solidifies or hardens. This method is predominantly employed in the production of brassieres, headgear, and contoured support components for structured garments.

### 10.2.1 Fusing

During the fusing process, a thermoplastic resin is applied to the underside of a fabric to facilitate its bonding with another fabric strip by subjecting it to heat and pressure at a specific temperature and duration in the fusing press. Temperature is usually between 110°C and 175°C, while 150°C is more prevalent (Tyler, 2008). The fusing time typically ranges from 10 to 20 seconds, with pressure being 2–3 $kg/cm^2$ for the desired effect. The temperature and pressure settings of the fusing press are vital for desired bond strength and appearance. Fusing enhances the garment's appeal as it helps strengthen and stabilize its crucial areas without compromising its draping quality and helps preserve the underlying fabric's crisp appearance. This technique imparts form, rigidity, and support to various garment components such as cuffs, collars, and waistbands. Over-fusing leads to a stiff, board-like appearance, whereas under-fusing results in a weak bond susceptible to disintegration after laundering or dry cleaning. It is crucial to conduct appropriate testing and quality assurance measures to ensure that each garment's fusing process is executed accurately and uniformly.

Using fusible interlinings can reduce production time, lower direct labor expenses, and replace complex basting operations in large areas and sewing in small areas. While some prefer traditional canvas interlinings for their superior softness and contour, fusible interlinings offer ease of use and consistent quality. Fusible interlinings provide alternative methods of garment construction, such as using slotted interlinings. To achieve desired stiffness, handle, or draping qualities through the fusing process, manufacturers must contemplate the selection of the base fabric, type, and quantity of the fusible resin that forms the bond. Other fabric properties, such as crease recovery and air permeability, may be adversely impacted. If the low permeability of the outer fabric is essential for the final garment, the selection of interlinings to maintain this product characteristic is crucial. The strength of the bond must be sufficient to withstand subsequent operations, wear, and resistance to washing, drying, ironing, dry cleaning, or solvents. Failure to achieve a complete and even bond or delamination could cause bubbling in the outer fabric. Careful attention must be paid when fusing interlinings to garments to avert issues like strike-through or strike-back. Strike-through can lead to resin dots on the garment and potential markings during high-temperature ironing. Strike-back may contaminate equipment and adhere to the garment lining during pressing. Certain fabrics, such as showerproof rainwear, are challenging to fuse due to their smooth, hard surface and showerproof finish. Fusing interlinings to garments should not induce thermal shrinkage in the outer fabric. During fusing, many fabrics may experience shrinkage due to the high temperature, potentially resulting in ill-fitting garments during manufacturing and inaccurate sizing post-production.

Some fusing presses have a pre-heat zone set at a lower temperature to prevent shrinkage. The heat generated from the fusing process can cause dye sublimation, resulting in discoloration of some fabric parts. Additionally, pile fabrics are susceptible to being crushed during fusing, leading to a different appearance between fused and unfused parts of the garment. Although showerproof materials can absorb water through the fused areas, the development of water-resistant interlinings has addressed this concern effectively. Appropriate selection of interlining for the outer fabric and the precise control of the fusing process to achieve a satisfactory appearance and performance are essential in garment manufacturing. Fusing operations hold greater significance than other aspects of clothing production, as their effectiveness cannot be assessed visually and necessitates specialized, destructive testing methods. To ensure successful fusing operations, closer and more continuous cooperation among material suppliers, production managers, and equipment manufacturers is required.

*The Fusing Process*: The fusing process employs a specialized press that applies heat and pressure over a designated period. This process causes the resin to change from a solid to a liquid state, requiring appropriate pressure to permeate the fibers of the outer fabric and fusible base cloth. Upon cooling, the resin hardens and bonds the two components together. The amount of time needed for the process varies depending on the type of fabric and resin used. The attributes of a fused laminate garment rely on three aspects besides the outer fabric: the interlining's base fabric, the type of fusible resin, and the pattern of resin application on the base cloth. The most prevalent techniques to apply the resin are dry dot printing, scatter coating, and paste coating. Each technique requires the utilization of specifically selected resin particle sizes, with powdered resin being used under dry dot printing and scatter coating utilizing the largest particles. The paste coating encompasses mixing fine resin powders with water and other substances to create a uniform paste, which is then applied to the base fabric through printing. The selection of the resin type and coating method for fused garments hinges on multiple factors, such as cost, fusing characteristics, compatibility with special equipment, durability during laundering and dry cleaning, and intended end use of the garment. Thus, the fusing process involves placing a single interlining piece with the resin side facing downward on the fabric and its right side facing upward. This method, known as single fusing, is deemed the safest, as it simplifies adjusting press conditions to attain the optimal temperature at the glue line. Nevertheless, various other configurations can present components to the fusing press and incorporate them into the garment. A few of these alternatives will be outlined to demonstrate the diverse approaches to constructing garments utilizing fusible interlinings. These include:

*Reverse Fusing*: This approach involves placing the outer fabric above the fusible interlining, a technique widely employed in the fusing of shirt and blouse collars.

*Sandwich Fusing*: This approach entails fusing two pairs of components together, generating two laminates. The two outer fabrics form the outermost layer of the sandwich, comprising four layers, while the two interlinings are situated between them.

*Double Fusing*: This technique involves fusing two types of interlining to the outer fabric in a single operation, commonly utilized in constructing shirt collars and men's jacket fronts.

### 10.2.2 Welding and Molding

The popularity of clothing with a seamless appearance, especially in the performance apparel sector, is increasing due to the benefits of greater comfort and durability combined with reduced bulk. The seamless appearance is achieved through two methods: seamless knitting and welding. Seamless knitting involves producing the whole garment in one piece on a knitting machine. At the same time, welding fuses together layers of fabric using heat and pressure, often reinforced by thermoplastic films or tapes. The welding process in garment production involves using a thermoplastic material to bond two fabrics. Unlike the typical method of fusing interlinings where heat is applied externally, the heat must be concentrated on the thermoplastic material. The selection of welding techniques relies on various factors such as the fabric type, intended application, and the required bond strength, and each method has its advantages and drawbacks. Different equipment like a flatbed press, rotary machine, stationary press, and traditional press like tray press can be used to bond seams (Tyler et al., 2012). Four main approaches through which welding can be done are (i) thermoplastic adhesive using heat and pressure, (ii) ultrasonic sound waves as a source of heat plus pressure, (iii) high-frequency radiation using pressure, and (iv) infra-red light radiation with the addition of pressure (Tyler et al., 2012).

One of the methods to achieve this objective includes *hot air welding* that directs a stream of controlled temperature and speeds hot air onto the thermoplastic material to achieve a consistent and strong bond. Consequently, the substrate surface melts, and a cold roller is used to direct the same as to wet the fibers. The molten material cools and forms a cohesive bond upon contact with a cold surface. The fabric can be a hot-melt adhesive if the constituent materials are thermoplastic. For fabrics with a higher melting point, another thermoplastic material having a lower melting temperature is introduced to the joints. Another bonding method, *hot bar welding*, utilizes a heated bar pressed against the thermoplastic material to melt and bond the fabrics. It is a rapid process when a bar is used or at a rate of a few meters per minute if rollers are used. When joining two similar layers, the outer layer of the material is melted during this technique (Jones, 2013).

*Thermal welding* helps bond the thermoplastic fabrics without requiring adhesives or other fastening techniques. The materials act as bonding agents, eliminating the need for additional adhesives to hold them together. The bonding resulting from this process is fast, durable, and lighter in weight, and reduces seam bulk while providing elasticity making it an excellent bonding process for undergarments in direct contact with the body. However, excessive heat through the fibers may result in fiber degradation and stiffness. The traditional thermal bonding process entails melting surfaces individually before joining them, potentially causing fiber degradation. Advanced thermal welding techniques, such as *dielectric, ultrasonic*, and *laser welding*, can activate thermoplastic adhesives; however, the emphasis is on joining substrates without additional adhesive material. These sophisticated thermal welding processes offer a more controlled and expeditious method of bonding, generating heat primarily at the joint interface. *Dielectric welding* is an apt technique for polarizable materials with short molecular chains, such as polyamide and polyester. This approach is employed for joining woven and nonwoven fabrics and can also be used to bond fabrics to polymeric films. Different coatings and adhesives, such as Polyvinyl Chloride (PVC), Polyurethane (PU), and polyester, can be utilized in dielectric welding. This process is commonly used to manufacture protective clothing, such as rainwear and waterproof garments. *Ultrasonic welding* uses high-frequency sound waves to melt the thermoplastic material and bond fabrics. Ultrasonic welding extensively adopted the technique of joining small plastic parts in substantial quantities. The procedure entails generating oscillations of one substrate against a static second substrate by employing an ultrasonic generator, which induces intense frictional heat between the two substrates, creating a molten weld zone. With the application of pressure and subsequent cooling, a strong bond is formed. Fabrics with 60%–65% thermoplastic content, such as polyester or nylon, are appropriate for ultrasonic welding as the weld quality is better in such fabrics. Usage of thermoplastic adhesive is required for fabrics with a lower thermoplastic level, with the additional strengthening of hot melt reinforcing tape to be applied over the face of the underside of the seam (Tyler et al., 2012). *Laser welding* involves a concentrated light beam traversing through one of the fabric layers, subsequently heating and melting the thermoplastic coating or adhesive film within the joint interface. The molten polymer is then subjected to pressure and cooled to create a strong bond. The melting only occurs at the joint interface, leaving the outer surfaces of the fabrics unaffected. Research indicates that laser welding can improve the automation of the seaming process, delivering exceptional performance, aesthetics, and singular step, thereby facilitating its application in textiles for airbags, furniture, and apparel.

The textile industry has, thus, witnessed various trends over the past decade, including advanced thermal joining processes like ultrasonic welding, adhesives like nonblocking hot melts and reactive "B"-staged adhesives, bonded multilayer laminates, flame-retardant adhesive systems, and high-strength contact cement. Hot-melt adhesives and moisture-curing PUs are replacing solvent- or water-based systems as the most prominent volume trends in the industry.

Molding: Molding involves permanently stretching or shrinking the fabric of a garment by subjecting it to heat until it begins to soften, forming it into the desired shape, and ultimately cooling it so that the new shape becomes permanent. This process induces a permanent deformation in the fabric to create the desired shape. Garments made through molding typically exhibit simplistic styles, as the technique eliminates a crucial element in creating visually appealing garments: strategic seam placement. This technique is frequently used in fabricating hats and standardized garment components, where shaping can be achieved cost effectively. Although molding generally removes all seams or joints, there are some cases where seams may still be present in the final product, depending on the material and mold design. For example, if the mold has intricate shapes or undercuts. In that case, generating multiple molds and assembling the final product from numerous pieces might be necessary, potentially resulting in seams or joints. In addition, some materials might shrink or warp slightly as they cool or cure, leading to minor gaps or seams in the final product. Achieving adequate adhesion to fibers and fabrics is possible when the material is dry; however, adhesive joint failures may transpire during post-care procedures, such as washing and dry-cleaning treatments (Holme, 1999). These challenges can often be mitigated through meticulous design and production techniques.

### 10.2.3 Application of Adhesives

Adhesives facilitate the bonding of disparate materials without necessitating sewing as a procedure. The technique can bond the materials; waterproof seams; attach trims, labels, or patches to the garments; and create seamless designs by joining various fabric layers. Adhesive tapes can seal the seams proving particularly advantageous in waterproof garments resulting in increased aesthetics and stronger seams. However, the technique influences the production cost of the garments and imposes design limitations in terms of reduction in the number of seams and control of the shape of the garment. Adhesives are accessible in different forms, such as tapes, liquids, and hot-melt adhesives. *Liquid adhesives* can be applied using spray or a roller, while tape adhesives can be pre-cut and directly attached to the garment. *Hot-melt adhesives*, which melt when heated, are utilized to manufacture outerwear and activewear. Adhesives curtail conventional sewing methods' cost and production time, offering design flexibility and superior finishing in the final garment. They allow design flexibility and a better finish in the end garment. Some limitations of adhesives include reduced durability, the necessity for periodic reapplication, and diminished strength, posing challenges in maintaining highly stressed garment areas. In summary, the usage of adhesives in garment manufacturing is a prevalent technique that provides numerous benefits in terms of flexibility in design and production efficiency. Nonetheless, careful consideration must be given to specific materials and their applications to ensure the desired durability and performance of the end product.

## 10.3 PRESSING EQUIPMENT: CONVENTIONAL AND MODERN

The pressing process elevates a garment's aesthetic appeal for sale, brand recognition, and customer satisfaction. Ill-presented garments can adversely affect a brand's reputation and diminish the product's retail value. Pressing garments eliminates manufacturing creases and wrinkles, clarifies pleats if present, and ensures consistency among various garment components. Furthermore, it stabilizes the garment to retain its shape and relaxes any stresses induced during manufacturing. Thus, the primary purpose of pressing is refreshing the garment post-production, creating creases as per the garment design, shaping the garment to conform to body contours, and finishing unfinished garment parts for further sewing. Functionally, pressing is a process that utilizes the application of pressure and heat for crease formation and to provide the required shape to the garment. The optimal temperature and pressure levels to prevent distortion and damage depend on the fabric composition. For example, chiffon or silk being delicate fabrics require lower pressure and temperature levels, and heavier fabrics like denim or wool may require higher levels for a smooth appearance. The utilization of steam during pressing introduces moisture, which aids in swelling fibers and stabilizing the garment's form. The moisture content of steam varies with pressure; higher-pressure steam contains less moisture. Depending on the pressing equipment, a vacuum may be applied post-pressing to draw cool air through the garment, reducing temperature and moisture content while enhancing shape retention. Different types of pressing equipment are available in the market depending upon the garment and its requirement.

Conventional pressing equipment includes steam electric irons with thermostat regulation. These irons are triangular, similar to a domestic iron, with parallel sides and a pointed tip for opening seams on trouser legs and sleeves. Steam is supplied from a boiler adjacent to the pressing unit or the factory's main steam supply. Industrial steam irons are sizable and potent devices engineered for commercial settings such as factories, predominantly utilized in the garment industry for pressing large orders. The pressing table is another piece of equipment with a large surface. It typically includes a built-in vacuum, making it suitable for pressing large textile products such as tablecloths, bedding, and curtains. Specialized machines like form finishers are primarily utilized to smooth out wrinkles in suits and jackets. Different types of pressing equipment are available to enhance the quality of garments produced. Some

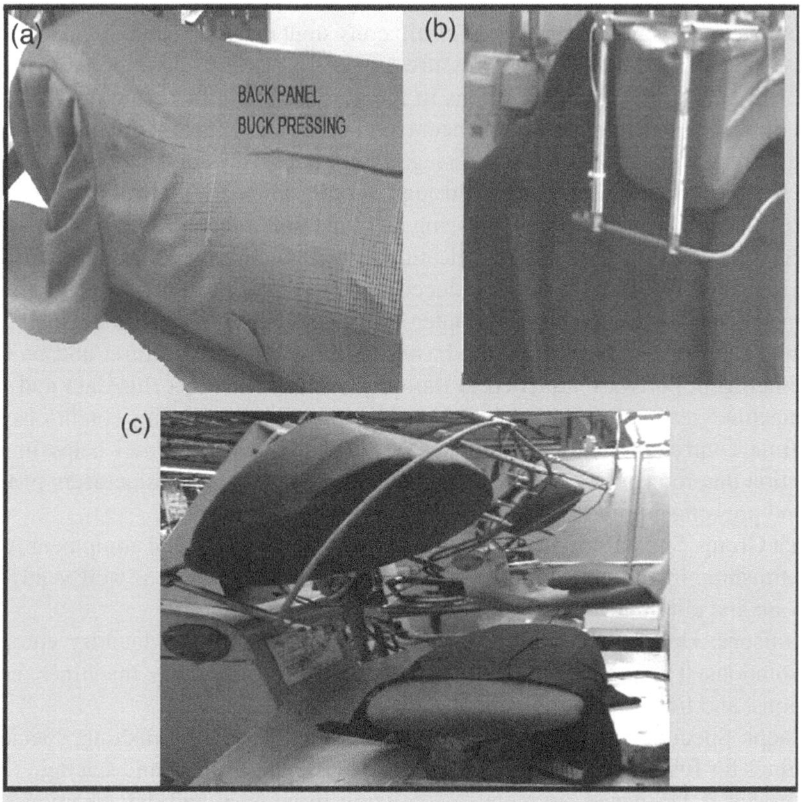


**FIGURE 10.1** Specialized buck presses for jackets and trousers: (a) A pressing buck for back panels in coats and jackets. The contoured surface of and precise heat application help in eliminating wrinkles in the back section of garment. (b) A buck press used for the shoulder area of the coats and jackets. Targeted heat and pressure are applied to mold the shoulder pads and fabric for precise and well-fitted contours. (c) Trouser pressing machine helps in applying controlled heat and pressure enabling sharp creases for professional finish. Topper press used to press trouser tops through heated plate that smoothens out wrinkles in the trouser tops, particularly around the pockets and the waistbands.

pressing devices can be designed with a programmed logic circuit to operate in varying cycles depending on the garment type, and others come with additional attachments, such as bucks or matching heads. Additional extensions can be attached to accommodate changes in garment shape. Examples of specialized equipment include a steam air finisher, carousel press, steam tunnel, and uniquely designed presses for jackets and trousers.

Specially designed machines are available for creasing and pleating garments. Creasing equipment presses the edges of clothing components to make sewing easier. Pleating machines generate a series of creases in a specific pattern or randomly. Blade-type and rotary-type machines can be utilized for rapid and accurate pleat creation. Some garments are manufactured as wash-and-wear or permanent-press garments by special finishing treatments that impart crease-resistant properties. Erroneous selection of parameters during pressing can lead to shrinkage, color loss, or fabric degradation. It is crucial to precisely control the variables in all pressing equipment to prevent variability and attain consistent results. Technological developments, such as accurate sensing, pressure monitoring, and strength of a cooling vacuum, can aid in mitigating variations in the garment manufacturing process. Apart from these, critical innovations of recent years, like intelligent software that allows process control and real-time monitoring, digital control systems like programmable logic controllers, and automation through robotic pressing and folding machines, have revolutionized the garment pressing and finishing industry. These developments enable quick identification and rectification of issues, improve efficiency resulting in improved quality and efficiency, reduced labor costs, and minimum human error, leading to consistent results. Computer-controlled pressing machines have actuators and sensors for data collection based on pressure, temperature, and moisture. The intelligent software attached to the machine can interpret this data and adjust the machine settings accordingly to ensure appropriate pressing conditions. This real-time control provides consistent quality of the garments and helps in waste reduction due to critical errors. Some of the world-famous manufacturers of finishing and pressing equipment are as follows:

Veit Group: Specializes in developing and producing industrial equipment for garment finishing including ironing, pressing, and fusing machines, as well as advanced software and automation systems.

Ramsons: Designs, manufactures, and distributes industrial laundry equipment and solutions. The product portfolio includes a range of laundry machines, custom solutions, and finishing equipment.

Macpi: Specializes in the industrial production of ironing, finishing, specialized machines for fusing and bonding fabrics, and automatic cutting and sewing.

Forenta: A US-based company specializing in laundry and dry-cleaning equipment for the garment care industry, offers laundry equipment, including steam presses, dry-cleaning machines, finishing equipment, laundry presses, etc.

Cissell: Manufactures and distributes a wide range of laundry equipment, including commercial dryers, washer-extractors, finishing equipment, flatwork ironers, etc.

Unipress: Their product portfolio includes commercial ironing machines, pressing machines, and finishing equipment, with custom solutions tailored to the specific needs of customers.

## 10.4 CARE LABELING OF CLOTHING

Regular refurbishment of apparel and textiles is necessary to preserve their appearance and functionality for reuse. Consumers, textile and apparel industries, and textile care organizations all share the responsibility of ensuring the desired functional properties of the products are retained. As consumers often lack the expertise to determine the appropriate care treatments for clothing, it falls upon apparel manufacturers to provide the necessary care instructions for their products. Care labels guide customers on the required cleaning procedures for a particular garment. These labels are visual representations that eliminate the need for translation into multiple languages, aiming to enable inexperienced customers to interpret and adhere to the recommended actions accurately. The parameters required to design appropriate care instructions depend on various interrelated factors like the pre-processing of the product, fiber content, fabric construction, and its end use. Care labels guide consumers on maintaining their clothing, covering aspects such as washing, bleaching, drying, ironing, and dry cleaning. By adhering to the instructions on care labels, consumers can ensure that the appearance and fit of their garments will be preserved through repeated cleanings. For manufacturers, care instructions could damage garments, resulting in complaints, expensive customer returns, and a tarnished reputation. Customer dissatisfaction, thus, can be circumvented by providing suitable care instructions. Garments that are easy to maintain are often preferred over those with complex or challenging-to-understand care procedures. The ultimate objective is communicating care and maintenance instructions effectively to the end customer. The country of origin on the care label directs the consumer to the place of manufacturing of the garment. To facilitate easy customer access, the care label should be permanently attached and positioned on the side or bottom of the garment.

A universal care labeling system is nonexistent; however, global commerce can be strengthened by reducing standardization and technical barriers by implementing a comprehensive care labeling system. The internationally renowned ASTM D5489-18 is a "standard guide for care symbols for instructions on textile products." This guide offers care instructions for textile products conveyed in pictorial format for universal comprehension. The Federal Trade Commission, USA, provides "The Care Labelling Rule" that provides manufacturers and importers with timely purchase instructions; the Rule mandates that textile apparel and goods manufacturers and importers furnish buyers with care instructions via care labels or other means. It also prohibits dishonest acts or practices that withhold standard care instructions and mandates precise language and symbols to convey care procedures properly. The European Union (EU) Textile Regulation stipulates that textile products sold in the EU must have care label instructions that are durable, clear, and easy to understand. JIS L0001–2014, based on ISO standards, are the care labeling standards introduced by the Japanese Standards Association. This system utilizes symbols to indicate care instructions, with each symbol representing a specific meaning, such as drying method, washing temperature, ironing temperature, etc. Since 1975, the International Association for Textile Care Labelling (GINETEX) has been the global authority regulating care labels. Its member countries include Belgium, France, Germany, England, Netherlands, Israel, Austria, Switzerland, and Spain. The organization's mission is to educate consumers about the appropriate care labeling of textiles using standardized care labeling symbols that are simple and universally understood regardless of language.

Primary challenges associated with care labels include more precise directions, the type of material used for label manufacturing, and overly restrictive instructions. To address these issues, manufacturers can provide clear care instructions using universal language or symbols. Incorporating extra details by manufacturers through fabric performance codes like social responsibility, environmental labels, and finishing information would further assist consumers in understanding the garment while making purchases. Fashion and textile trade fair organizer Première Vision devised fabric performance codes comprising a comprehensive set of sustainable development indicators. These codes evaluate multiple aspects of sustainable development, including responsible manufacturing processes, social responsibility, environmentally friendly materials, and fair labor practices. The objective is to advocate sustainable methodologies and enhance customer transparency concerning products available in the market.

Sustainable care labels address the growing demand from environmentally conscious consumers supporting global initiatives to reduce ecological impacts and encourage responsible consumption. Eco-friendly labels made from biodegradable materials like recycled paper aid in minimizing environmental harm by decreasing waste generation and conserving natural resources. Recycled labels are made from post-consumer waste materials like plastic, paper, or textiles, which are transformed into new products instead of being sent to landfills, resulting in lower environmental impact than extracting natural resources. These labels are made from renewable resources, including organic cotton or recycled polyester. Additionally, eco-friendly inks free from heavy metals and harmful chemicals are used for printing. Biodegradable labels decompose naturally over time and are manufactured from bioplastics or papers derived from plant-based materials such as sugarcane, bamboo, or corn. Design techniques, like using smaller label sizes, can also contribute to creating eco-friendly labels by minimizing the amount of adhesive required.

## 10.5 GARMENT PACKAGING

The primary purposes of packaging revolve around distribution and merchandising. Distribution includes the efficient transportation of apparel while maintaining product quality, whereas merchandising focuses on enhancing the visual appeal of garments to attract customers. Both functions safeguard the durability and distinctive characteristics of various styles throughout their journey from production to retail. Paper, plastic, and cardboard are usually the materials used for garment packaging. For protective garment bags, plastic materials are used, whereas cardboard and paper are used for boxes, labels, and envelopes. Branding or labeling on garment packages is achieved through methods such as screen printing, digital printing, or offset printing. Garment packaging comes in diverse shapes and sizes, tailored to the specific clothing items being packaged, such as suits in box-shaped containers, T-shirts in flat envelopes, and dresses in large bags. Based on the type of packaging, the closure methods can be a ziplock closure for plastic bags, string ties, or adhesive tapes for closing envelopes and boxes. The durability of garment packaging during storage and shipping can be improved using foam inserts, bubble wrap, and reinforced edges. Environmentally friendly packaging alternatives include biodegradable materials, recycled material packaging, and reduced packaging solutions.

### 10.5.1 Conventional Modes of Packaging

The secure distribution of goods within the value supply chains makes packaging crucial. As a significant consumer of materials, the packaging industry accounts for approximately 40% of plastics and 50% of paper usage in Europe (Coelho et al., 2020). Although packaging serves an essential purpose, it has various environmental implications and constitutes a substantial expense in the current supply system. Conventional packaging typically comprises boxes, cartons, bags, cases, crates, and wraps. One of the most common forms of conventional packaging for garments includes plastic bags, as they are transparent, lightweight, and easy to handle, and can be closed using a ziplock closure or with a heat-sealing machine.

In contrast, paper bags present an eco-friendlier alternative to plastic bags due to their biodegradable nature. However, they lack durability and offer less protection to the enclosed garments. Packaging materials such as paper and cardboard are susceptible to moisture absorption and chemical attacks. To counteract these vulnerabilities, coatings of specific polymers are applied as barriers, enhancing both material strength and chemical resistance simultaneously. Large quantities of garments are packaged in cardboard boxes due to their stackable nature, sturdiness, and ease of sealing with adhesive tapes. Designed for convenient transportation, they are recyclable but comparatively more expensive than paper or plastic bags. Lightweight garments, like T-shirts, are often shipped in flat envelopes made of paper or cardboard, which are lightweight, affordable, and recyclable. Crates and cases are typically constructed from wood, while cartons and boxes are often made of plastic or cardboard. Wrapping materials and bags typically consist of plastic or paper film. Distinct from other packaging types, cases are completely enclosed, devoid of openings on any side. Crates exhibit gaps between the wooden boards on their sides or ends. Boxes are characterized by separate covers, while cartons lack separate enclosures. Another form of conventional packaging includes garment labels that provide branding and other customized information about the brand. They can be made of either synthetic material or paper that can be attached to the garment with either sewing or adhesive. Glass, while recyclable and easily repurposed for applications like concrete aggregates, possesses notable drawbacks, such as its heavy weight and fragility, making it prone to breakage under extreme temperatures or improper handling.

### 10.5.2 Drawbacks Associated with Conventional Packaging

Environmental issues have prompted a growing awareness among various societal groups regarding the significance of environmental sustainability. The environmental impact of conventional garment packaging methods is significant, as they are often composed of non-biodegradable materials that take years to decompose. These materials contribute to pollution and landfills that further harm wildlife and ecosystems. Also, manufacturing these materials is resource intensive, requiring substantial water and energy. Paper products like cartons and cardboard demand large quantities of wood pulp, water, and energy, while plastic packaging relies on non-renewable petroleum products. Conventional garment packaging practices might not offer adequate protection against environmental factors like dust, moisture, or UV radiation, potentially damaging the product during storage and shipping and leading to customer returns and dissatisfaction.

Moreover, they must be designed for efficient space utilization, resulting in excessive packaging and increased shipping costs. For example, large cardboard boxes lead to space wastage in shipping containers or trucks. The conventional garment packaging methods are a time-consuming, labor-intensive, costly process, resulting in higher shipping and handling fees for customers that can negatively impact sales and profitability. Distribution is a crucial factor in the environmental consequences of packaging and needs more effective solutions to reduce its environmental impact (Escursell et al., 2021). The transportation of these packaging materials contributes to carbon emissions, as their manufacturing process involves the usage of heavy machinery and equipment that generate greenhouse gases. Transportation to large distances further adds to the emissions generated by the process. Improper disposal of these packaging materials, especially plastic and foam, threatens wildlife and the ecosystem. The materials can be ingested by animals or disintegrated foam into smaller pieces is hazardous to aquatic life. The prevalent packaging materials derived from non-renewable materials like plastic must be circumvented under all circumstances (Escursell et al., 2021). The use of unnecessary and problematic plastic packaging materials should be eliminated through the use of innovative and eco-friendly solutions. The ultimate goal of packaging materials should be to prevent any material from becoming waste.

### 10.5.3 Introduction to Sustainable Packaging and Eco-Friendly Labels

Sustainable packaging has garnered considerable attention from consumers and retailers in recent years due to mounting concerns surrounding environmental sustainability. As an integral component, sustainable packaging can substantially lessen the fashion industry's environmental impact. Enhancing sustainability across the packaging value chain can be achieved by facilitating the collection and sorting of packaging materials for recycling, reuse, composting, and waste-to-energy conversion while ensuring the proper disposal and processing of sorted packaging. Moreover, sustainability can be improved by sourcing materials more sustainably, curbing the usage of materials and resources, and still maintaining the necessary functions of packaging. Sustainable packaging uses design techniques and materials that are environmentally friendly and mitigates adverse impacts on Earth. Such practices include the utilization of biodegradable materials that decompose safely and rapidly, as well as using recycled materials, which reduces the demand for new resources. Bioplastics derived from sugarcane and cornstarch offer another example: they decompose naturally without releasing harmful chemicals into the environment. Using recycled paper for packaging leads to waste minimization and diminishes the necessity for extracting new resources. Sustainable packaging can also be implemented through thoughtful design techniques; for instance, decreasing packaging material helps reduce waste and minimize transportation emissions. Efficient manufacturing processes prioritizing energy conservation and waste minimization aid in diminishing the environmental impact of the packaging production process. Incorporating green packaging as a competitive strategy has become increasingly vital for businesses in the modern era. Eco-friendly packaging is crucial to curtailing the environmental impact of packaging waste. This can be achieved by optimizing packaging to reduce its size relative to its contents, maximizing the amount of recycled material in the packages, and utilizing more sustainable materials in packaging design. Additional research is required to explore novel packaging materials derived from cellulose-based and renewable sources. Enhancing distribution processes and adopting innovative production methods, such as 3D printing and additive manufacturing, may contribute to optimizing package shape and size, resulting in more eco-friendly production by reducing $CO_2$ emissions (Escursell et al., 2021).

*Eco-Friendly Labels*: Eco-labels promote environmental sustainability by encouraging consumers to make environmentally conscious purchasing choices. These eco-friendly labels communicate the use of sustainable methods and materials to consumers, enabling them to make better-informed decisions. The assumption is that if consumers are provided with appropriate label information, they will be more likely to make sustainable purchasing decisions. The idea behind eco-labels is to promote environmental sustainability by encouraging consumers to make more sustainable purchasing choices. Eco-friendly labels convey sustainable methods and materials to the end consumer, helping them make more informed purchasing decisions. They can help reduce the environmental impact by using sustainable materials and practices and conserving natural resources and waste minimization. Consumers tend to support organizations committed to promoting sustainability and reducing environmental damage. The labels indicate that the product made from sustainably sourced materials is energy efficient, without harmful chemicals. Eco-friendly labels indicate certified products toward environmentally responsible practices generating verified product environmental claims. Some prevalent eco-friendly labels that help consumers find environmentally friendly products are EU eco Label, Organic, Cradle to cradle, Earth check, Forest Stewardship Council, Energy Star, etc. Not all labels serve the same purpose regarding sustainable practices; some may focus on fair trade, energy efficiency, human safety, the absence of harmful chemicals, sustainable design, or raw materials.

## 10.6 CONCLUSION

Fusing is a well-established process for interlining attachment; welding is increasingly used in lingerie and sportswear, while adhesive use is still considered niche. Molding is commonly applied in creating headwear, bras, and structured garments. The textile industry has, thus, witnessed various trends over the past decade, including advanced thermal joining processes like ultrasonic welding and adhesives like nonblocking hot melts. Technological developments, such as accurate sensing, pressure monitoring, and measuring the strength of a cooling vacuum, can aid in mitigating variations in the garment manufacturing process. Apart from these, critical innovations of recent years, like intelligent software that allows process control and real-time monitoring, digital control systems like programmable logic controllers, and automation through robotic pressing and folding machines, have revolutionized the garment pressing and finishing industry. The importance of garment care labels, packaging, and sustainable practices in the fashion industry has grown due to increased environmental awareness. Care labels inform consumers about proper garment maintenance, while packaging serves distribution and merchandising functions. The industry is moving toward eco-friendly labels, sustainable packaging materials, and design techniques that reduce waste, conserve resources, and minimize environmental impact. Eco-labels further enable consumers to make informed purchasing decisions by highlighting sustainable product methods and materials, fostering a more environmentally conscious market.

## REFERENCES

**Books and Journals:**

Coelho, P. M., Corona, B., ten Klooster, R., & Worrell, E. (2020). Sustainability of reusable packaging–current situation and trends. *Resources, Conservation and Recycling: X, 6* (March), 100037. https://doi.org/10.1016/j.rcrx.2020.100037

Escursell, S., Llorach-Massana, P., & Roncero, M. B. (2021). Sustainability in e-commerce packaging: A review. *Journal of Cleaner Production, 280*, 124314. https://doi.org/10.1016/j.jclepro.2020.124314

Holme, I. (1999). Adhesion to textile fibres and fabrics. *International Journal of Adhesion and Adhesives, 19*(6), 455–463. https://doi.org/10.1016/S0143-7496(99)00025-1

Jones, I. (2013). The use of heat sealing, hot air and hot wedge to join textile materials. *Joining Textiles: Principles and Applications*, 355–373. https://doi.org/10.1533/9780857093967.3.355

Petrie, E. M. (2015). Alternative fabric-joining technologies. In *Garment Manufacturing Technology*. Elsevier Ltd. https://doi.org/10.1016/B978-1-78242-232-7.00013-8

Tyler, D. J. (2008). *Carr & Latham's Technology of Clothing Manufacture*. Blackwell Publishing. https://www.amazon.co.uk/Carr-Lathams-Technology-Clothing-Manufacture/dp/1405161981/ref=sr_1_1?s=books&ie=UTF8&qid=1332101048&sr=1-1

Tyler, D., Mitchell, A., & Gill, S. (2012). Recent advances in garment manufacturing technology: Joining techniques, 3D body scanning and garment design. In *The Global Textile and Clothing Industry: Technological Advances and Future Challenges*. Woodhead Publishing Limited. https://doi.org/10.1533/9780857095626.131

# 11 Evaluation Techniques, Quality Control, and Garment Costing

*Sujit Shrikrushnarao Gulhane*

## 11.1 OBJECTIVE EVALUATION

Objective evaluation methods in the apparel industry involve the use of quantifiable and measurable criteria to assess the quality of garments. These criteria often pertain to the garment's physical attributes, such as fabric strength, seam strength, color fastness, and dimensional stability. Here are some key objective evaluation techniques.

### 11.1.1 Fabric Strength Testing

Fabric strength testing is a crucial aspect of evaluating the quality and durability of textiles. It involves assessing a fabric's ability to withstand mechanical stresses such as stretching, tearing, or abrasion. Several standardized methods exist for fabric strength testing, each focusing on different aspects of strength.

#### 11.1.1.1 Tensile Strength Testing (ASTM D5034-21)

Tensile strength testing measures a fabric's resistance to being stretched or pulled apart. It involves cutting fabric samples into standardized strips, clamping them, and applying a steadily increasing force until the fabric breaks. The maximum force endured before rupture is recorded (ASTM D5034-21, 2021).

#### 11.1.1.2 Elmendorf Tear Strength Testing (ASTM D1424-21)

The Elmendorf tear test is a variation of tear strength testing. It measures the force required to propagate a single tear in a fabric sample that has been slit along one edge. Children's clothing may be subject to rough handling and tearing. The Elmendorf tear test ensures that such garments can withstand tearing, making them more durable. Tear strength testing assesses a fabric's resistance to tearing. It involves cutting a fabric sample into a specific shape, creating a small initial cut, and then applying force to propagate the tear. The force required to tear the fabric completely is measured. Fabrics used for outdoor tents need to resist tearing when exposed to wind or accidental damage. Tear strength testing ensures that the fabric used in tent construction can endure such stresses without tearing apart (ASTM D1424-21, 2021).

DOI: 10.1201/9781003432739-11

#### 11.1.1.3 Grab Test (ASTM D5035-11)

The grab test assesses the fabric's tensile strength, similar to tensile strength testing. However, instead of cutting fabric strips, a rectangular fabric sample is used, and force is applied to the fabric's shorter sides until it ruptures. The fabrics used for applications like parachutes rely on the fabric's strength to safely slow down and land a person or object. The grab test ensures that parachute materials can withstand the forces involved in deployment and descent (ASTM D5035-11, 2019).

#### 11.1.1.4 Ball Burst Test (ASTM D3787-16)

The ball burst test measures a fabric's resistance to localized pressure. A steel ball is pressed against the fabric until it bursts through. The test provides insights into a fabric's resistance to punctures and impacts. Fabrics used in protective clothing, such as firefighter gear, need to resist punctures from sharp objects or falling debris. The ball burst test helps ensure the fabric's ability to protect wearers (ASTM D3787-16, 2020).

#### 11.1.1.5 Abrasion Resistance Testing (ASTM D4966-22)

Abrasion resistance testing assesses how well a fabric can withstand wear and abrasion. It typically involves rubbing the fabric against an abrasive surface under controlled conditions and measuring the number of cycles required to cause visible damage. The abrasion resistance of fabrics used in applications such as work gloves needs to endure friction and abrasion when used in rugged environments. Abrasion resistance testing helps ensure that these gloves remain intact and functional after extended use. These standardized fabric strength testing methods are essential for selecting materials suitable for specific applications and ensuring product quality and safety. By conducting these tests, manufacturers can confidently produce textiles and products that meet the required strength standards and perform as expected in various real-world scenarios (ASTM D4966-22, 2022).

### 11.1.2 Colorfastness Testing

It is a crucial aspect of assessing how well the color of a fabric holds up when exposed to various external factors. There are standardized methods for conducting colorfastness testing, and these tests are essential in different applications across the textile and apparel industry. The various types of colorfastness testing procedures are described in detail below.

#### 11.1.2.1 Washing Colorfastness Testing (ISO 105-C06)

Washing colorfastness testing evaluates how well a fabric's color withstands multiple washing cycles. The test involves placing fabric swatches with specified detergent and water in a washing machine and subjecting them to repeated wash and rinse cycles. Afterward, the swatches are assessed for color change and staining. The garments such as t-shirts are subjected to frequent washing. Washing colorfastness testing ensures that the colors on a printed or dyed t-shirt remain vibrant and do not fade or bleed significantly after multiple washes (ISO 105-C06, 2010).

#### 11.1.2.2 Lightfastness Testing (ISO 105-B02)

Lightfastness testing assesses a fabric's resistance to fading when exposed to artificial light, particularly ultraviolet (UV) light. Fabric samples are exposed to controlled light conditions for a specified duration, and color changes are evaluated using color fastness standards. Fabrics used in outdoor furniture upholstery must resist fading when exposed to sunlight. Lightfastness testing helps determine which materials can maintain their color vibrancy despite prolonged UV exposure (ISO 105-B02, 1994).

#### 11.1.2.3 Rubbing or Crocking Colorfastness Testing (ISO 105-X12)

Rubbing colorfastness testing assesses a fabric's resistance to color transfer or fading when subjected to friction. In this test, a piece of white fabric is rubbed against the tested fabric, and any color transfer is evaluated. The garments such as denim jeans are subject to constant rubbing and friction, which can cause dye transfer. Rubbing colorfastness testing ensures that the dye used in denim remains intact, preventing the jeans from staining other fabrics or losing their color due to everyday wear (ISO 105 X12, 2016).

#### 11.1.2.4 Perspiration Colorfastness Testing (ISO 105-E04)

Perspiration colorfastness testing assesses how well a fabric's color withstands exposure to artificial perspiration. Fabric samples are subjected to a solution that simulates human sweat, and color changes are evaluated. Athletic sportswear, like jerseys and leggings, needs to maintain their color even when exposed to sweat during physical activities. Perspiration colorfastness testing ensures that the garments don't lose their color or develop unsightly stains (ISO 105-E04, 2013).

#### 11.1.2.5 Waterproof and Water-Repellent Colorfastness Testing (ISO 4920)

These tests evaluate how well a fabric's color withstands exposure to water and moisture. Methods vary depending on whether the fabric is intended to be waterproof or water-repellent. Color changes and water resistance are assessed. The special garments like raincoats need to remain colorful and appealing despite prolonged exposure to rain and moisture. Waterproof and water-repellent colorfastness testing ensures that the raincoat's color doesn't wash away or fade when wet (ISO 4920, 2012).

### 11.1.3 Dimensional Stability Testing (AATCC Test Method 135, ISO 5077)

It is a critical evaluation method used to assess how well a fabric maintains its size and shape after being subjected to various environmental conditions, including washing and mechanical stress. Standardized testing methods help manufacturers and quality control professionals ensure that textiles and garments meet size and shape requirements. Dimensional stability testing evaluates a fabric's ability to retain its original dimensions, including length, width, and shape, after undergoing specific environmental conditions. The test typically involves the following steps:

*Sample Preparation:* Fabric samples are carefully cut and measured according to standardized dimensions.
*Conditioning:* The samples are conditioned in a controlled environment with specified temperature and humidity levels. This step ensures that the fabric is in an equilibrium state before testing.
*Test Procedures:* Fabric samples are subjected to various stressors, such as washing, dry cleaning, or exposure to moisture and heat. The specific test conditions depend on the intended application and the fabric type.
*Measurement:* After the test procedures are completed, the samples are re-measured to determine any changes in length, width, or shape. Any dimensional changes are recorded and analyzed.
*Calculations:* Dimensional changes are often expressed as a percentage of shrinkage or expansion relative to the original dimensions.

Dimensional stability testing is essential for various textile and garment applications to ensure that products maintain their size and shape under real-world conditions. The curtains and drapes need to maintain their length and shape for proper window coverage. Dimensional stability testing ensures that they do not shrink or elongate when cleaned or exposed to humidity. Automotive upholstery, including seat covers and headliners, should remain dimensionally stable to provide comfort and aesthetic appeal to vehicle interiors. Dimensional stability testing ensures that the upholstery fabric retains its shape and fit over time (AATCC TM 135, 2018; ISO 5077, 2022).

### 11.1.4 Garment Construction Assessment

Seam strength testing (ASTM D1683, ISO 13935-1) is a critical evaluation method used to assess the strength and durability of seams in textiles and garments. Standardized testing methods help ensure that the seams can withstand the stresses and strains they will encounter during normal use. Seam strength testing evaluates the ability of a seam to resist separation or rupture when subjected to tension or pulling forces. The test typically involves the following steps:

*Sample Preparation:* Fabric samples are selected with the seam of interest, which could be a flat seam, overlapped seam, or any other relevant seam type.
*Sample Mounting:* The fabric sample is securely mounted in the grips of a universal testing machine. The seam is positioned along the path of the applied force.
*Force Application:* A steadily increasing force is applied to the fabric in a direction perpendicular to the seam. The force continues to be applied until the seam ruptures or fails.
*Measurement:* The maximum force applied to the seam before rupture is recorded as the seam strength. This is typically reported in units such as Newtons or pounds-force.

*Calculations:* Seam strength may be expressed as the maximum force sustained before failure or as the force per unit width of the seam, depending on the specific testing standard.

Seam strength testing is essential for various textile and garment applications to ensure that products can withstand the mechanical stresses they will encounter during use. The garments such as jeans typically have multiple seams, especially in high-stress areas like the crotch and knees. Seam strength testing ensures that the seams of denim jeans can endure the stresses associated with everyday wear.

Seam strength testing is a crucial quality control measure in the textile and apparel industry. By conducting these tests according to standardized methods, manufacturers can guarantee that the seams in their products are strong and durable, providing customers with reliable and long-lasting textile and garment items. This is especially important in applications where seam failure could result in safety risks or reduced product performance (ASTM D1683/D1683M-22, 2022; ISO 13935-1, 1999).

### 11.1.5 Garment Fit Testing

Garment fit testing is a critical evaluation process used to assess how well a garment's design, size, and shape match the intended wearer's body dimensions and proportions. Fit testing helps ensure that the garment provides comfort, mobility, and a flattering appearance to the wearer. While there isn't a single standardized method for fit testing, it typically involves several steps and evaluations. The primary goal of garment fit testing is to ensure that a garment fits the wearer's body comfortably, conforms to its shape, and allows for ease of movement. The process typically involves a series of evaluations and adjustments to optimize the fit. Fit testing aims to address aspects such as:

*Sizing:* Determining the appropriate size for the wearer based on body measurements, such as chest, waist, hip, inseam, and height.

*Ease:* Ensuring that the garment provides the right amount of ease (extra space or allowance) in critical areas, such as the underarm, crotch, and knee, to allow for comfortable movement.

*Proportion:* Evaluating how well the garment's design and construction align with the wearer's body proportions, considering factors like torso length, arm length, and leg length.

*Drape:* Assessing how the fabric drapes on the body, ensuring it falls naturally and flatters the wearer's shape.

*Comfort:* Verifying that the garment's fit does not cause discomfort, such as tightness, constriction, or chafing, during activities relevant to its intended use.

#### 11.1.5.1 Steps in Garment Fit Testing

The steps involved in garment fit testing may vary depending on the brand, product type, and target audience, but typically include the following:

*Wearer Selection:* Identifying individuals who represent the target audience and whose body measurements align with the intended garment size.

*Garment Sizing:* Fitting wearers with garments in various sizes to determine the best fit for each individual.
*Evaluations:* Wearers provide feedback on comfort, mobility, and overall fit of the garment. Fit testers may assess aspects like freedom of movement, comfort in different positions (standing, sitting, walking), and any pressure points or areas of discomfort.
*Adjustments:* Garments may be adjusted and redesigned based on feedback and evaluations to improve the fit and overall wearability.
*Iterative Process:* Fit testing may involve multiple iterations and adjustments until the desired fit is achieved.

Fit testing is crucial for delivering garments that meet customer expectations for comfort, style, and performance. It helps manufacturers and designers refine their products to cater to a diverse range of body shapes and sizes while ensuring that the clothing meets its intended functional and aesthetic purposes (Liu et al., 2022; Petrova and Ashdown, 2012).

### 11.1.6 Button and Fastener Testing

Garment button and fastener testing is a critical evaluation process used to assess the functionality, durability, and safety of buttons, zippers, snaps, and other fasteners used in clothing and apparel. Standardized testing methods help ensure that these components meet quality and safety standards. The primary goal of garment button and fastener testing is to assess the performance of closures and fastening components in clothing and apparel. This testing helps ensure that buttons, zippers, snaps, hooks, and other fasteners function as intended (e.g., buttons can be securely fastened and unfastened, zippers open and close smoothly), withstand repeated use without breaking or malfunctioning, do not pose a safety hazard (e.g., buttons do not easily detach, preventing choking hazards for children's clothing), and meet any applicable regulatory or industry standards for quality and safety. Standard methods for garment button and fastener testing are discussed below.

#### 11.1.6.1 Button Strength Testing (ASTM D4846)

This test assesses the strength of buttons by determining the force required to pull a button through a fabric. A button pull tester or a tensile testing machine equipped with the necessary fixtures to hold and apply loads to the button. The gradually increasing pulling force is applied perpendicular to the surface of the button. The force is applied continuously until the button breaks or fails. The maximum force applied just before the failure is recorded. This force is often reported in units such as Newtons or pounds-force. The primary objective of this test is to ensure that buttons can withstand the stress they may encounter during normal use without breaking or failing (ASTM D4846-96, 2021).

#### 11.1.6.2 Zipper Strength Testing (ASTM D2061)

Zipper strength testing evaluates the ability of a zipper to withstand the forces applied during normal use, such as opening and closing. Button Attachment Strength

Testing: This test assesses the strength of the attachment between buttons and the fabric. It involves measuring the force required to detach a button from the fabric (ASTM D2061-07, 2021).

#### 11.1.6.3 Zipper Endurance Testing (ASTM D2062)

This test evaluates the zipper's ability to endure repeated opening and closing cycles without failure. Snaps and Fastener Durability Testing: For snaps, hooks, and other fasteners, tests are conducted to evaluate their durability, including how well they withstand repeated fastening and unfastening. Garment button and fastener testing is an essential quality control measure in the textile and apparel industry. These tests help manufacturers and designers ensure that the fasteners in their products are durable, functional, safe, and compliant with industry standards and regulations. Garment fastener testing ultimately contributes to the overall quality and safety of clothing and apparel items (ASTM D2062-03, 2021).

## 11.2 SUBJECTIVE EVALUATION

While objective evaluation provides crucial data, subjective evaluation relies on human judgment to assess the overall aesthetic and tactile qualities of a garment. Fashion is, after all, a sensory experience. Here are some key aspects of subjective evaluation:

### 11.2.1 Design and Style

Garment design and style evaluation is a fundamental aspect of the apparel industry, encompassing the assessment of a garment's aesthetics, functionality, and overall design. While there are no specific standardized tests for design and style, this evaluation involves visual and qualitative assessments to ensure that garments meet aesthetic, market, and brand expectations. The primary goal of garment design and style evaluation is to assess how well a garment's design aligns with its intended purpose, target market, brand identity, and current fashion trends.

#### 11.2.1.1 Aspects of Design and Style Evaluation

*Aesthetics:* Evaluating the visual appeal, symmetry, color coordination, and overall attractiveness of the garment.

*Functionality:* Assessing whether the design allows for ease of movement, comfort, and functionality related to the garment's intended use (e.g., sportswear, workwear, formal attire).

*Market Relevance:* Ensuring that the design aligns with the preferences and needs of the target market or audience, whether it's children, teenagers, adults, or a specific demographic.

*Brand Identity:* Verifying that the design reflects the brand's image and values, maintaining brand consistency and recognition.

*Trend Analysis:* Keeping up with current fashion trends and market demands, ensuring that the garment is in line with the latest styles and consumer expectations.

#### 11.2.1.2 Methods for Garment Design and Style Evaluation

While there are no standardized tests, garment design and style evaluation typically involves the following methods:

*Visual Inspection:* Experts or designers visually assess the garment's overall design, details, proportions, and color combinations to ensure they meet aesthetic and brand standards.

*Wearable Prototypes:* Creating prototypes or samples of the garment to evaluate how it looks, feels, and functions when worn. This helps assess factors like fit, comfort, and functionality.

*Focus Groups and Surveys:* Collecting feedback from focus groups or surveys to gauge consumer preferences, opinions, and reactions to the garment's design and style.

*Comparison with Competing Products:* Analyzing how the garment compares with similar products in the market in terms of design, style, and functionality.

*Trend Analysis:* Keeping abreast of current fashion trends and market research to ensure that the garment design aligns with the latest consumer preferences and industry standards.

#### 11.2.1.3 Applications and Examples of Garment Design and Style Evaluation

Garment design and style evaluation applies to a wide range of clothing and apparel items. Here are some examples:

*Casual T-Shirts:* Casual T-shirts come in various designs and styles, from graphic prints to minimalist aesthetics. Evaluation ensures that the chosen design resonates with the target audience's preferences and aligns with the brand's image.

*Athletic Wear:* Athletic wear, including sports jerseys and leggings, needs to combine style with functionality. Evaluation considers both the garment's appearance and its performance during physical activities.

*Formal Wear:* Formal wear, such as suits and evening gowns, requires careful design and style evaluation to ensure that it meets the standards of elegance, sophistication, and brand identity.

*Children's Clothing:* Children's clothing designs should be age appropriate, fun, and comfortable. Evaluation ensures that the designs align with both children's preferences and parental expectations.

*Outerwear:* Outerwear garments, such as coats and jackets, require a balance between style and warmth. Evaluation ensures that the design appeals to consumers while providing adequate protection from the elements.

*Fashion Collections:* Fashion designers create entire collections that need to be cohesive in style, reflecting a particular theme or concept. Evaluation ensures that the collection's pieces work together to convey the desired message or image.

Garment design and style evaluation plays a vital role in the textile and apparel industry. While there are no standardized tests, careful assessment of aesthetics, functionality, market relevance, and brand identity helps designers and manufacturers create garments that meet consumer expectations and market demands. Successful design and style evaluation results in products that are not only visually appealing but also resonate with the intended audience and effectively represent the brand (Eckman et al., 1990).

## 11.2.2 Color Harmony

Garment color harmony evaluation is a crucial process in the fashion and apparel industry, ensuring that the colors used in a garment work well together, create an appealing visual composition, and resonate with the intended audience. While there are no specific standardized tests for color harmony, it involves visual assessments and qualitative analysis. The basic objective of garment color harmony evaluation is to assess how well the colors used in a garment complement each other and create a visually pleasing composition.

### 11.2.2.1 Aspects of Garment Color Harmony Evaluation

*Color Combinations:* Evaluating the combinations of colors used in the garment, whether they are complementary, analogous, monochromatic, or triadic, and whether they create the desired visual impact.

*Contrast:* Assessing the level of contrast between colors, which can affect the garment's visual balance and emphasis on specific areas or details.

*Aesthetics:* Determining whether the chosen colors align with the overall aesthetic and design intent of the garment.

*Market Appeal:* Ensuring that the color palette resonates with the preferences and expectations of the target market or audience, taking into account cultural and regional differences.

*Brand Identity:* Verifying that the color choices align with the brand's image and values, maintaining brand consistency and recognition.

### 11.2.2.2 Methods for Garment Color Harmony Evaluation

Garment color harmony evaluation relies on visual assessments and qualitative analysis, as there are no specific standardized tests. Common methods include:

*Visual Inspection:* Experts or designers visually assess the garment's color composition, combinations, and contrast to ensure they meet aesthetic and brand standards.

*Color Swatching:* Using color swatches or samples to analyze how different colors work together in the garment's context and lighting conditions.

*Focus Groups and Surveys:* Collecting feedback from focus groups or surveys to gauge consumer reactions to the garment's color choices and overall visual appeal.

*Comparison with Competing Products:* Analyzing how the garment's color palette compares with similar products in the market to identify trends and preferences.

*Use of Color Theory:* Applying principles of color theory, such as the color wheel, to evaluate the harmony, contrast, and balance of colors in the garment.

#### 11.2.2.3 Applications and Examples of Garment Color Harmony Evaluation

Garment color harmony evaluation applies to a wide range of clothing and apparel items. Here are some examples:

*Casual Apparel:* Casual clothing, such as T-shirts, shorts, and casual dresses, often aims for a relaxed and comfortable look. Evaluation ensures that the color combinations are visually appealing and resonate with the casual style.

*Athletic and Sportswear:* Sportswear and activewear often use vibrant and energetic color palettes to convey a sense of movement and activity. Evaluation ensures that these colors work harmoniously and contribute to the intended athletic aesthetics.

*Formal and Eveningwear:* Formal and eveningwear requires careful color selection to convey elegance and sophistication. Evaluation assesses whether the color choices align with the formality of the garments and create a visually pleasing composition.

*Children's Clothing:* Children's clothing often incorporates bright and playful colors. Evaluation ensures that the colors are not only visually appealing to children but also align with the brand's identity.

*Outerwear and Seasonal Collections:* Seasonal collections, including outerwear for winter or swimwear for summer, may use seasonal color palettes. Evaluation assesses whether the colors evoke the intended season and appeal to consumers.

*Fashion Runway Collections:* Fashion designers create runway collections that often push creative boundaries with color. Evaluation ensures that the color choices contribute to the collection's artistic vision and impact.

Garment color harmony evaluation is a critical aspect of fashion and apparel design. While there are no standardized tests, it involves careful analysis of color combinations, contrast, aesthetics, market appeal, and brand identity to create garments that are visually appealing and resonate with consumers. Successful color harmony evaluation results in products that not only look good but also effectively communicate the desired style, mood, and brand image (Shi et al., 2016).

### 11.2.3 Texture and Handfeel

Garment texture and handfeel evaluation is a crucial process in the fashion and apparel industry, focusing on the tactile aspects of clothing. This evaluation assesses the texture, surface feel, and overall tactile qualities of fabrics used in garments. While there are no specific standardized tests, it involves tactile assessments and qualitative analysis. The garment texture and handfeel evaluation is to assess how the fabric's texture and surface qualities contribute to the overall sensory experience when wearing the garment.

#### 11.2.3.1 Aspects of Garment Texture and Handfeel Evaluation

*Texture:* Evaluating the surface texture of the fabric, including its smoothness, roughness, weave, and patterns. It assesses how the texture aligns with the intended garment design and purpose.

*Handfeel:* Assessing the tactile qualities of the fabric when touched and manipulated. Handfeel includes attributes like softness, stiffness, warmth, coolness, and the sensation against the skin.

*Comfort:* Determining whether the fabric's texture and handfeel contribute to the overall comfort of the garment when worn for extended periods.

*Aesthetics:* Verifying that the texture and handfeel align with the garment's aesthetic and design intent, enhancing its visual and tactile appeal.

*Market Appeal:* Ensuring that the texture and handfeel resonate with the preferences and expectations of the target market or audience.

#### 11.2.3.2 Methods for Garment Texture and Handfeel Evaluation

Garment texture and handfeel evaluation primarily relies on tactile assessments and qualitative analysis, as there are no specific standardized tests. Common methods include:

*Touch and Feel:* Experts or designers touch, feel, and manipulate the fabric to assess its texture, smoothness, softness, stiffness, and other tactile qualities.

*Comparison with Competing Products:* Analyzing how the fabric's texture and handfeel compare with similar products in the market, identifying unique selling points or areas for improvement.

*Wearable Prototypes:* Creating prototypes or samples of the garment to evaluate how the fabric's texture and handfeel contribute to the wearer's comfort and overall sensory experience.

*Consumer Surveys and Focus Groups:* Collecting feedback from consumers through surveys or focus groups to gauge their reactions to the fabric's texture and handfeel, including preferences and comfort assessments.

*Expert Panels:* Assembling expert panels or sensory evaluation teams to provide qualitative assessments of the fabric's tactile qualities and its suitability for the intended purpose.

#### 11.2.3.3 Applications and Examples of Garment Texture and Handfeel Evaluation

Garment texture and handfeel evaluation applies to various clothing and apparel items. Here are some examples:

*T-Shirts and Casual Tops:* For everyday wear, comfortable and soft textures are often preferred. Evaluation ensures that the fabric's texture and handfeel align with the casual and relaxed style of these garments.

*Lingerie and Intimate Apparel:* Intimate apparel requires soft, smooth, and comfortable textures that feel gentle against the skin. Evaluation focuses on ensuring that the fabric enhances the wearer's comfort.

*Formal Attire:* Formal wear, such as suits and dresses, often uses fabrics with specific textures and handfeel to convey a sense of luxury and sophistication. Evaluation ensures that the fabric contributes to the garment's elegance.

*Outerwear:* Outerwear garments, such as coats and jackets, may use textured or insulated fabrics to provide warmth and protection. Evaluation assesses how the fabric's texture and handfeel contribute to functionality and comfort.

*Athletic and Sportswear:* Sportswear requires fabrics with moisture-wicking, cooling, or insulating properties. Evaluation focuses on how the fabric's texture and handfeel enhance performance and comfort during physical activities.

*Accessories:* Accessories like scarves, gloves, and hats rely on texture and handfeel to provide warmth and style. Evaluation ensures that the tactile qualities of these items meet consumer expectations.

Garment texture and handfeel evaluation plays a vital role in fashion and apparel design. While there are no standardized tests, careful analysis of texture, handfeel, comfort, aesthetics, and market appeal helps designers and manufacturers create garments that offer a positive sensory experience to consumers, enhancing overall wearability and satisfaction. Successful texture and handfeel evaluation results in products that not only look good but also feel comfortable and pleasant to wear (Fan et al., 2004).

### 11.2.4 Comfort and Wearability

Garment comfort assessment is a crucial process in the fashion and apparel industry, focusing on the comfort and wearability of clothing. This evaluation assesses how comfortable a garment is when worn, taking into account factors like fit, fabric, construction, and overall design. While there are no specific standardized tests, comfort assessment involves a combination of objective measurements and subjective evaluations. Garment comfort assessment is to evaluate how well a garment provides comfort to the wearer during various activities and in different environmental conditions.

#### 11.2.4.1 Aspects of Garment Comfort and Wearability Assessment

*Fit and Sizing:* Assessing whether the garment fits well, provides ease of movement, and avoids constriction or tightness.

*Fabric Properties:* Evaluating the fabric's breathability, moisture-wicking ability, softness, and overall feel against the skin.

*Seam and Construction Quality:* Examining the construction of the garment, including seam placement, stitch type, and reinforcement, to ensure it doesn't cause discomfort or chafing.

*Functionality:* Determining whether the design and features of the garment align with its intended use (e.g., sportswear, workwear, loungewear).

*Temperature Regulation:* Assessing how well the garment manages temperature and moisture, keeping the wearer comfortable in different weather conditions.

*Aesthetics:* Verifying that comfort considerations do not compromise the garment's overall aesthetic and design intent.

*Market Appeal:* Ensuring that the comfort level resonates with the preferences and expectations of the target market or audience.

#### 11.2.4.2 Methods for Garment Comfort Assessment

Garment comfort assessment involves a combination of objective measurements and subjective evaluations, and it often includes the following methods:

*Fit and Sizing Analysis:* Using standardized measurements to evaluate how well the garment's sizing and fit match the intended size chart.
*Wearable Prototypes:* Creating wearable prototypes or samples of the garment to assess how it fits, feels, and functions when worn for an extended period.
*Moisture Management Testing:* Conducting tests to evaluate how well the fabric manages moisture, such as moisture-wicking tests for activewear or moisture retention tests for cold-weather garments.
*Thermal Comfort Testing:* Assessing how the garment manages temperature, including heat retention and insulation, to ensure comfort in different weather conditions.
*Biomechanical Analysis:* Employing motion capture technology and biomechanical analysis to evaluate how well the garment allows for ease of movement during various activities.
*Subjective Surveys and Focus Groups:* Collecting feedback from wearers through surveys or focus groups to gauge their comfort, including fit, fabric feel, and overall satisfaction.
*Expert Panels:* Assembling expert panels to provide qualitative assessments of the garment's comfort attributes and its suitability for the intended purpose.

Comfort assessment of garments is essential for fashion and apparel design. It involves evaluating fit, fabric, construction, functionality, and wearability to create garments that offer a comfortable and pleasant experience to consumers. Successful comfort assessment results in products that not only look good but also provide an optimal level of comfort and satisfaction when worn (Tadesse et al., 2021).

## 11.3 QUALITY CONTROL AND QUALITY ASSURANCE IN THE APPAREL INDUSTRY

The quality of clothing is influenced not solely by the quality of the materials employed during its production, but also by variables dictated by the technology and parameters of the clothing manufacturing process. The decision to utilize specific processes and parameters can ultimately impact the overall quality of the garments. In garment manufacturing fabric stands as the fundamental raw material in the apparel industry. The primary objective for clothing manufacturers lies in transforming flat fabrics into shell structures that conform to the contours of the human body. This conversion of fabric into apparel entails a range of procedures, including spreading, pattern crafting, cutting, stitching, bonding, ironing, and packaging. The choice and utilization of these processes and parameters can ultimately impact the clothing's quality.

### 11.3.1 Process Control in Fabric Spreading

Spreading is the process of layering material on top of one another, creating a fabric stack or "lay." This lay can comprise a single fabric ply or numerous plies, ranging from just a few to potentially hundreds. During spreading, it's essential to consider various critical factors, including ply alignment, ply tension, bowing, splicing, and static electricity.

*Ply Alignment and Tension:* Ensuring that fabric plies are aligned correctly in terms of both width and length is crucial. Any discrepancies here can lead to wasted material during precision cutting, as excess trimming will be required to match the narrowest and shortest plies. Additionally, maintaining proper tension in the spread fabric is vital to avoid issues with contraction or expansion after cutting.

*Bowing:* Bowing refers to the deviation of filling yarn from a straight line across the fabric width. This distortion can introduce imbalances in fabric stresses, resulting in uneven ply tightness or slackness. These issues can lead to components being smaller or larger than intended, with the added risk of distortion during laundering or dry cleaning.

*Splicing:* Splicing involves overlapping two ends of fabric within a ply. The degree of overlap is critical—too little can result in incomplete pattern sections, while excessive overlap leads to unnecessary fabric wastage. Proper attention to splicing is vital for optimizing material usage.

*Static Electricity:* The development of static electricity in the fabric can disrupt the spreading process, resulting in incomplete pattern sections. Remedies include increasing humidity in the cutting room or implementing static eliminators to mitigate this issue (Vilumsone-Nemes, 2015).

### 11.3.2 Process Control in Pattern Making

The pattern-making process introduces various potential defects that can significantly affect pattern quality. These defects include:

*Pattern Alignment:* Correct alignment of pattern pieces with the fabric's grain is fundamental. The grain line of pattern pieces should align parallel to the warp in woven fabrics or the wales in knitted fabrics. In cases of bias cutting, where the grain lines typically form a 45° angle to the warp, precision is essential to ensure proper draping.

*Mixed Pattern Parts:* Incorrectly labeled pattern parts can lead to the incorrect assembly of parts with the wrong size or shape.

*Line Definition:* Issues such as excessively thick chalk lines, indistinctly printed lines, or incomplete powdering of the fabric stack can result in cutting inaccuracies.

*Marker Manipulation:* At times, markers may not fully utilize pattern edges, or patterns may be adjusted to save fabric, potentially leading to missing sections at the margins of pattern pieces.

*Marker Width:* When markers are excessively wide, components at the lay's edges may suffer from incomplete cutting, negatively affecting the final product's quality (Viļumsone-Nemes, 2012).

### 11.3.3 Process Control in Cutting

Cutting quality is a cornerstone of overall product quality and the efficiency of the construction process. Several defects can arise during cutting, including:

*Frayed Edges:* Fraying can obstruct the cutting process, leading to knife blockages or fabric damage. The extent of fraying depends on fabric construction and the quality of the cutting tools.

*Fuzzy, Ragged, or Serrated Edges:* Subpar cutting tools can produce edges that hinder sewing and compromise sewing quality. Issues such as burrs, chips, or dulled knife edges contribute to this problem.

*Ply-to-Ply Fusion:* Adjacent fabric plies sticking together can impede sewing and cause operational difficulties. Preventing fusion involves ensuring the right knife speed, maintaining sharp blades, using wax paper between plies, and lubricating the cutting blade.

*Single-Edge Fusion:* In some cases, a single ply's cut yarn ends may fuse together, forming a hard, brittle rim on the cut edges. While this can prevent fraying, it can make sewing manipulation challenging and create uncomfortable seams for consumers.

*Pattern Precision:* Misshaped or distorted pattern perimeters as a result of under- or overcutting are often due to poor manual control of the cutting machine and inaccuracies in marker alignment. To ensure precise patterns, it's crucial to verify markers before cutting, employ tensionless spreading, and allow the fabric time to relax. After cutting, verifying the top, bottom, and middle plies against the pattern is also essential.

*Notches:* The size and placement of notches should be carefully managed to prevent sewing challenges and ensure proper alignment during stitching.

*Drilling:* Drilling should be conducted with care to avoid issues such as holes that are too small or excessive heat generation, which can lead to fabric fusion. In some cases, marking fluid may be necessary to ensure drill holes are adequately marked for further processing while remaining easily removable if required (Viļumsone-Nemes, 2012).

### 11.3.4 Process Control in Sewing

In the realm of apparel production, the challenge of sewing damage is a prevalent concern, often leading to minor aesthetic imperfections. This issue relates to the actual creation of stitches, giving rise to challenges such as poor seam appearance and reduced seam performance. Stitch formation defects can manifest as skip or slipped stitches, staggered stitches, unbalanced stitches, variable stitch density, and needle, bobbin, or looper thread breakage.

Among these, skip or slipped stitches occur when the machine's hook or looper fails to catch the loop in the needle thread. In lock stitching, this results in a minor gap in the seam or less-than-ideal topstitching but doesn't compromise seam integrity.

However, in chain stitching, it creates a vulnerability that can lead to unraveling during garment use. Thread properties, particularly elongation and recovery, are crucial in preventing skip or slip stitches. Other contributing factors may include bent needles, incorrect needle size or type, improper thread tension, and issues related to material control.

To address these problems effectively, timely adjustments and replacements are required, such as checking and rectifying timing, selecting the right needle, ensuring proper thread tension, and enhancing material control. In the context of process control in sewing, these categories provide valuable insights into identifying, addressing, and preventing issues that can affect garment quality, appearance, and performance. Effective control measures encompass a range of actions, from selecting the right equipment and materials to conducting timely maintenance and adjustments in the sewing process. By focusing on these areas, manufacturers can optimize their sewing operations and enhance the overall quality of their products. The details of some major sewing defects are discussed below.

#### 11.3.4.1 Uneven Stitches

In the realm of sewing, addressing uneven and staggered stitches is crucial for ensuring the quality and durability of the final product. Unbalanced stitches in lock stitching can compromise the potential stretch of seams, particularly in knitted fabrics, often resulting in undesirable cracking. To rectify this issue, adjustments to bobbin tension are essential, ensuring that a full bobbin smoothly slides down the thread when held by its end. Careful fine-tuning of needle thread tension is also necessary to prevent thread breakage. In cases where maximum seam stretch is required, precise adjustments to both needle and bobbin thread tension may be needed, ultimately contributing to balanced and resilient seams.

#### 11.3.4.2 Variable Stitch Density

Variable stitch density poses another significant challenge in sewing, often stemming from insufficient foot pressure, especially in drop feed systems, which can lead to uneven fabric feeding through the machine. This problem is more pronounced when dealing with materials featuring sticky or slippery surfaces. To combat this issue, it is imperative to ensure that the foot pressure is sufficient for uniform feeding. In instances involving challenging fabrics and specific sewing scenarios, specialized feed systems become indispensable for achieving consistent feeding of all material plies. Operators also play a role in maintaining consistent stitch density by maintaining a constant sewing speed and avoiding abrupt changes in stitching pace. In sewing, addressing unbalanced stitches and variable stitch density necessitates a combination of technical adjustments, equipment suitability, and operator proficiency to ensure consistent and high-quality stitching results.

#### 11.3.4.3 Seam Pucker

In the world of apparel production, the appearance of the garments holds paramount importance, and seam pucker can significantly impact their visual appeal. While eliminating seam pucker entirely is a near-impossible feat, it is widely acknowledged that accepting a minor degree of pucker is a common practice.

Seam pucker emerges due to various factors, including structural jamming, differential feed, sewing thread tension, sewing thread shrinkage, fabric shrinkage, and pattern mismatch.

##### *11.3.4.3.1 The Impact of Seam Pucker*

Seam pucker manifests as a wrinkled appearance along the seam, exerting a substantial influence on the overall aesthetics of the apparel. Recognized as a sewability challenge for nearly seven decades, seam pucker stands as a vital parameter for quality control in the apparel manufacturing industry. It typically arises from the improper selection of sewing parameters and material characteristics, leading to unevenness in the fabrics being stitched together and compromising their visual appeal.

In severe cases, seam pucker can resemble a wave, originating from the seam and extending throughout the entire garment. Even in milder instances, the presence of a wave is noticeable and far from desirable, making garments with pronounced seam puckers unwelcome among consumers.

##### *11.3.4.3.2 Identifying Causes and Solutions for Seam Pucker*

*Structural Jamming:* This type of pucker is observed in tightly constructed fabrics that offer limited space for sewing threads. Solutions include opting for thinner threads, smaller needle sizes, needle plates, and presser feet with smaller holes, reducing stitch density, and stitching in a direction that accommodates the fabric's yarns.

*Differential Feed:* Addressing feeding pucker requires using an appropriate feed dog, adjusting presser foot pressure, and controlling sewing speed to minimize the accumulation of fabric and resulting pucker.

*Sewing Thread Tension:* Tension pucker arises when sewing thread is subjected to high tension. Solutions include reducing thread tension, selecting threads with reduced extensibility, and ensuring proper feed timing.

*Sewing Thread Shrinkage:* This issue can be mitigated by choosing threads with good lubrication or opting for synthetic threads. The choice of thread type plays a crucial role in preventing pucker.

*Fabric Shrinkage:* To combat fabric-related pucker, refinishing fabrics to reduce shrinkage effects or using synthetic threads can be effective.

*Mismatched Patterns:* Ensuring that fabric patterns match in length is vital to prevent puckering caused by discrepancies in pattern alignment (Choudhary et al., 2018).

#### 11.3.4.4 Thermal Damage

When it comes to sewing, fabric can suffer damage due to thermal and mechanical factors, falling into two distinct categories. Thermal problems relate to the impact of heat during sewing operations and encompass various causes and solutions. One prominent issue in sewing is the heating of the needle, which, in turn, affects the material and thread used in the stitching process. Needle heating results from the friction between the needle and the rapidly moving thread as it passes through the needle eye. This heat concentration can reach temperatures ranging from 300°C to 350°C. The consequences of high needle temperatures can be significant, especially with the prevalent use of thermoplastic fabrics and sewing threads, such as nylon and polyester. These materials may soften, melt, and break when exposed to high needle temperatures, leading to disruptions in the stitching process and reduced production efficiency. Furthermore, softened polymer from either the sewing thread or the fabric can accumulate in the needle eye, potentially causing clogs and thread breakage. Thermal damage to the fabric can also impact its properties, performance, and appearance.

##### 11.3.4.4.1 Causes and Solutions for Thermal Damage

*Excessive Sewing Speed:* Reducing sewing speed can mitigate the impact of high temperatures.

*Inappropriate Needle Point:* Selecting needles with sharper or finer ball points can help.

*Needle Size Too Large:* Reducing needle size can lower temperatures.

*Inappropriate Fabric Density:* Ensuring the fabric density is appropriate can prevent thermal damage.

*Inadequate Fabric Finishing:* Increasing the content of softeners on the fabric can counteract thermal damage.

*Inappropriate Sewing Thread:* Choosing the right sewing thread is crucial to minimize thermal damage.

##### 11.3.4.4.2 Factors Affect Needle Temperature during Sewing

*Sewing Speed:* Higher sewing speeds lead to higher needle temperatures, with equilibrium being reached for specific combinations of variables.

*Needle Size:* Larger needle sizes require more force for fabric penetration, resulting in higher temperatures.

*Needle-Point Shape and Surface Finish:* Needle-point shape and surface finish can impact heat generation, with heavy-ball needles typically operating at higher temperatures.

*Presser Foot and Throat Plate:* These machine parts influence fabric restraint, with throat plate hole size being a critical variable. Adjusting the pressure of the presser foot can also affect needle temperature (Jayakumari et al., 2022).

#### 11.3.4.5 Mechanical Damage in Sewing

In the realm of stitching, cloth can suffer mechanical damage in contrast to thermal damage. Mechanical damage, often characterized by fragmented or broken fibers, filaments, or yarns, becomes apparent either immediately after stitching or sometimes only after the product undergoes tension, stress, strain, deformation, or multiple cleaning cycles. While woven fabrics can sometimes hide this type of damage under the sewing thread, knitted structures tend to exhibit more severe visual effects. The impact of mechanical damage extends beyond the immediate seam area and is deemed unacceptable due to its negative implications on both aesthetics and performance.

##### 11.3.4.5.1 Causes and Solutions for Mechanical Damage

*Needle Size Too Large:* Reducing the needle size can alleviate this issue.

*Inadequate Fabric Finishing:* Increasing the softener content on the fabric can help mitigate mechanical damage.

*Inappropriate Needle Point:* Choosing needles with sharper or finer ball points can address this problem.

*Damaged Needle Point and Throat Plate:* Replacement of damaged parts is essential.

*Mismatch between Needle Size and Throat Plate Hole:* Reducing the fit's tightness can prevent damage (Palaniappan, 2020).

## 11.3.5 Process Control in Fusing Operations

Fusible interlinings play a crucial role in mitigating seam pucker issues. These interlinings consist of a base fabric layered with a thermoplastic adhesive resin, which, when subjected to heat and pressure, can bond with another fabric. While fusible interlinings have significantly enhanced efficiency and quality in the apparel industry, selecting the right interlining for a specific shell fabric remains a challenge. Incorrect handling during fusing can lead to undesirable consequences such as alterations in shell fabric color, surface appearance, shrinkage, stretching, or stiffening.

To ensure the final garment's optimal appearance and drape, it's imperative that the chosen interlining complements the shell fabric. However, fusible interlinings come in various compositions, fabric structures, and adhesive types, making the selection process complex. While many companies offer guidelines for selecting fusible interlinings, the conventional approach often relies on trial and error or past experience. Collaborative efforts between manufacturers and interlining suppliers, alongside pre-production testing, can significantly reduce issues in interlining selection and usage.

### 11.3.5.1 Strike-Through and Strike-Back in Fusing

The fusing process involves several parameters, including fusing method, fusing machine, and fusing conditions (temperature, time, pressure, and cooling). Inadequate selection or adjustment of these parameters can result in fusing problems like strike-through, strike-back, color changes, and distortion. Strike-through manifests when adhesive resin appears on the exterior of the fused fabric, while strike-back involves adhesive resin appearing on the non-adhesive side of the fusible interlining.

#### *11.3.5.1.1 Causes and Solutions for Strike-Through*

*Prolonged Fusing Time:* Reduce it, especially for thinner fabrics, and perform testing before bulk production.

*Excessive Fusing Temperature:* Decrease the temperature, and conduct pre-production fusing tests.

*High Fusing Pressure:* Follow the machine manufacturer's pressure guidelines. Reduce pressure for lightweight and loosely constructed outer fabrics.

*Incompatible Fusible Interlining (Excessive Adhesive):* Avoid using interlinings with excessive adhesive. Reduce the size and weight of adhesive dots. Test before bulk production.

*Inappropriate Fusing Method for Thin and Loosely Constructed Fabrics:* Use frame fusing and perform tests before bulk production.

*11.3.5.1.2 Causes and Solutions for Strike-Back*

*Inappropriate Fusible Interlining:* Avoid lightweight substrate materials and interlinings with excessive adhesive. Reduce adhesive dot size. Test before bulk production.

*Prolonged Fusing Time:* Reduce it and conduct pre-production tests.

*Excessive Fusing Temperature:* Lower the temperature, and utilize temperature measurement strips for optimization.

*High Fusing Pressure:* Follow machine manufacturers' instructions for pressure settings. Reduce pressure for lightweight fabrics.

#### 11.3.5.2 Color Changes during Fusing

Improper fusing can lead to temporary or permanent color changes, including shine, glazing, or discoloration in fabrics. Such color variations stem from the effects of heat on certain dyes.

*11.3.5.2.1* Causes and Solutions for Color Changes during Fusing

*Temporary Color Changes in Synthetic Fabrics after Heat Application:* Regulate pressing temperature, pressure, and time to avoid overheating. Allow fused fabrics to cool to regain their color.

*Excessive Temperature, Pressure during Fusing:* Cover the fusing material with a cloth during heat pressing. Select a multivariable adhesive requiring lower temperature, pressure, and time.

#### 11.3.5.3 Fusing Distortion

Fusing distortion occurs when panels become distorted during the fusing process. Distorted panels cannot be corrected and must be discarded as waste. Factors contributing to fusing distortion include inadequate fusing conditions, improper panel handling, fabric instability, and incorrect fusing directions.

*11.3.5.3.1 Causes and Solutions for Fusing Distortion*

*Poor Fusing Machine Settings and Conditions:* Reduce fusing press speed and pressure, establish optimum machine settings in consultation with the interlining supplier, and allow time for cooling before handling.

*Incorrect Panel Orientation:* Change panel orientation to prevent pressure roller catch. Pre-set panels through hand-ironing, or use a flatbed fusing press if available.

*Unstable Fabric/Substrate Properties:* Use stable fabrics with high shear rigidity, and handle fabrics with care during the fusing process.

*Wrong Fusing Direction:* Fuse interlining and outer fabrics in the same direction to ensure consistent dimensional changes (Thilagavathi and Viju, 2012).

### 11.3.6 Process Control in Pressing Operations

Pressing serves as a final opportunity to refine the shape of apparel, smoothing out creases, and enhancing overall appearance. It plays a vital role in preserving the garment's shape and fit. A well-executed pressing job can significantly influence the impression it makes on potential customers, potentially allowing for higher pricing. Proper pressing should result in an impeccably smooth and unblemished look, free from shine, scorch marks, melting, clamp imprints, or water stains. There should be no unintended creases, pleats, pressed wrinkles, or hems mistakenly pressed on the wrong side. For instance, pants and shorts' seams and side seams should be pressed seam-on-seam to create a clean center crease.

Fabrics must be engineered to exhibit optimal pressing characteristics, and garments should undergo pressing under appropriate conditions.

#### 11.3.6.1 Preparation and Handling

Before pressing, it's essential to remove all tacking threads and pins to prevent marking the garment. To avoid scorching, melting, hardening, or shriveling, it's advisable to test the iron's heat (temperature) before pressing, ensuring it suits the garment's fabric. The heat test should be performed on a spare piece of the actual fabric to determine the right temperature. When positioning a garment on a pressing machine, it's crucial to maintain the fabric's grain alignment to prevent unnecessary creases. All seams should be kept in a straight line. To prevent shine or iron marks on the right side, the pressing should be done on the garment's wrong side. If necessary, place a strip of paper under seam turnings to shield the right side of the fabric or apparel from marks.

#### 11.3.6.2 Pressing Conditions

Pressing encompasses five main components: heat (temperature), moisture (often as steam), pressure, time, and equipment. Depending on factors like the garment's volume, fashion, and structure, pressing can be carried out using industrial or household irons (with or without steam) or heavy presses. Industrial flat irons similar to those used at home are typically employed for low-volume, high-fashion, or highly structured apparel. Conversely, high-volume, ready-to-wear apparel typically undergoes pressing using presses or pressing machines such as flatbed, buck presses, form presses, and upright presses, known for their efficiency. The garment is usually placed over the machine, where steam and pressure are applied automatically, followed by vacuum extraction of excess moisture. The ideal pressing conditions vary depending on fabric types and apparel variations. Different fabrics possess distinct properties, necessitating tailored pressing conditions. For instance, microfiber apparel has low heat resistance and can be challenging to press without causing shine, marks, or glazing. To address this, microfiber or extremely lightweight fabrics are often pressed from the inside using foam finishes. If pressing from the fabric's "right side" is necessary, the temperature must be very low, with perfectly dry steam, as synthetics cannot tolerate excess moisture. Pressing duration and ambient relative humidity during preconditioning and post-conditioning phases also significantly influence pressing performance (Thilagavathi and Viju, 2012).

### 11.3.7 Storage and Packaging of Garments

Following pressing, garments should be wrinkle free, creaseless, and maintain their shape for storage and delivery to customers. While the industry offers manual or semi-automatic packaging machines, careless or improper storage and packaging can negatively impact the final product's appearance.

#### 11.3.7.1 Garment Storage

Pressed apparel should be stored in a cool, dry location with ample space to prevent overcrowding, which can lead to wrinkles and hinder air circulation, possibly causing moisture buildup and mildew. Additionally, overcrowding may result in excessive heat, leading to plastic bags bonding with the fabric. The storage area should remain as clean as possible to safeguard against dust, dirt, atmospheric pollutants, and pests like moths that could harm the apparel. During storage, garments should not experience strain or movement that could weaken the fibers and cause breakage. To avoid pressing sharp folds, it is recommended to fold knitted or stretched apparel, storing them in plastic bags. Nylon hosiery should also be kept in plastic bags to prevent snags, while sweaters should be shielded from lint contamination in plastic bags. Wool-containing clothing should be stored in a moth-free or mothproof environment.

#### 11.3.7.2 Vacuum Packing for Garments

Tailored jackets, in particular, benefit from vacuum packing, which reduces storage space requirements and prevents creases from reappearing during storage and transportation. Vacuum packing is also environmentally friendly and reusable. This method, available in various sizes for different types of apparel, protects against dirt, moisture, and insects due to its waterproof and airtight properties (Thilagavathi and Viju, 2012).

## 11.4 GARMENT COSTING

Garment costing involves determining the total cost of producing a garment, including materials, labor, overhead, and profit margin. Accurate costing is essential for setting competitive prices and ensuring profitability.

### 11.4.1 Material Cost

Garment material cost estimation is a critical component of garment costing in the fashion and apparel industry. It involves assessing the costs associated with the materials used to produce a garment, including fabric, trims, and other components (Magreba Takebira, 2016). Accurate material cost estimation is essential for pricing decisions, cost control, and overall profitability. Garment material cost estimation is to determine the total cost of the materials needed to produce a garment. This includes calculating the cost of fabric, lining, interlining, buttons, zippers, threads, labels, and any other materials used in the construction.

#### 11.4.1.1 Factors of Material Cost Estimation

*Material Type and Quality:* The cost of different materials varies based on their type (e.g., silk, cotton, polyester) and quality (e.g., high-end, luxury, basic).

*Material Consumption:* Estimating the quantity of each material required for each garment based on its design, size, and style.

*Material Price:* Determining the unit price of each material based on sourcing and supplier negotiations.

*Material Waste:* Accounting for any waste or excess material that may be generated during the cutting and sewing processes.

*Trims and Accessories:* Calculating the cost of trims (buttons, zippers, etc.) and accessories (labels, hangtags) that contribute to the overall material cost.

#### 11.4.1.2 Methods for Garment Material Cost Estimation

Garment material cost estimation is typically done using a combination of methods and tools, including:

*Bill of Materials (BOM):* Creating a detailed BOM that lists all the materials required for each garment. The BOM includes material descriptions, quantities, unit prices, and total costs.

*Sample Making and Prototyping:* Creating garment prototypes to determine the actual material consumption and identify any inefficiencies or excess material.

*Sourcing and Supplier Negotiations:* Establishing relationships with suppliers and negotiating prices for bulk purchasing of materials.

*Historical Data:* Analyzing historical data on material costs for similar garments to estimate costs for new designs.

*Software Tools:* Using specialized software and spreadsheets to calculate material costs based on the BOM and other relevant data.

### 11.4.2 Labor Cost

Garment manufacturing labor cost estimation is a crucial aspect of the apparel production process, determining the cost associated with the labor required to make a garment. Accurate labor cost estimation is essential for pricing decisions, production planning, and cost control. While there is no single standardized method, the estimation process involves various factors and calculations.

#### 11.4.2.1 Factors of Garment Manufacturing Labor Cost

The garment manufacturing labor cost includes calculating the labor cost for various production processes, such as cutting, sewing, pressing, and finishing. Labor cost estimation takes into account various factors, including:

*Labor Type:* Identifying the types of labor involved, including skilled (e.g., master tailors) and unskilled (e.g., machine operators) labor.

*Labor Rate:* Determining the hourly or piece-rate wages for each category of labor, which can vary based on location and skill level.

*Production Time:* Estimating the time required for each production process, considering factors like garment complexity, design, and machinery efficiency.

*Efficiency and Productivity:* Assessing the efficiency and productivity of the production team, including any breaks, downtime, or interruptions.

*Overhead Costs:* Accounting for indirect labor costs, such as supervision, administration, and maintenance of machinery.

#### 11.4.2.2 Methods for Garment Manufacturing Labor Cost Estimation

Labor cost estimation in garment manufacturing typically involves the following methods and considerations:

*Time and Motion Studies:* Conducting time and motion studies to analyze and record the time spent on each production process. This data is used to calculate labor costs.

*Standard Minute Value (SMV):* Using the SMV system, which assigns a standard time value to each task or operation involved in garment production. The SMV is multiplied by the labor rate to calculate labor costs.

*Piece Rate:* For tasks where output can be easily quantified (e.g., attaching buttons), labor costs are calculated based on the number of pieces completed multiplied by the piece rate.

*Labor Rate Negotiations:* Negotiating labor rates with workers or labor unions based on factors like experience, skill level, and regional wage standards.

*Historical Data:* Analyzing historical data on labor costs for similar garments or production processes to estimate costs for new designs.

### 11.4.3 Overhead Costing and Profit Margin Estimation

It is a vital aspect of the apparel production process, as it involves calculating the indirect costs associated with manufacturing garments and determining the profit margin for each product. Accurate overhead costing and profit margin estimation are essential for pricing decisions, cost control, and overall profitability. Garment manufacturing overhead costing is to allocate indirect costs that cannot be directly attributed to a specific product. These indirect costs include expenses like factory rent, utilities, maintenance, machinery depreciation, administrative salaries, and quality control expenses. Accurate overhead costing is crucial for determining the true cost of production for each garment. Profit margin estimation, on the other hand, involves calculating the amount of profit that can be earned from each garment after considering both direct costs (such as material and labor costs) and indirect costs (overhead costs). It helps in setting the selling price and ensures that the business is profitable.

#### 11.4.3.1 Methods for Garment Manufacturing Overhead Costing and Profit Margin Estimation

Overhead costing and profit margin estimation in garment manufacturing typically involve (Islam Rajib et al., 2023) the following methods and considerations:

*Overhead Allocation:* Allocating overhead costs to each garment by using appropriate allocation bases. For example, overhead costs can be allocated based on machine hours, labor hours, or square footage of production space.

*Cost Centers:* Categorizing overhead costs into different cost centers based on their nature (e.g., production, administration, quality control) to track and allocate expenses more accurately.

*Historical Data:* Analyzing historical data on overhead expenses to estimate the average indirect cost associated with manufacturing a garment.

*Activity-Based Costing (ABC):* Using ABC methods to allocate overhead costs more precisely by identifying the specific activities that drive indirect costs and allocating those costs accordingly.

*Profit Margin Formula:* Calculating the profit margin by subtracting the total cost (including overhead) from the selling price. The profit margin is usually expressed as a percentage of the selling price.

## 11.5 CONCLUSION

This chapter has provided a comprehensive insight into the fundamental pillars of success within the dynamic fashion and apparel manufacturing industry, and the intricate relationship between quality and cost-effectiveness, recognizing their pivotal roles in shaping the industry's destiny.

Through the exploration of evaluation techniques, we learned how crucial it is to employ both objective and subjective methods to assess apparel quality accurately. This ensures that consumer expectations are not just met but exceeded, cementing the industry's reputation for excellence.

Moving on to quality control and assurance, we discovered the vital importance of not only identifying defects but also establishing robust protocols to rectify them. This commitment to quality assurance ensures that every garment leaving the production line upholds the highest standards, fostering customer trust and loyalty.

Lastly, our exploration of garment costing has shed light on the financial underpinnings of apparel production. Understanding the intricate cost structures involved enables industry stakeholders to make informed decisions, optimize resource allocation, and ultimately determine the viability of production ventures.

In essence, this chapter serves as a crucial resource for professionals, scholars, and enthusiasts within the fashion and apparel manufacturing sector. It emphasizes the significance of continually balancing quality and cost-effectiveness, ensuring the industry's continued success in a constantly evolving landscape.

## REFERENCES

AATCC TM 135, 2018. *Dimensional Changes of Fabrics after Home Laundering*. American Association of Textile Chemists and Colorists.

ASTM D1424-21, 2021. *Standard Test Method for Tearing Strength of Fabrics by Falling-Pendulum (Elmendorf-Type) Apparatus*. ASTM International.

ASTM D1683/D1683M-22, 2022. *Standard Test Method for Failure in Sewn Seams of Woven Fabrics*. ASTM International.

ASTM D2061-07, 2021. *Standard Test Methods for Strength Tests for Zippers*. ASTM International.

ASTM D2062-03, 2021. *Standard Test Methods for Operability of Zippers*. ASTM International.

ASTM D3787-16, 2020. *Standard Test Method for Bursting Strength of Textiles—Constant-Rate-of-Traverse (CRT) Ball Burst Test*. ASTM International.
ASTM D4846-96, 2021. *Standard Test Method for Resistance to Unsnapping of Snap Fasteners*. ASTM International.
ASTM D4966-22, 2022. *Standard Test Method for Abrasion Resistance of Textile Fabrics*. ASTM International.
ASTM D5034-21, 2021. *Standard Test Method for Breaking Strength and Elongation of Textile Fabrics (Grab Test)*. ASTM International.
ASTM D5035-11, 2019. *Standard Test Method for Breaking Force and Elongation of Textile Fabrics (Strip Method)*. ASTM International.
Choudhary, A.K., Sikka, M.P., Bansal, P., 2018. The Study of Sewing Damages and Defects in Garments. *Research Journal of Textile and Apparel* 22.
Eckman, M., Damhorst, M.L., Kadolph, S.J., 1990. Toward a Model of the In-Store Purchase Decision Process: Consumer Use of Criteria for Evaluating Women's Apparel. *Clothing and Textiles Research Journal* 8.
Fan, J., Yu, W., Hunter, L., 2004. *Clothing Appearance and Fit: Science and Technology, Clothing Appearance and Fit: Science and Technology*. Woodhead Publishing.
Islam Rajib, Md M., Parvez, Md M.H., Islam, Md S., Ahmed, T., Islam, Md R., 2023. Complete Garment Costing with Major Cost Breakdown. *Journal of Textile Science and Technology* 09.
ISO 105-B02, 1994. *Textiles–Tests for Color Fastness, Part B02: Color Fastness to Artificial Light*. International Organization for Standardization.
ISO 105-C06, 2010. *Textiles Tests for Colour Fastness Part C06: Colour Fastness to Domestic and Commercial Laundering*. International Organization for Standardization.
ISO 105-E04, 2013. *Textiles—Tests for Colour Fastness—Part E04: Colour Fastness to Perspiration*. International Organization for Standardization.
ISO 105 X12, 2016. *Textiles–Tests for Colour Fastness: Part X12: Colour Fastness to Rubbing*. International Organization for Standardization.
ISO 13935-1, 1999. *Textiles-Seam Tensile Properties of Fabrics and Made-Up Textile Articles-Part 1*. International Organization for Standardization.
ISO 4920, 2012. *Textile Fabrics–Determination of Resistance to Surface Wetting (Spray Test)*. International Organization for Standardization.
ISO 5077, 2022. *Textiles — Determination of Dimensional Change in Washing and Drying*. International Organization for Standardization.
Jayakumari, M., Rajalakshmi, M., Priyalatha, S., Kapoor, V., Sampath, M.B., Suganth, G., Prakash, C., 2022. Investigation of Change in Sewing Needle Temperature: Part I: Influence of Fabric Types and Sewing Parameters. *Journal of Testing and Evaluation* 50.
Liu, K., Wu, H., Zhu, C., Wang, J., Zeng, X., Tao, X., Bruniaux, P., 2022. An Evaluation of Garment Fit to Improve Customer Body Fit of Fashion Design Clothing. *International Journal of Advanced Manufacturing Technology* 120.
Magreba Takebira, U., 2016. Proficient Costing Procedure of a Garment from Yarn to Consumer. *American Scientific Research Journal for Engineering* 19.
Palaniappan, G., 2020. Analysis of Sewing Defects and Control Measures for Apparel Industry. *International Journal of Polymer and Textile Engineering* 7.
Petrova, A., Ashdown, S.P., 2012. Comparison of Garment Sizing Systems. *Clothing and Textiles Research Journal* 30.
Shi, C., Huang, M., Li, Z., Liu, Y., Wang, H., 2016. Evaluation and Optimization of Color Harmony Model with Context of Patch and Garment. *Guangxue Xuebao/Acta Optica Sinica* 36.
Tadesse, M.G., Loghin, C., Dulgheriu, I., Loghin, E., 2021. Comfort Evaluation of Wearable Functional Textiles. *Materials* 14.

Thilagavathi, G., Viju, S., 2012. Process Control in Apparel Manufacturing. In: Majumdar, A., Das, A., Alagirusamy, R. and Kothari, V.K. (Eds.), *Process Control in Textile Manufacturing*. Woodhead Publishing.

Viļumsone-Nemes, I., 2012. *Industrial Cutting of Textile Materials, Industrial Cutting of Textile Materials*. Woodhead Publishing.

Vilumsone-Nemes, I., 2015. Fabric Spreading and Cutting. In: Nayak, R. and Padhye, R. (Eds.), *Garment Manufacturing Technology*. Woodhead Publishing.

# 12 Artificial Intelligence and Robotics

## *A Wave of Transformation for Garment Industry*

*Amit Madhu*

### 12.1 ARTIFICIAL INTELLIGENCE AND ROBOTICS REBOOTING THE GARMENT INDUSTRY

Artificial intelligence (AI) has transformed the clothing business in this information technology era. For better problem-solving and decision-making, AI employs conventional mathematical and logical models. AI is being employed more and more in the supply chain, sewing floor, production planning, pattern development, and design of clothing. The clothing industry uses a variety of AI methods, including neural networks, expert systems, fuzzy logic, evolution strategies, genetic algorithms, artificial immune systems, and multiagent systems. AI may be used to accomplish important goals including raising productivity, cutting production costs, and enhancing quality [1–3].

The possible applications of AI in the garment industry are quite diverse and range from design assistance systems to fashion recommendation systems through sensory evaluation, quality control, smart tracking systems, fashion forecasting, fashion e-marketing, and decision-making in supply chain management. Since they can handle complex and multicorrelated data and dependencies relating to sensory quality attributes, consumer behaviour, designer knowledge, and so forth human factors, these systems are specifically suited for the garment industry. Robotics are now used to move parts and materials around a factory or to speed up manufacturing, which reduces the need for human labour and increases production efficiency [2,4–6].

In essence, there are several ways AI may help the fashion business in all its facets. Let's go through the important ones [3,5,7].

- Future trend forecasting and prediction using AI are extremely beneficial for fashion firms. It aids in their ability to adapt to quick changes in this unstable economy.
- Algorithms powered by AI assist fashion firms in automating labour-intensive, regular processes. It comprises assessing client behaviour patterns, colour matching, quality control examinations, and other things.

 DOI: 10.1201/9781003432739-12

- Customer service: Chatbots powered by AI are a huge help to customer care personnel.
- Management of inventories and the supply chain is made possible by the use of AI technologies by businesses. They are incredibly useful in setting up the delivery of goods in accordance with demand forecasting.

### 12.1.1 Machine Learning

AI's machine learning field enables computers to learn and develop on their own without being supervised learning. Making computer programmes that can acquire data and gain knowledge on their own is the focus of machine learning. In order to uncover patterns in data and make better judgements moving forward based on the examples we offer, the learning process activates with observations or data, such as examples, first-hand experience, or learning. The basic goal is for computers to acquire knowledge entirely on their own, without help from humans, and to change how they behave as a consequence. Depending on the learning method used to forecast the output, machine learning can be supervised, semi-supervised, unsupervised, or reinforced. By employing trend forecasting, improving the quality and design of new items, and using smart maintenance techniques, machine learning may help fashion firms run their businesses more successfully [2,8,9].

One of the many approaches to include machine learning in systems for forecasting in the fashion industry is to anticipate fashion trends using AI. Making precise demand predictions and identifying patterns in consumer purchasing behaviour are made possible by machine learning. It is quickly rising to the top of the list of methods used in the fashion sector to predict trends. Machine learning-driven fashion forecasting uses big data and predictive analysis to anticipate sales, consumer analysis, and future demand for new goods based on an algorithm of style, trend, and seasonality. The underlying idea behind market and sales forecasting is the utilisation of data from prior consumer interactions with the company's commercial environment, such as searches and purchases. AI then draws inferences about, for instance, seasonality in sales and fluctuations in demand. As a result, merchants may utilise AI prediction to foresee the demand for a specific product at a given moment. Overall sales are increased and stock is managed more effectively as a result [5,8].

Smart algorithms must include machine learning since it essentially guarantees that AI evolves over time in response to new information. In order to ensure precise and accurate collection planning and purchasing, machine learning-driven projections are continuously updated in real time. The goal of trend forecasting is to enhance demand planning so that businesses may decrease resource costs and waste from surplus commodities.

Additionally, machine learning may enhance a product's functionality and aesthetics. Machine learning is similar to AI in that it may find flaws and errors before finalising the product. The information is subsequently given to the design team for assessment and revision [2,6]. By anticipating and assessing a machine's performance, availability, and longevity, machine learning helps businesses prevent expensive downtime. Machine learning may be used to plan maintenance inspections and notify operators before a mechanical malfunction by analysing operational data from the past and present.

The well-known corporations are utilising cutting-edge machine learning algorithms to estimate their local sales more precisely and anticipate changes in worldwide patterns. Machine learning is used by H&M to study local consumer behaviour, various retailers' financial gains, the quantity and variety of items sold, and many other variables. These are standardised and taken into account with care. While ASOS uses machine learning to more accurately predict the lifetime value of customers. Even slight changes in this indicator could be a sign of broad behavioural pattern changes. In this approach, the business will always be aware of the yearly spending preferences of its clients. Such data enables the marketing industry to decide more effectively on the whole sales plan [4,5,9].

The use of chatbots by fashion firms to understand more about their customer needs and intended purchases is one of the primary ways AI and machine learning are used in the fashion industry. An online shopper seeking for a new shoe or clothing, for instance, may easily communicate with an intelligent agent via a website or mobile app. Burberry, Tommy Hilfiger, and Levi's are just a few of the well-known fashion brands that use chatbots. Even small-time fashion businesses are already adopting machine learning algorithms to comprehend rapidly changing consumer demands and expectations, outside of giant e-commerce merchants or brands like Amazon. For instance, Stitch Fix, a web-based styling firm, uses machine learning algorithms to enhance supply chain management and client experiences [3,5,8,9].

In the past several years, there has been an incredible rise in the use of machine learning in the garment sector. More recent research on machine learning claims that advances in using deep learning for fashion challenges have been fuelled by the availability of large-scale datasets like Deep Fashion. One of the most recent applications uses a fashion design framework that successfully synthesises realistic pictures using generative adversarial training. The unique feature of the suggested system is that it generates pictures of new objects that are compatible with a particular query item rather than proposing already-existing items from the dataset. The framework builds an automated model of the user's sense of style before creating an item for a specified query item [5,7,10].

Among other uses, machine learning algorithms may estimate inventory demand accurately, cutting down on waste and the cost of unsold inventory. In other words, machine learning for fashion is changing the way that fashion firms create their items, promote them, and deliver them to customers.

### 12.1.2 Digital Factories

As automation technology advances and becomes more sophisticated, the apparel industry starts to integrate these technologies and replace human labour with mechanisation in various production-related steps. In the 20th century, automation has advanced from basic cutting and sewing to more specialised and varied garment manufacturing machinery, such as pre-sewing, sewing, and post-sewing equipment.

In the manufacturing of clothing, there are several automated processes. These include application of radio frequency identification in automation, fabric inspection; application of computer-aided design (CAD) and computer-aided manufacturing (CAM) in fabric spreading, cutting, sewing, pressing, and material handling [7,11].

#### 12.1.2.1 Fabric Inspection Automation

Previously, fabric inspections were done manually, which frequently made it difficult to precisely identify problems. The fabric inspection procedure is made more effective with the usage of automated technology. The most challenging textile processes to automate is fabric examination. For automatic fabric inspection, a variety of techniques, including statistical, spectral, and model-based approaches, can be used. In each of these techniques, the fabric picture is modified by a modelling tool or piece of software to get details about the degree of the fabric problem. The found flaws are automatically noted on the fabric, and if there are too many of them in one fabric lot, it is discarded. Three automated optical fabric inspection systems are BarcoVision's Cyclops, Zellweger Uster's Fabriscan, and Elbit Vision System's I-Tex [11,12].

#### 12.1.2.2 Automatic CAD and CAM

Traditionally, cloth design was done by hand, in order to spend a lot of time and capture the concept on paper. Today, however, clothing is made by companies that employ CAD and CAM. Computer programme called CAD is used to develop 3D designs for garments, and CAD then delivers those designs to CAM. Based on the data, CAM oversees and controls the production process [12].

#### 12.1.2.3 Spreading of Fabric Automatically

Many years ago, spreading fabric was a laborious process that required employees to spread the fabric by hand. In automated fabric spreading, the task is completed swiftly by maintaining the length-width in accordance with the operator's directions. The spreader automatically splices and spreads fabric from new roll exactly where the previous fabric roll exhausts. An efficient and high-quality sensor is employed to detect faults, and the flawed cloth is automatically cut. Counting the ply number, and when the designated ply number is reached, the machine automatically shuts off. This machine can spread any kind of cloth. Less labour is needed, which takes less time. However, operating this equipment requires a professional operator and is highly expensive [11].

#### 12.1.2.4 Fabric Cutting Automation

In automatic fabric cutting, the patterns are cut more precisely and smoothly. According to the pattern's design for the garment, it is now recorded directly in the computer memory without being printed out, and directly feed to the cutting machine to automatically and properly cut numerous layers of cloth together. Somewhere else, a laser is likewise used to cut things. When compared to manually or operator-controlled devices, the usage of automatic cutting machines has lowered both the number of labour and the amount of time. Bullmer Cut, Eastman CNC, and Tukatech CNC are some of the modern automatic cutting machines [7,10].

#### 12.1.2.5 Automation in Sewing

Industrial robots are developed now to handle the fabric during sewing operations, eliminating the need for human labour. This will allow for the sewing process to be automated. These automated machines make seams in a manner akin to that of conventional sewing machines. Robotic sewing machines can create and switch between a variety of sewing stitches (including chain, overlock, and double lock) easily.

Sewbo, a robot created by Zornow can autonomously handle fabric components during stitching, is one such example. Zornow created the "Sewbo" robot, which can sew a T-shirt from beginning to end. This accomplishment was a significant step toward obtaining total outfit automation at 100%. It is possible to programme the robot for a certain size and look. The robot must be reprogrammed whenever the size or kind of clothing changes.

LOWRY SewBot is a unique kind of robot created specifically for the garment business, and it was created by the American firm "Softwear Automation". These SewBots were created utilising cutting-edge industrial 4.0 revolution technologies, like sophisticated robotics or computer vision, which can analyse and work with cloth much like a person. SewBots can complete a variety of garment production operations, including fabric cutting, sewing, waking, labelling, and inspection, all under the direction of a single touch screen. In a practical example, SewBot increases the production of T-shirts by 71% compared to human sewing line [2,3,6].

Recently, a 3D sewing technology was created by Philipp Moll GmbH & Co. that could automatically construct 3D seams. The robotic arm swiftly scans fabric pieces using a laser scanner, stitches them using pre-designed patterns, and snips the thread ends. 3D sewing robotic arm was also built in China. The sewing of automobile interiors is presently done with 3D robotic arms. Airbag materials, car seat coverings, and apparel (trousers, jackets, and shirts) may all be created using 3D stitching technology. With the aid of this 3D technology, higher-quality, more productive sewing items may be produced. Additionally, 3D stitching technology lowers labour expenses. Robotic 3D sewing technology can open up new sewing possibilities and produce top-notch clothing [7,12].

There are currently a number of cutting-edge technologies on the market, including machine for cutting and twisting collar points, cuff and collar shaping press, automatic velcro attaching machines for jackets, microcomputer-based button holing machine, automatic button feeder and attaching machine, program controlled single-needle lock stitch machine and automatic thread trimming, automatic pocket and sleeve creasing machine, automatic label cutting, shirt pressers, jacket finishers, pressing robots for shirts, and automatic poly bagging machine.

Even in the 21st century, the majority of clothing industries still require a significant amount of labour to perform several steps in the production process, including cutting, sorting, and transportation between other steps. Some automated devices can carry out a few linkages, such as using sewing machines to stitch cut parts together [11]. Therefore, a mix of a sizable work force and some automated gear is still required to produce clothing on an assembly line. According to the machine vision hypothesis, cutting components can be automated using industrial robot technology, eliminating the need for manual labour. The robots' ability to properly determine the size and form of the cutting item allows them to cut it in accordance with scanning results.

Automatic cutting sorting requires the classification of several parts that have been mixed together as well as the division of the elements needed for a ready-made garment. In actuality, AI and Internet of Things technologies enable automated sorting. With this technique, cloth rolls are allocated QR codes and then joined together. The grasping robots may scan a QR code on an item of clothing during sorting to read its information and the area in which it should be placed, then leave it with other clothes before moving it to the next stage. When compared to manual sorting, this technique significantly reduces labour and time expenses [6,11].

Although the sewing process can be greatly simplified by today's sewing machines, the operators must still regulate the cutting parts to ensure that they travel through the sewing machines uniformly and smoothly. Through programmable intelligent robots, we can fully automate this procedure. All that is required to get the robot to function properly is to upload the mobile cutting instruction to it in the form of programming [1,7].

### 12.1.3 Robots for Material Handling and Manufacturing

In order to boost production and reduce labour costs, material handling is crucial in the textile and apparel sectors. The majority of material handling in the production of clothing entails moving fabric, cut components, trimmings, and completed garments from one work station to another. Automated handling of materials entails the use of the proper tools, machinery, or even robots to move the materials.

Robotics, particularly in the creation of clothing, is a unique field of research that calls for knowledge in the fields of mechanical engineering, machine learning, and fashion. Automation and robotics are not the same thing. Software alone can automate processes; no actual robotic gear is necessary. Despite being frequently utilised for work automation, robotics may also be employed for other purposes. The main objective of automation is to have a machine carry out activities automatically [3,7]. Robotics often takes a more flexible approach with the aim of producing solutions in an unperfect reality. The difference is often that robots may be employed for a variety of difficult operations, not just one. They can detect and respond to signals from the outside environment. On the other hand, a pocket setting, which can only set pockets, may be mentioned in the context of industrial automation approaches. In certain production environments, if something should be configured improperly, the machine will not function. Part of what distinguishes a robot from other forms of automation is its autonomy to react depending on perceiving the outside world [11,12].

Beyond the production floor, robots are employed in the fashion business. Already, picking and packaging tasks are being handled by robots in warehouses all around the world. Billions of dollars have been spent on robotic automated infrastructure by companies like Tencent, Alibaba, and Amazon to handle the repetitious process of discovering inventory products and packing them into boxes to dispatch to clients. A well-known illustration of robotic warehouse automation is Kiva Systems. Amazon bought them in 2012, changing their name to Amazon Robotics. Major companies in the fashion business, such as Saks Fifth Avenue, Gilt Groupe, The Gap, and others, were using them prior to the takeover [8,9].

Similar businesses like Kiva Systems exist today, including AutoStore, Dematic, and OPEX. These robots are often seen as belonging to the school of thinking that holds that people should work with robots rather than being replaced by them. Collaborative robotics is the term used to describe this concept. Industrial robots have often been big, metal robots. As a result, they were harmful to people since they might easily accidentally hit someone who was in front of one of them. To prevent this, certain soft robotic collaborative robots are constructed of soft materials. Others could be smaller, lighter, and less powerful, and have better sensing systems [2,3,12].

### 12.1.4 Manufacturing Software for Customer Relationship Management

Customer relationship management (CRM) means to create long-lasting customer connections and great customer experiences by gathering, managing, and utilising data wisely. Appropriately managed data that are collected from all client interactions assist businesses in developing targeted marketing replies, coming up with fresh ideas, customising products and services, and ultimately providing superior value to customers and a competitive edge in the market. The amount, pace, diversity, and processing power of data have increased in the digital era, resulting in new technological solutions, such as the development of AI methods [5,6]. AI is the capacity of a machine to precisely read a bulk of data, learn from that data, and use those learnings to achieve specific tasks and targets.

AI technology advancements benefit both businesses who create CRM systems and businesses that utilise CRM. These advancements are now necessary to succeed in the CRM environment. In fact, new CRM features like personality insight services, chatbot services, website morphing, image and facial recognition technologies, and programmatic advertising demand a significant amount of data to be crunched in real time, without AI advancements which is nearly impossible to implement [6,13].

Academics assert that AI is the next step toward a unique and competent administration of client interactions, in addition to its importance in the commercial sector. Application of AI-based solutions to CRM empowers businesses to better assimilate customer data, increasing their capacity to anticipate plan and take advantage of imminent opportunities. CRM is the output of ongoing evolution and integration of marketing ideas with newly available data, technologies, and organisational forms.

### 12.1.5 Fashion on Demand

Due to their short lifespan and high volatility, which are caused by the constantly shifting fashion trend, it is harder to predict consumer demand for stylish goods. Fashion sales forecasting has been successfully carried out using a variety of AI-based hybrid methodologies. Basically, AI considerably simplifies decision-making in the area of trends forecast in the fashion sector [5,9].

The idea of mass customisation is the process of creating products and services to satisfy unique consumer requirements. Due to its widespread use, mass customisation has become a common technique across most sectors. Customers frequently struggle to obtain appropriate clothing in the fashion sector that meets their demands. Sometimes they may feel uneasy wearing dresses due to the colours, styles, patterns, or any other factors. Therefore, numerous studies recommended a method of user-driven customisation where the buyer may design his or her own garment.

#### 12.1.5.1 Customer Experience That Is Customised

Customers benefit from a distinctive experience that is secured by AI-driven fashion software. It encompasses the following options: virtual changing rooms, personalised offers, and digital assistance. Fashion firms frequently deploy an AI-driven function that lets customers virtually try on clothing using their phone's camera, which gives better idea to them whether an apparel fits or not.

Virtual assistants like Google Home or Amazon Alexa may be familiar to you. Instead of painstakingly looking for stylish things, customers may speak with one. These helpers might be quite beneficial and save a lot of time. The analysis of client queries is a strength of clever algorithms that allows the creation of individualised product based upon the customer's preferred size and colour suggestions. These are more methods for using AI to personalise and interactively enhance the shopping experience [1,4,5].

Researchers and practitioners now have access to a data mine owing to social media platforms and e-commerce sites in recent years. Images, words, hashtags, likes, shares, and comments all reveal details on the current activity of users. To understand user opinions and the future of fashion, several fashion firms monitor customer dialogues in the Web 2.0 environment. On social media platforms like Instagram, certain design firms and trend-forecasting businesses post their most recent collections in order to gauge consumer interest [10]. Picture-sharing websites like Pinterest, Tumblr, and Instagram are assisting in facilitating the flow of fashion knowledge by keeping brand customers abreast of the most recent style information. Additionally, individuals are adopting fashion icons extremely, which democratises the industry.

Artificially intelligent chatbots are designed to understand human language and are even able to respond appropriately to inputs that are not pre-defined. The majority of chatbots fall into one of two fundamental categories: artificially intelligent and programmed. Only a predetermined set of rules may be followed by scripted/programmed chatbots. While, more recently, chatbots have the ability to include visuals in a discussion along with text. The image-based characteristics and visual search are made possible by computer vision algorithms. While many businesses produce chatbots for everyday usage, some focus exclusively on developing chatbots for retail applications. Because a general-purpose chatbot could become perplexed while talking fashion or retail with a consumer, these chatbot services are more likely to assist fashion merchants.

Since adopting new technology in 2018, Alibaba, a global leader in fashion retail, has changed consumer buying. The business unveiled in-store features with the opening of their first FashionAI store, including Bluetooth chips built into each and every item of merchandise, smart clothing tags, and intelligent mirrors. Customers now have access to AI-driven fashion recommendations tailored to their personal interests due to these technologies. Additionally, FashionAI data may be connected with the business' smartphone app owing to omnichannel technology, creating a smooth and uniform user experience [6].

Tommy Hilfiger was the first to launch the project "Reimagine Retail", which gives fashion designers access to AI-based design capabilities. In order to create customised apparel, fashion students might proficient a diversity of technological skills including natural language processing or computer vision. With the use of AI, fashion students might get knowledge from thousands of photos that are directly relevant to the industry, enhancing their creativity and speeding up lead times for the industry [5,6].

The online shopping experience has been completely transformed by Amazon's AI-powered product suggestion engine. An AI-empowered fashion design algorithm introduced by Amazon can create clothing by duplicating the design elements of several in-demand garments and applying them to a brand-new article of clothing. Amazon's Echo Look fashion assistant, which can offer tailored recommendations based on machine learning, enables its second use case [3,6].

An augmented reality (AR)–based full-featured clothing customisation system was proposed by recent study done in 2021. This system incorporates the installation of the material library for the selection of styles, colours, patterns, and fabrics for the users to pick. Users may further customise the system by fitting themselves into virtual clothing in addition to the custom designing process. The system makes use of a somatosensory virtual changing room, OpenGL 3D graphics, and Azure Kinect somatosensory technology. The user may add several designs to the fabric, and when satisfied, he can try a virtual fitting. The system is made up of useful components. The pattern, colour, fabric, and style selections are part of the design sub-module. The virtual fitting sub-module uses the Azure Kinect somatosensory camera to record the depth data stream of human bones, enabling bone-tracking processing and virtual try-on of 3D clothes with accurate positioning on the human body. Then, as part of the rendering process, it selects the best tools for producing clothing textures, lighting, and other elements so that viewers may view virtual clothes with more life-like results. The hand-made module is used and necessitates hand-made by qualified experts [11,14].

### 12.1.6 AR in Apparel Design

Physical and digital shopping will be reconnected via AR. Because of its interaction, virtual reality (VR) and AR appear to be the newest hot technology in fashion retail and fashion events. This new strategy known as Seamless Retail, which means to build experiences with any type of touchpoint, such as media, websites, and catwalks, goes beyond just browsing social media, e-commerce sites, and stores. Trip is the essential concept behind this new fashion movement, not just in the sense of a fun shopping experience but also in the sense of a true aesthetic experience that even includes our sensibilities. As fashion brands grow and become more numerous every day, there is fierce rivalry to win over customers. VR and AR appear to be the next way to make online experience unique, set a brand apart from rivals, and convey compelling and immersive stories [3,15].

In addition, fashion consists of two core elements: an impalpable element and material element. Impalpable elements refer to the cultural significance of fashion and the capacity of fashion to incarnate desires and imagination. The material element refers to the physicality of fashion items and addresses materials and fabrics, seams, and other quality issues. VR and AR can do far more and elicit additional physical and sensory sensations than internet stores with photographs of apparel can, which can only transmit the impalpable side of fashion of a particular piece of clothing. We may characterise VR and AR as creating a sensual and aesthetic experience.

A VR environment will encourage the physical and multisensory interaction of fashion-related trade businesses, shopping service content to customers, accurate and realistic product presentation, and facilitating a real purchasing capacity. A thorough examination of the type of experience given to clients as they shop for fashion items utilising VR has not yet been conducted. Many retail firms have already used AR technologies in their applications and in-store interactions. Businesses who use AR technologies stand to gain greatly.

Specifically, the COVID pandemic accelerated the concept of AR and VR for e-commerce websites, because of that fashion customers had to shop online. Not only would seeing a product in VR be helpful for size testing, but it can also be used to exhibit things in a more realistic way, encouraging buyers to purchase more and with fewer uncertainties. The GAP "DressingRoom" app, which uses AR to let customers try on clothes wherever they are using a Google Tango-enabled device, allows users to create an avatar based on their body size and type and check how various items of clothing will look on them from various angles. If they like what they see, they can also purchase the item directly from the app [3,15].

The COVID epidemic has caused several changes, forcing fashion week and other events to reconsider their format. During the epidemic, the customary catwalks, photo shootings, and mass events were not possible. The biggest events need to be reinvented, and the fashion industry chose to use technology to accomplish so. London Fashion Week, the first all-digital fashion week in Europe, utilised internet platforms for new designs, podcasts, virtual showrooms, and short films. However, digital fashion weeks lacked a sensory approach, whereas VR and AR may simulate intense involvement and engagement [3].

The ORDRE firm offers services to construct a cutting-edge VR fashion show. The business can give customers inside the retail network VR headsets so they may view shows and fashion parades from the front row from anywhere in the world. The 360° virtual showroom that ORDRE offers enables fashion businesses to develop collections in a realistic way. Customers may move the 3D models to zoom, spin, and examine every aspect of the clothing in addition to viewing fabric and silhouette information [12]. Customers can genuinely live and enjoy the collection and clothing because of this interaction and realism.

VR and AR in particular have the potential to become a widely used technology in the last stages of the value chain, particularly for online retail concerns like virtual fit and the gamification of the shopping experience. It might be claimed that VR and AR may represent the internet shopping platforms of the future and that they will be quite inventive in their attempts to replicate a sensory purchasing experience [1,14].

Virtual try-on is a component of AR in fashion retailing that simulates the look of clothing product combinations on a human form using customised or non-personalised virtual models. Since the "Magic Mirror" idea is the widely utilised virtual try-on solutions, it may be applied in actual clothes stores. A big screen is used by "Magic Mirror" to show the user several clothing while emphasising hair and makeup simulations. In this approach, a touch mechanism is used for the interaction [15]. In order to provide customers more realistic try-before-you-buy clothes buying experience, a more recent study recommended a more sophisticated virtual trial room incorporating Internet of Things (IOT) mirrors.

VR fitting rooms may be used for fashion gamification. In the research "Fashion Island", a prototype of a VR fitting room was evaluated by young fashion clients, where members of the group as avatars could point, observe, and wear clothes in a calm setting similar to that of an island in open water. Obsess, a company, is able to design VR stores that are accessible even from a phone. Obsess can transform virtual spaces into stores and showrooms for customers, but the firm also provides AR marketing experiences like pop-up AR stores in public places. The list of Obsess' clienteles is extensive, and the company offers high-quality service [3,13].

Designer labels aren't far behind and have begun to present limited-edition collections and provide in-person experiences. One of Australia's top bridal couture manufacturers, Suzanne Harward, has launched the Suzanne Harward Virtual View smartphone application, which makes shopping for bridal gowns for brides a simple and alluring experience. The app uses innovative AR technology to let users examine unique bridal couture collections on "life-size models" at their convenience, making for a pleasant and interesting experience. Intricate features, superior fabric, and a 360° view of the model allow brides to zoom in and visualise how each gown will look in real life. This opens the door for more confident and educated choice, which has been shown to be beneficial to both retailers and consumers [7,14,15].

Similar to this, Wanna Kicks allows customers to try on sneakers virtually, which seems usually like consumers do in physical stores, and they believe that AR try-on may carry away the difference between online and offline shopping. By 2022, 3.5 billion individuals, or "44%" of the world's population, will be utilising AR, according to a Digi-Capital prediction [15].

## REFERENCES

1. Noor, A., Saeed, M. A., Ullah, T., Uddin, Z. and Khan, R. M. W. U. (2021), "A Review of Artificial Intelligence Applications in Apparel Industry", *The Journal of the Textile Institute*. https://doi.org/10.1080/00405000.2021.1880088.
2. Abd Jelil, R. (2018), "Review of Artificial Intelligence Applications in Garment Manufacturing", in *Artificial Intelligence for Fashion Industry in the Big Data Era*, Springer Series in Fashion Business, Thomassey, S. and Zeng, X. (eds), 97–123. Singapore: Springer, ISBN 978-981-13-0080-6.
3. Silvestri, B. (2020), "The Future of Fashion: How the Quest for Digitization and the Use of Artificial Intelligence and Extended Reality Will Reshape the Fashion Industry after COVID-19", *ZoneModa Journal*, Vol. 10, Issue 2, 61–73. https://doi.org/10.6092/issn.2611-0563/11803.

4. Luce, L. (2019), *Artificial Intelligence for Fashion: How AI Is Revolutionizing the Fashion Industry*, Apress, eBook ISBN: 978-1-4842-3931-5. https://doi.org/10.1007/978-1-4842-3931-5.
5. Banerjee, S. S., Mohapatra, S. and Saha, G. (2021), "Developing a Framework of Artificial Intelligence for Fashion Forecasting and Validating with a Case Study", *International Journal of Enterprise Network Management, Inderscience Enterprises Ltd*, Vol. 12, Issue 2, 165–180.
6. Lu, Y. (2021), "Application of Artificial Intelligence in Clothing Production", Advances in Social Science, Education and Humanities Research, Vol. 638, *Proceedings of the 2021 International Conference on Public Art and Human Development* (ICPAHD 2021), 364–367, DOI:10.2991/assehr.k.220110.071.
7. Akram, S. V., Malik, P. K., Singh, R., Gehlot, A., Juyal, A., Ghafoor, K. Z. and Shrestha, S. (2022), "Implementation of Digitalized Technologies for Fashion Industry 4.0: Opportunities and Challenges", *Scientific Programming*, Vol. 2022, 17 pages. https://doi.org/10.1155/2022/7523246.
8. Chang, A. A., Cynthia, D., Ramadhan, J. F., Adnan, Z. K. S., Kanigoro, B. and Irwansyah, E. (2021), "Fashion Trend Forecasting Using Machine Learning Techniques: A Review", in *Data Science and Intelligent Systems. CoMeSySo 2021. Lecture Notes in Networks and Systems*, Vol. 231, Silhavy, R., Silhavy, P. and Prokopova, Z. (eds). Cham: Springer. https://doi.org/10.1007/978-3-030-90321-3_5.
9. Chen, I. F. and Lu, C. J. (2021), "Demand Forecasting for Multichannel Fashion Retailers by Integrating Clustering and Machine Learning Algorithms", *Processes*, Vol. 9, 1578. https://doi.org/10.3390/pr9091578.
10. Gu, X., Gao, F., Tan, M. and Peng, P. (2020), "Fashion Analysis and Understanding with Artificial Intelligence", *Information Processing & Management*, Vol. 57, Issue 5. https://doi.org/10.1016/j.ipm.2020.102276.
11. Sikka, M. P., Sarkar, A. and Garg, S. (2022), "Artificial Intelligence (AI) in Textile Industry Operational Modernization", *Research Journal of Textile and Apparel*. https://doi.org/10.1108/RJTA-04-2021-0046.
12. Jindal, H. and Kaur, S. (2021), "Robotics and Automation in Textile Industry", *International Journal of Scientific Research in Science, Engineering and Technology*, Vol. 8, Issue 3, 40–45. https://doi.org/10.32628/IJSRSET21839.
13. Mohammadi, S. O. and Kalhor, A. (2021), "Smart Fashion: A Review of AI Applications in the Fashion & Apparel Industry", https://doi.org/10.48550/arXiv.2111.00905.
14. Jayamini, C., Sandamini, A., Pannala, T., Kumarasinghe, P., Perera, D. and Karunanayaka, K. (2021), "The Use of Augmented Reality to Deliver Enhanced User Experiences in Fashion Industry", *Presented at: IFIP Conference on Human-Computer Interaction*, Bari, Italy, 30 Aug–3 Sep 2021. Proceedings of the Multisensory Augmented Reality Workshop, Interact. Lecture Notes in Computer Science. Vol.12936, Springer Verlag. https://ris.utwente.nl/ws/files/276566325/Karunanayaka2021multisensory.pdf.
15. Mohammadi, S. O. and Kalhor, A. (2021), "Smart Fashion: A Review of AI Applications in Virtual Try-on & Fashion Synthesis", *Journal of Artificial Intelligence and Capsule Networks*, Vol. 3, Issue 4, 284. https://doi.org/10.36548/jaicn.2021.4.002.

# 13 Innovative Approaches and Key Trends in Smart Textiles

*Yamini Jhanji Dhir*

## 13.1 INTRODUCTION

Smart textiles are state-of-the-art textiles designed and developed with innovative and emerging technologies to offer enhanced functionality over and above the traditional textiles. Smart textiles exhibit the ability of interacting, responding and adapting in accordance with the changing environment or wearer inputs. The smart textiles feature smartness owing to multiple functions, namely sensing, feedback, information recognition and accumulation, response, self-diagnosis, self-repair and adaptation, all accomplished by them (Figure 13.1).

The integration of technology into fabrics has paved the way for a myriad of smart applications in textiles, fashion, healthcare and sports. Designers are in the quest to experiment with smart fabrics to create functional yet stylish ensembles. The smart

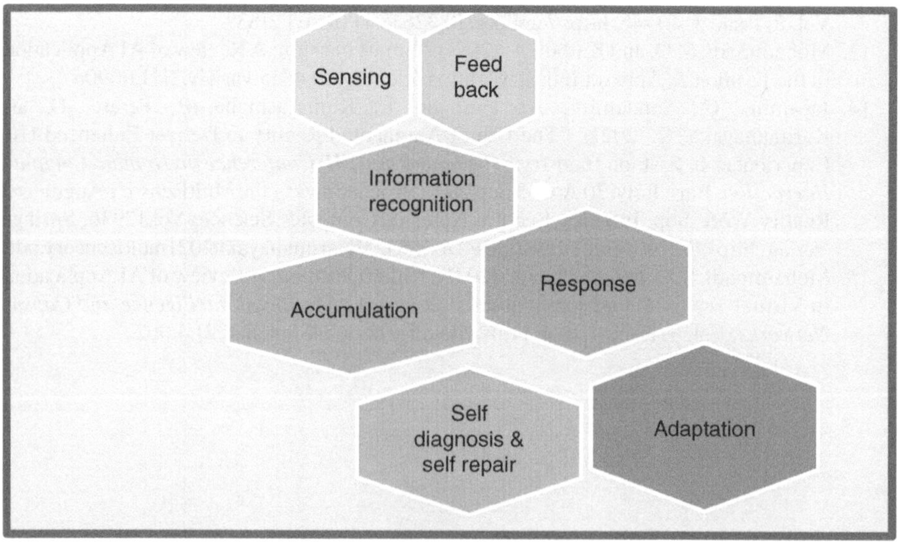


**FIGURE 13.1** Functions of smart textiles.



DOI: 10.1201/9781003432739-13

wear responding to music and lighting up, jackets with built-in navigation and sportswear monitoring wearer's physiological parameters are testament of colossal potential of new genre of smart clothing that is not just visually stunning but is practical too. The new generation of smart textiles are designed and developed utilizing a myriad of fibres, yarns and fabrics featuring exceptional mechanical, chemical, electrical and thermal performances. The integration of electronic components, nano materials, phase change materials and shape memory materials capable of responding to external stimuli enables the development of a range of smart textiles showcasing a spectrum of functionalities (1–3).

Some state-of-the-art high-tech garments that are transforming the whole concept of clothing include anxiety detection clothing and attires featuring facial concealing attributes.

Designers in the recent times attempt to test the limits of clothing by incorporation of innovative and established technologies to create unconventional attires possessing the ability to expand as per wearer's changing physiologies, change colour, show agility and can even show notifications akin to smartphones. It can thus be recapitulated that the rapidly expanding sector of smart textiles offers radical solutions for multitude of domains and resonates the meeting point of fashion and technology (4–6).

## 13.2 KEY TRENDS IN SMART TEXTILES

Key trends pivotal in shaping the realm of smart textiles are enlisted below:

i. Integration of sensors and actuators

   The integration of sensors and actuators aids in development of smart ensembles that are sensitive and adaptive to external stimuli. The intricate interaction between fabric and technology so achieved not just broadens the application spectrum of smart textiles but is beneficial for the user too. The data pertaining to user's surroundings and ambient conditions can be accumulated through varied sensors like temperature, humidity and proximity sensors. Likewise, actuators are essential in smart textile systems to utilize data obtained from sensors and offer dynamic response such as changing the shape, colour or temperature of the clothing.

ii. Development of novel fabrics

   Novel materials intended for specialized functionalities are conductive, stretchable and breathable and thus outshine in contrast to their conventional counterparts. The novel fabrics have emerged as key trend in smart textile domain as they not only serve as perfect substrate for embedding electrical components but also are comfortable and safe for the users. Hence, novel fabrics play a pivotal role in developing flexible and efficient smart textile solutions suited to the needs of myriad of industrial setups.

iii. Internet of Things (IoT) integration

   The integration of smart textiles and IoT is synonym with convergence of textiles and the digital world broadening the application spectrum of smart textiles from interactive textiles reacting to smart home surroundings to health monitoring synchronized with wearables. The integration of IoT with smart textiles not only enables smooth communication with numerous systems and gadgets but also offers user-friendly smart interface. The data exchange and command interpretation by smart textiles has simplified considerably by IoT integration.

iv. Use of artificial intelligence (AI)

AI plays a significant role in the field of smart textiles since data evaluation from embedded sensors and prediction of wearer behaviour can be easily accomplished by AI. The user interaction with smart textiles can be increased based on inputs received from AI. The features can be customized to individual tastes thereby enhancing personalized user experiences through AI data processing.

v. Increasing demand for wearable technology

As more and more population is committed to healthy lifestyle options, each individual is in the quest to track and monitor physiological parameters to avert any medical exigency. Accordingly, there has been an upsurge in the demand of wearable electronics that can effectively monitor various health parameters and ensue early detection of any underlying medical condition.

vi. Sustainable approaches for smart textiles

The world-wide clamour for adoption of sustainable approaches in textile supply chain has impacted the design and development of smart textiles as well. The rising awareness towards eco-friendly raw materials and production techniques fosters the stakeholders associated with smart textile domain to adopt sustainable practices in design and development of smart wear.

vii. Widespread acceptance in healthcare

Smart textiles are gaining traction in healthcare sector for monitoring patient's vital parameters, administering medication, wound treatment and early detection of diseases (7–10).

## 13.3 CLASSIFICATION OF SMART TEXTILES

Smart textiles are classified on the basis of their response to the external stimuli and are generally categorized as passive smart, active smart and ultra smart wear.

Passive smart textiles often referred to as first generation of smart textiles have the ability to sense external factors; however, they cannot adapt or change themselves as per perceived information. The fabrics primarily focus on monitoring and data collection rather than any active change in their properties. Passive smart materials are developed utilizing conductive, responsive textiles and nanomaterials.

*Passive materials* although are not as dynamic in their responses compared to active and ultra smart counterparts; however, their sensing and data collection capabilities provide valuable insights for a gamut of applications ranging from UV protection, air quality sensing and environmental monitoring to military wear, protective clothing and healthcare. Likewise, apparels designed for antimicrobial protection, plasma-treated clothing and optical-sensor-embedded fabrics are some examples of first generation of smart materials.

Undoubtedly, they showcase functionality beyond their traditional counterparts but cannot outshine the active smart materials in terms of sensing and adaption. A classic example of passive smart material is a cooling fabric that aids in temperature regulation but fails to provide cool sensation to wearer. The fabric structure simply enables quick liquid evaporation.

*Active smart textiles* exhibit the ability of sensing and responding to external conditions or stimuli. These textiles comprise of high-tech materials like shape memory alloys, electroactive polymers and thermochromic dyes, enabling them to change their colour, shape or stiffness in accordance with the external stimuli. They possess the ability of memorizing shape and structure along with retaining the chameleon effect and regulating consistent temperature. Some other smart features attributed to second generation of smart textiles include resistance to water ingress, vapour absorption, heating fabric of clothing and heat storage. The versatility and prompt responsiveness of active smart fabrics to the external factors make them potential candidates in interactive user experience, thermal regulation and adaptive clothing.

*Ultra smart textiles*, the third generation of smart textiles, are class apart with sensing, reacting and adaptation capabilities. The ability of ultra smart textiles to review number of data types, make forecasts and adapt to ambient conditions without the need of predetermined tuning make them outshine in contrast to the first two generations of smart textiles. The built-in microcomputer in the smart wear enables them to function like human brains. Spacesuits, sportswear, musical jackets and wearable computers are some examples of ultra smart textiles.

Another classification of smart textiles can be based on the specific textile composites and other high-tech, functional materials utilized for their design and development. There have been relentless collaborative efforts from researchers and manufacturers for integration of cutting-edge technologies, like the incorporation of conductive materials, shape memory polymers, phase change materials, chromic materials and nanomaterials into textile structures for enhanced functionality and versatility of smart textiles (Figure 13.2).

The accomplishment of designated function in a manual or programmed condition by smart textiles is achievable by three components: sensors, actuators and controlling units. The sensors serve as peripheral nervous system for signal detection and thus sensors are essential component for passive smart textiles. The detected signal is acted upon either directly or through a central control unit and thus active smart textiles for their seamless performance require both sensors and actuators. A central processing unit with cognition, reasoning and activating capacities are vital for ultra smart textiles. A robust smart textile system can thus be built by the incorporation of multiple functions like sensing, actuation, powering, communication and data processing. Nevertheless, the most sought after garment properties like flexibility, comfort and washability cannot be trade off with incorporation of smart features (11–13).

The succeeding section discusses some of the high-tech garments that are perfect amalgamation of smart and aesthetic features.

*Smart knitwear by the name Manifesto Collection* launched by Italian Startup Capable is pragmatically developed using AI algorithms to safeguard wearer's biometric data by blocking facial recognition software.

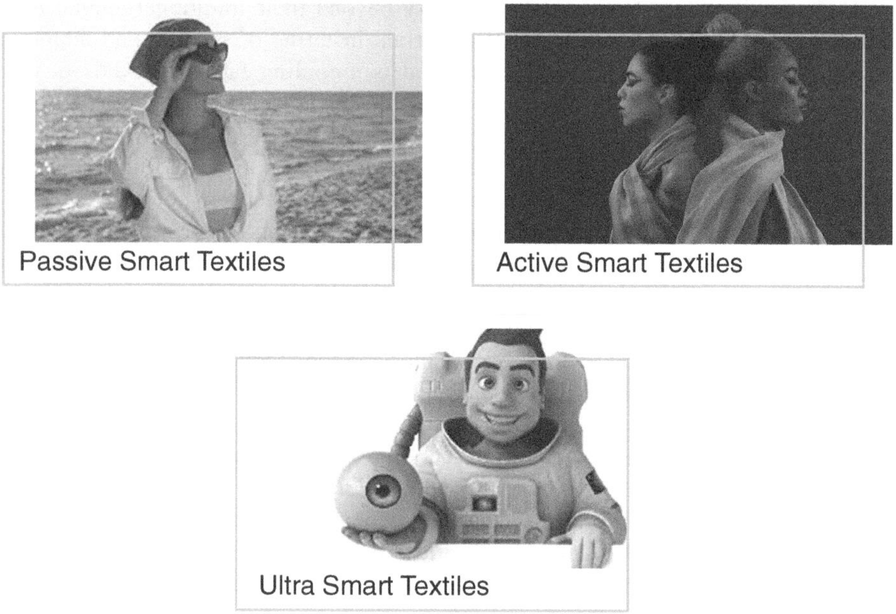


**FIGURE 13.2** Classification of smart textiles.

A pair of polysensory garments also referred to as *Emotional Clothing* created by Polish designer Iga Węglinska are capable of changing colour in response to changing mood and stress levels of the wearer through the inbuilt sensors in the clothing.

*Kinetic garments* – Soft assembly created by Pola Demianiuk with built-in inflatable and expandable elements can curl around the wearer's body. The movement of clothing is credited to soft robotics primarily focusing on technologies mimicking living organisms.

*Wearable technologies of Petit Pli* basically deal in expandable clothing that can stretch as per the wearer's need, particularly pregnant women and kids. The stretchable material utilized for designing clothing enables them to comfortably fit the pregnant women as they approach the full term or as the kids grow rapidly.

*An assortment of robotic dresses* composed of silicone, organza and glass, based on the theme Flowing Water, Standing Time, features rippling, expansion and contraction similar to living organisms. Ying Gao, the creative mind behind the designing of dress, started with an intent to create a smart-sensor-embedded ensemble capable of responding to changing colours in the vicinity.

*A prototype jacket with concealed solar panels* to harvest solar energy was designed by researchers at Aalto University. The jacket is believed to serve as energy autonomous product capable of giving power to wearable devices such as humidity and temperature measuring sensors (14, 15).

Laura Deschl fascinated with yoga and benefits of acupuncture created knitwear referred to as *Healing Imprint*. The ensemble was stitched to create a grid-like structure for insertion of small massage balls on specific acupuncture parts on body like feet, hands, head and body.

*Down-Less Down Jacket* features a typical puffer jacket silhouette; however, it is hollow between the transparent layers unlike the traditional ones that contain feathers between the layers. The jacket keeps the wearer warm for extended periods owing to the infrared-retaining technology in the jacket. The feathers and synthetic materials used for warmth retention in traditional down jacket have been replaced in the smart jacket by Solament material capable of converting the near infrared light into heat thereby instantly warming the wearer (15–19).

The concept of 3D knitting develops three-dimensional shapes unlike the conventional two-dimensional fabrics produced on circular and flat bed knitting machines thereby eliminating the additional processes involved in cutting and sewing room. The concept has been used by researchers to produce a *4D-knit dress* composed of heat-activated yarns. The shape and fit of the tube-shaped dress can instantly be altered by heat application utilizing a robot arm. The dress offers customization as per wearer's style and fit preferences and portray unique silhouettes ranging from a form fitting sheath dress to a voluminous bubble dress. The dress not only promotes minimalistic approach but is also an adaptable and more sustainable alternative in contrast to traditional garments developed by a series of cutting and sewing operations thereby minimizing the cutting and sewing room wastage (Figure 13.3).

The heat-activated yarn used in the making of 4D dress is composed of nylon mixed with a soft blend of viscose and polyester yarn.

The fabric so knitted from the yarn is heat responsive and shrinks with the knitted structure allowing for different behaviours across the garment (20, 21).

## 13.4 INTRODUCTION TO WEARABLE ELECTRONICS

Wearable electronics also referred to as E-textiles are a crucial segment of smart textiles and are intended for myriads of applications owing to their sensing, reacting and adaptation properties in response to external stimuli.

Wearable electronics are versatile textiles and garnering massive popularity owing to potential applications in sportswear, defence, training assessment, medical procedures, health monitoring, rehabilitation, fashion arena and even as global positioning system. The upsurge in demand for wearable sensors in recent times can be attributed to miniaturization of electronic devices, affordability, ergonomics, smartphone connectivity and easy use and care features.

A higher degree of integration and user comfort in wearable electronics is ensued by development and incorporation of flexible and stretchable electronic circuits. A fibre-based flexible textile structure offers unique advantages over a conventional rigid and bulky electronic device with former showcasing the ability of accommodating complex and severe deformations with high stability. Furthermore, textiles are soft, lightweight and breathable ensuing that the embedded electronics closely matches the body contours and are comfortable for the users. The electronic components can be used on textile substrate in numerous modalities such as centrally controlled sensors, actuators, circuit tree added conductive stitched or embroidered fabric, planer electric yarn, data transferring devices, conductive ink printing, and textile bluetooth antenna.

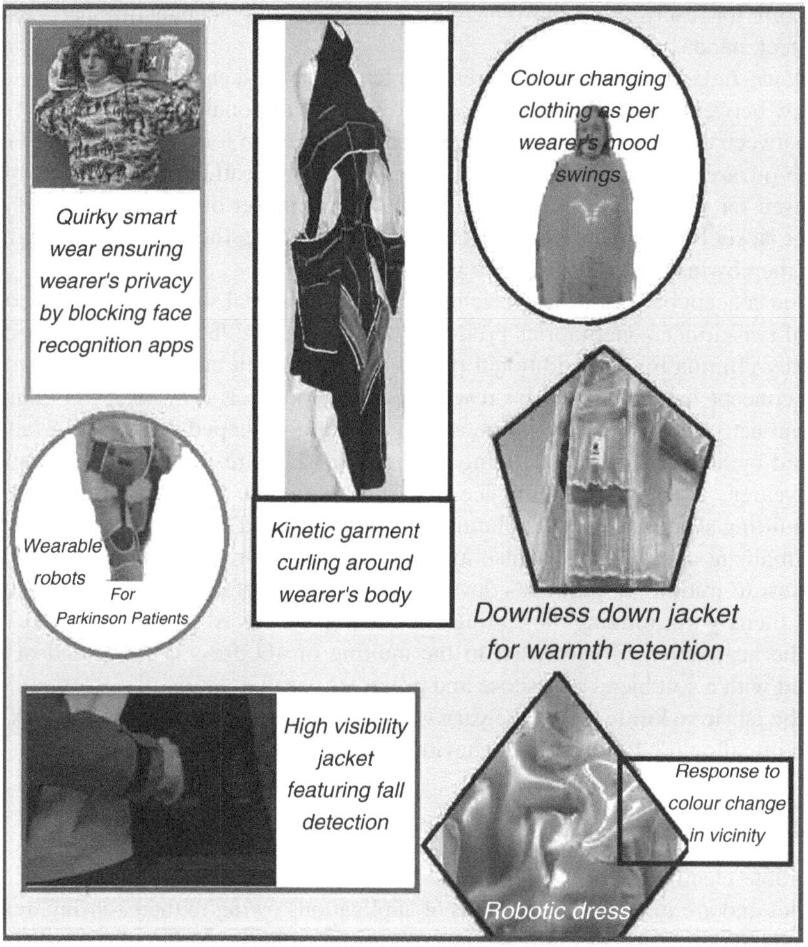


**FIGURE 13.3** Key trends in smart textiles.

The materials generally used to impart conductivity in wearable electronics include single and multi-walled nano carbon tubes and Ag, Cu and Au nanoparticles in sandwiched, coated, shelled, tri-lobal or twisted configurations. Integration of electronic devices and algorithms in E-textiles enables gathering, processing and digitization of data pertaining to human body motion in dynamic, real-time situations for a range of healthcare facilities such as wound healing, electrotherapy and improved blood circulation (22–24).

A range of wearable electronics such as intelligent vest, headbands for facial Electromyography (EMG) and knee sleeves are embraced by users during and after medical operational procedures to avert any health hazards. A trailblazing innovation: Biodegradable smart shirt equipped with data management-based information applications is being successfully utilized for monitoring the patient's vital body parameters such as heartbeat and palpitation rate. Apart from tracking of vital signs, the user's posture and mobility patterns can also be analyzed in real time by the sensors. The sensors capable of detecting changes in pressure, strain and body positioning are integrated in textiles to obtain comprehensive data pertaining to wearer's activity levels and sleep quality. Accordingly, the data can serve as a useful guide for users to embrace healthy lifestyle choices along with improvement in physical rehabilitation and posture.

Integration of electronics in military wear enables detection of enemies and biochemical threats. Apart from functional clothing, wearable electronics are also redefining the fashion ensembles by inclusion of embroidery, printing and other surface embellishment techniques which in first glance appear to be aesthetic appreciation for the clothing but have a concealed electronic component to render unconventional functionalities like geo-navigation, playing music and mobile connect. Phillips Electronics and Levi Strauss have collaborated for fabricating a smart jacket featuring embroidered textile keypad for handsfree driving, mobile connection, answering calls, playing music and geo-navigation.

One approach to develop wearable electronics is by integration of electrical and mechanical components within knitted structures by intermeshing conductive yarns into loops thereby eliminating the need of attaching or glueing electric components over textile substrates. The layout composed of a planar design of an antenna, RFID tag or any other electronic structure is converted into a CAD knitting program. Along with CAD program, stiches represented by grid are prepared. The CAD specifications of the prototype to be developed are exported to the knitting machines to knit conductive fabrics as per CAD specifications. The conductive knit structures so obtained are converted into smart clothing to exhibit varied smart features like monitoring wearer's vital parameters (25, 26).

The following section presents some of the trailblazing development in wearable electronics tailored as per the application area.

## 13.4.1 Latest Developments in Wearable Electronics

### 13.4.1.1 Sensor-Embedded Automatic Urine Detecting Diapers

The latest innovation in wearable electronics is a diaper with embedded film-based sensor prototyped by Toray Industries. One of its kind of diaper not only conforms to diaper contours and offers additional comfort but also serves as a notification alert for care givers regarding change time of the damp diaper. The task of caregivers becomes less cumbersome as it enables them to monitor diaper changing times in real time with a PC, smartphone or tablet. The diaper change time can be accurately predicted and optimized owing to automatic urine detection thereby decreasing diaper consumption. Moreover, the automatic detection of urination timing and real-time monitoring also enables caregivers to encourage the patients to use the washrooms independently.

The sensor embedded in the diaper utilizes a semiconductive carbon nanotube composite-based flexible film with semiconductor circuits mounted above it. Semiconductor circuits are directly printed on the film offering design flexibility and economical production of small lots. Furthermore, the film-based sensor guarantees durability and peeling resistance unlike their silicon counterparts.

#### 13.4.1.2 High-Visibility Parka Jacket with Fall Detection System

PYKRS, a Spanish startup specializing in smart clothing, has recently launched a smart jacket that is a perfect blend of comfort, safety and style for urban commuters.

The state-of-the-art jacket is specifically designed for pedestrians and two-wheeler commuters to overcome the challenges such as bad weather conditions, poor visibility and traffic faced by them while on the go.

The parka jacket features an electroluminescent lighting system covering the jacket's upper torso and offers thrice the visibility compared to any other reflective material. The inbuilt heating system with carbon fibre element in the jacket enables the heating of chest and back area of jacket rapidly.

Another salient feature of the jacket is the sensor and GPS-enabled fall detection system thereby preventing the users from sudden fall or accident raising alarm by sending users' location to their emergency contacts.

A smart light emitting fibre developed by group of researchers in Singapore is expected to address the issues associated with silicon-based textile systems. The fibre showcases the ability to generate electricity, has wireless visual–digital interface thereby eliminating the necessity of extra chips or batteries.

#### 13.4.1.3 Touch-Enabled Embroidery-Based Sensor

A fabric-based sensor capable of controlling electronic devices by touch has been developed merging machine learning and three-dimensional embroidery techniques.

The device comprises of the embroidered pressure sensor and a microchip for data processing and distribution. The sensor is triboelectric featuring the ability to power itself using the electric charge generated as a result of friction between the multiple layers of sensor. The device was versatile, offering multiple functions like controlling video games and music-playing application on mobile through different gestures on the sensor.

#### 13.4.1.4 Smart Shirt for Visually Impaired Individuals

A smart shirt system with inbuilt ultrasonic sensors is developed for visually impaired individuals. Apart from sensor, the shirt is equipped with vibration motors, power supplies and microcontroller. The ultrasonic sensor in the smart shirt senses the ambient conditions and detects the obstacles while actuators guide the user via feedback process interpreted in a signal-processing unit. The smart shirt can serve as a boon for visually impaired people since position of obstacle within the detection rate can be accurately identified.

#### 13.4.1.5 Hue-Changing Jacket

Firefly jacket by Vollenbak, a UK-based clothing brand, possesses the ability to change its hue in response to sunlight and thus offers the advantage of rendering two looks – copper orange indoors and yellow-green in darker vicinities.

#### 13.4.1.6 Nanodiamond Jacket for Thermal Regulation

Nanodiamond coated on cotton fabric-based smart textiles developed at RMIT University, Melbourne, shows reduction in temperature by 2°C–3°C thereby offering rapid cooling effect to the wearer.

#### 13.4.1.7 Wearable Robot for Parkinson's Patient

A soft, wearable robotic device is designed to offer longer stride to patients suffering from Parkinson's disease. The robotic garment is worn around patient's hips and thighs, giving a gentle push to hips as the legs swing thus enabling the wearer to walk without freezing.

#### 13.4.1.8 Temperature-Sensing Modular Sportswear

A modular E-textile sportswear capable of recording skin temperature via inbuilt temperature is gaining traction among sports enthusiasts. The modular design enables convenient disassembling of the system for repair or recycling. Moreover, separate fabrication of electronic and textile components and integration at the final assembly stage simplifies the manufacturing process considerably. (24–27).

## 13.5 SMART TEXTILES FOR HEALTH MONITORING

Smart textiles are transforming the healthcare sector via seamless integration into daily lives for monitoring and enhancing well-being of individuals. A gamut of smart devices and accessories ranging from smartwatches to fitness trackers, smart glasses and smart footwear play a significant role in tracking vital health parameters, activity level, sleep patterns and preventive care.

Early detection and customized treatment plans are plausible because of real-time data collection. The role of wearables in managing and monitoring health parameters and insights on physiological parameters become all the more crucial in chronic conditions. Wearable sensor technology enables dynamic sensing of vital physiological parameters such as heart rate, blood pressure and oxygen rate, respiration rate, skin temperature and physical and brain-related activities. The real-time data related to ambient conditions and physiological parameters can be obtained from body sensor networks via wireless connectivity.

The new genre of sensors offers a wear and forget functionality on account of their design ergonomics, lightweight, wearer comfort, easy detachment, washing and after care convenience to users. The early diagnosis and monitoring the course of the disease and varied biological signals depend greatly on the ergonomics of wearable sensors. The durability and performance of wearable sensors can thus be increased considerably by designing them utilizing mechanical and bio compatible nanomaterials with an ability to mimic biofunctions.

Wearable biosensors have witnessed remarkable advancement as far as their monitoring capabilities are concerned. First-generation wearable biosensors, namely smart watches, footwear and headphones etc., are capable of monitoring wearer's biophysical aspects like activity level, heart rate and body temperature. Second-generation wearable biosensors are capable of biochemical and biophysical monitoring with applications spanning from health and wellness monitoring to disease detection and management. This category comprises tattoos, dermal patches, contact lens etc. and relies on invasive methods like microneedles and injectable devices.

Another classification of biosensors is based on their monitoring mechanism. Accordingly, biosensors can be categorized as temperature, pressure, optoelectronics, ultrasound, electrochemistry and magnetic field-based biosensors.

The fabrication of wearable sensor irrespective of their monitoring mechanism should take into consideration the biocompatibility of the sensor with the wearer's skin.

The sensors being worn on the skin should not only be comfortable but also adaptable, flexible and precise in their sensing abilities. The most suitable and versatile sensors are expected to conform to the body contours and thus generally utilize skin-inspired and patterned mesh designs in their construction. Hydrogels, textiles and paper serve as an effective biosensor owing to their flexible, stretchable and breathable substrates. The materials thus find widespread acceptance for development of wearable biosensors with effective monitoring of pH and sweat biomarkers (28, 29).

It can thus be recapitulated that the technological and design advancements in smart textiles and wearable biosensors have revolutionized healthcare sector by promotion of preventive strategies, cost reduction and improvement in overall health outcomes. The succeeding section discusses some trailblazing applications of wearable electronics deployed for health monitoring.

### 13.5.1 Application of Wearable Sensors in Health Monitoring

#### 13.5.1.1 Wearable Body Sensor Network for Monitoring Patient's Vital Parameters

Wearable body sensor network consists of wearable or implantable biosensors inside patient's body. A cluster of sensors carefully positioned on each organ for which vital parameters ought to be recorded send their readings as electromagnetic radio-frequency waves to group leader to be further communicated to central controller. The patient can thus receive their data on patients' personal home server via central controller that uses wireless protocol for data transmission.

#### 13.5.1.2 Conductive Ink-Based Sensors for Electrocardiogram Monitoring

Electrocardiogram (ECG) of patients can be monitored by textile-based sensor specifically tailored by inclusion of 12 ink-based dry electrodes, 5 respiratory strain sensors and an accelerometer. An adhesive layer, layer of conductive ink, binder, solvent and thickener comprise the electrodes used for monitoring. The anterior surface of the garment is covered with the respiratory sensors composed of an elastic ribbon impregnated with conductive ink. The elastic ribbon features an electrical connector at each end and a cover made of compression fabric.

#### 13.5.1.3 Monitoring of Oedema-Affected Patients by Smart Stocking

Washable stockings being prepped for clinical trials and validation are effective in measuring and monitoring of variations in leg volume of oedema-affected patients prone to fluid accumulation or swelling in the lower limbs.

#### 13.5.1.4 Smart Caps for Epilepsy Treatment

Woven materials with electrically conductive yarns are being used to develop brain caps that can sense brain activity enabling treatment of epilepsy by assimilation, storage and transmission of digital information.

#### 13.5.1.5 Conformable Suit for Monitoring Wearer's Physiological Parameters

E-textile-based conformable suit capable of wireless sensing and distribution is being utilized for monitoring the body's physiological parameters such as human skin surface temperature distribution, heart rate and respiration. A customized fabric capable of integrating with myriad of sensor-integrated circuits (ics) and interconnections in the form of flexible-stretchable electronic strips is utilized for designing the state-of-the-art suit. The two-layer industrial flexible printed circuit board is deployed to develop sensor ics and interconnections. The electronic strips are woven into the channels or pockets of the garment.

*Triboelectric all textile sensor array (TATSA) for dynamic and simultaneous monitoring pulse and respiratory signals.*

TATSA is a textile-based sensor that can effectively monitor pulse and respiratory signals dynamically and can transmit the physiological signals wirelessly to the intelligent mobile terminal application (app) for continual health monitoring.

A smart cardigan knitted fabric comprising of conductive yarn and commercial nylon yarn serves as a substrate for stitching the sensor array (TATSA). The conductive yarn utilized for knitting fabric is composed of stainless steel and terylene yarns that serve as conducting electrode. The salient feature of textile-based sensor is its ability of seamless monitoring without compromising the inherent handle and tactile properties of textile materials with an additional ability to withstand deformation.

#### 13.5.1.6 Life Belt Jacket for Mothers and Babies Inside Womb

Jackets composed of embedded sensors are being deployed to monitor blood pressure and ECG of mother and foetus (Figure 13.4). The jacket is designed with two modules – embedded sensors and electrodes in the jacket and handheld device to record the body parameters on tablet pc. The innovative smart jacket ensures sustained data recording and continuous monitoring. The timely diagnosis is thus possible as the jacket can continuously record physiological parameters and transmit to the healthcare professionals via a communication module.

#### 13.5.1.7 Pregsense for Monitoring Foetus Movement

A wearable sensor-integrated belt, Pregsense, can be draped around mother's tummy to monitor infant's pulse, kicks and position inside the uterus. The heart rates of both mother and growing foetus can be recorded in a smartphone to be later stored on a cloud-based database for doctors to keep a track of growth and development of baby inside the womb.

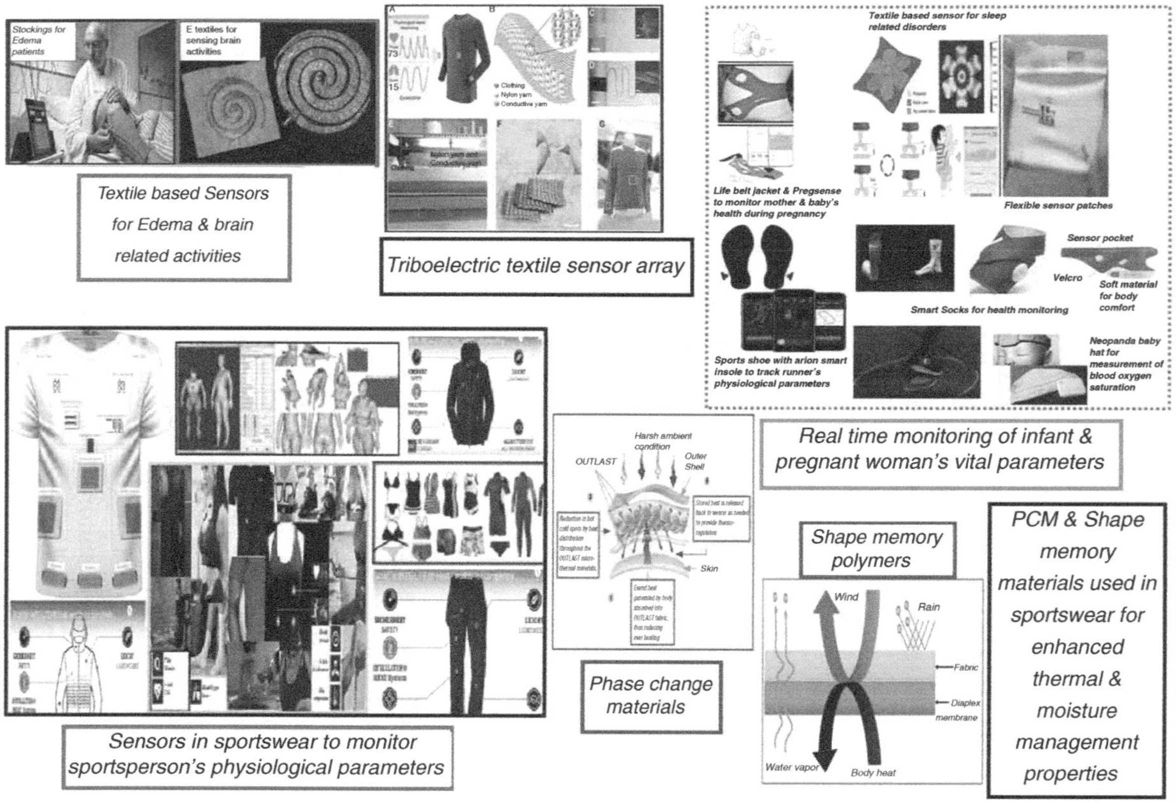


**FIGURE 13.4** Application of wearable sensors for health monitoring.

#### 13.5.1.8 Sensor-Embedded E-Bra for Real-Time Monitoring of Heart Rate and Blood Pressure

The inner garment for women, brassier, has been smartly designed with embedded sensor for real-time monitoring of heart rate and blood pressure (systolic and diastolic) along with data plotting on laptop and smartphone display.

#### 13.5.1.9 Textile-Based Sensor for Sleep-Related Disorder

A textile-based sensor along with wireless biomonitoring system (WBS) is considered to be a promising candidate for diagnosis of sleep-related disorder (obstructive sleep apnoea-hypopnea syndrome). The sensor and WBS system exhibit the capability to process the recorded human pulse wave signals thereby transmitting the information wirelessly and eventually displaying the health data of patient via an app interface in the smartphone.

#### 13.5.1.10 Flexible Sensor Patches for Post-Surgical Procedures

Wireless sensor patches comprising stretchable substrates flaunt of miniaturization and conformability and can be successfully deployed for unobtrusive motion sensing and recording. The flexible electronic-based sensor can sense patient's body temperature and adjust the temperature in accordance with patients' requirements. The sensor serves as a boon for post-surgical operations for patients who feel chill after the process.

#### 13.5.1.11 PUP Smart Socks for Fall Detection

PUP socks composed of conductive yarns feature a textile-based sensor underneath the foot for detection of falls and the balance issues. The socks are equipped with detachable housing including batteries, bluetooth and computer.

#### 13.5.1.12 Owlet Smart Sock for Monitoring Infant's Heart Rate

Owlet Smart Sock based on pulse oximetry technology is an effective smart textile to monitor infant's heart rate, sleeping and breathing patterns. The socks are chargeable via a base station and can be synchronized with smartphone for real-time data delivery (27–29).

#### 13.5.1.13 Sports Shoes with Arion Smart Insole

Arion smart insole sports shoes are regarded the future of running wearables since the footwear maximizes the runner's performance by measurement of interaction between foot and ground via pressure sensors embedded in the ultra-thin insole. Additionally, foot pods on the shoe's exteriors can effectively measure the runner's body movement. The insole and pods of smart footwear can be synched up to user-friendly app thereby serving as personal trainer for the sportsperson.

#### 13.5.1.14 Orthotic Footwear for Stress Relaxation in Ageing Feet

Nike, a popular sportswear brand, has come up with a trailblazing concept of orthotic features to relieve the ageing feet of stress by inclusion of polyurethane capsules filled with pressurized gas in the insole of footwear. The shock diffusion inserts equipped in the footwear ensure impact absorption thereby offering cushioning effect and comfort to the wearer.

## 13.6 FUTURE PROSPECTS OF SMART TEXTILES

The development of smart textiles in the near future holds humongous scope with lots of innovations already redefining and shaping the domain of smart textiles.

In present times, research in smart textiles is primarily targeted towards functional attributes with little attention paid to comfort, washability, folding ability and performance during and after use. Also, there is an ardent need of product diversification as far as smart features and functions are concerned. Concentration of multiple functions in a single product can go a long way in enhancing the diversification of single-product functions.

Adoption of sustainable principles during design and development of smart prototype can result in safe and eco-friendly product. Wearable electronics, in particular, should not pose any potential harm to the user and should be pragmatically developed considering energy conservation and environmental concerns during production process.

Affordable smart wear will ensure widespread acceptance and usage by consumers. Thus, advanced functionalities at affordable prices will strike the right chord among consumers prompting them to embrace the smart clothing.

## 13.7 CHALLENGES POSED BY WEARABLE ELECTRONICS IN HEALTHCARE

The promising avenues of wearable electronics in healthcare cannot be undermined; however, some challenges ought to be addressed in order to harness the full potential of this innovative and emerging domain of smart textiles. A persistent hurdle towards massive adoption of wearable electronics is related to washing, durability and comfort aspects of electronic components integrated into textiles. Furthermore, the power sources currently used for wearable electronics are far from being sustainable, flexible and lightweight. Interpretation and reliability in health settings demands standardized protocols for data collection and transmission. Also, storage and transmission of sensitive health data from wearable electronics should meet the privacy and security norms. Undoubtedly, wearable electronics can revolutionize patient monitoring and healthcare delivery if the challenges posed by them during designing and monitoring are well addressed.

## 13.8 SPORTSWEAR

Apart from healthcare, smart textiles are also transforming the world of sports with professionals and amateur sportsperson as well as individuals with non-sports background embrace professional and leisure sports attires as per their physical activity. Active and leisure sportswear laced with sensors, actuators and responsive capabilities can adapt themselves as per sportsperson's needs and preferences. The merger of technology and fashion resonates an era of smart textiles that can offer remarkable functionalities from temperature and moisture management to colour, phase and shape changing phenomena thus offering a perfect blend of functionality and aesthetics in the sports attire.

Sportswear is a crucial clothing category popular among sportsperson and fitness freaks that find application in professional sports, athletics and leisure activities.

Sports-specific clothing fulfils the wearer's requirements of comfort, dry feel next to skin and safety and protection from external elements. Sportswear and wearable electronics constitute an indispensable segment of smart textiles. Consequently, the field of sportswear has witnessed tremendous innovations and has emerged as one of the most promising avenues for further research exploration. Furthermore, designers, manufacturers and retailers have been joining hands to launch design collection intended for tech savvy and fitness buff consumers who aspire for functionality, aesthetics and smart features in their ensembles.

The sportswear industry is taken by storm by path breaking innovations as far as procurement of raw materials like high-performance and specialty fibres, yarns and engineering of fabric structures like double-layered, elastane and breathable fabrics are concerned. Furthermore, smart functionalities like antimicrobial, antistatic, anti-odour properties; monitoring sportsperson's physiological parameters; incorporation of smart materials like phase change materials and shape memory polymers; wearable sensors; tracking performance record of sportsperson and incorporation of smart technologies like smart coatings, nano technology and wearable electronics are engineered into sportswear.

The classification and requirements of sportswear vary as per the sportsman's level of physical activity, ambient conditions and specific sport. Accordingly, the sports and fitness aficionados prefer including active wear, leisurewear and athleisure in their closets. The unconventional and latest sportswear is often designed and tailored to offer functional, aesthetic and smart features in one ensemble meeting all the stated expectations of consumers (30, 31).

Athleisure is another key and captivating trend in sports arena with a perfect amalgamation of fashion, culture and urban lifestyle thereby presenting a sporty, quirky sports ensemble.

The smart sportswear ensemble is increasingly being incorporated with sensors exhibiting the capability of measuring body functions and biomechanics such as heart rate, breathing, lung capacity and perspiration or the muscle and joint movement. Smart sportswear so designed with integrated sensors can portray several salient and smart features like timely monitoring of any health ailment, reduction or elimination of injuries and prediction of injuries before hands as a result of AI-based analytics.

The succeeding sections shall pragmatically cover the characteristics, raw materials and varied innovations in the sportswear domain.

### 13.8.1 Characteristics of Sportswear and Athleisure

Sportswear demands various performance characteristics depending on multitude of factors; however, generalized performance features for any sportswear irrespective of type of sports include thermo-physiological and psychological comfort, lightweight, agility to wearer, fit, stretchability and easy-care properties.

The requirements for sportswear generally are based on a number of factors enlisted below:

a. Sport type – team or individual sport/professional or amateur sport
b. Level of physical activity – active/passive

c. Ambient conditions – indoor/outdoor sport
d. Specific functionality – wind proofing, antimicrobial etc.

### 13.8.2 Innovative Raw Materials for Sportswear

The utilization of high-performance fibres such as Coolmax®, Thermolite® and Thermocool® in performance and active wear results in increased surface area, better wicking and moisture management and in turn dry, cooler microclimate to wearer. Thermolite® is particularly suitable for cold-weather sportswear due to fabric's exceptionally high thermal insulation and moisture transmission properties. Winter sportswear comprising hollow core fibres possesses the ability to trap higher volumes of static air and thus provide enhanced warmth and wearer comfort without any additional weight or bulk unlike conventional fleece fabrics. The clothing is thus gaining popularity among sportspersons indulging in outdoor, winter sports like ice skating, mountaineering etc.

Dryarn, an innovative sustainable fibre from Aquafil, is a recyclable polypropylene microfibre and a preferred choice for developing sportswear fabrics that are sustainable, soft, anti-bacterial, lightweight, quick drying and comfortable and exhibit high thermoregulatory capacity.

Sportwool®, a two-layered moisture management fabric featuring wool on the inner side and synthetic fibre on the outer side, and Field sensor TM® with brushed inner side are other options for winter sportswear. Field Sensor, high-performance fabric from Toray, is a multilayered structure suitable for varied sports applications. The excellent moisture management properties and wick ability for rapid liquid transmission from next to skin to outer layer can be attributed to the fabric's specially engineered structure with distinct inner and outer layers composed of coarser denier yarn and fine denier hydrophobic polyester yarn in a mesh construction, respectively.

Additionally, thermoregulating materials like phase change or latent heat storage materials capable of sensing varying ambient conditions and responding by changing their phase are increasingly finding application in sportswear where sportsperson is exposed to prolonged, drastic environmental conditions. Outlast technology is involved in the development of microencapsulated Phase Change Materials (PCM)-coated fabric intended for sportswear and other smart textile applications. The sportswear developed with PCM-treated fabrics provides thermal balance and maintains constant body temperature to wearer by absorbing excess body heat at elevated temperatures due to metabolic heat production and releasing it as the temperature drops down during cooling.

The potential of shape memory polymers to obtain effective thermal and moisture management properties was first explored in sailor suit designed for Swedish sailors. The suit based on membrane technology employed waterproof, windproof and breathable Diaplex membrane. The smart membrane can sense the changing ambient conditions and respond by changing its shape, memorizing the original shape and returning to the original, memorized shape accordingly. The membrane undergoes Micro-Brownian motion as it senses elevated temperature thereby creating micro-pores for heat and moisture transmission through the membrane (Figure 13.4).

### 13.8.3 Innovative Wearable Sensors and AI-Based Technologies for Sportswear

The concept of sportswear design and development has been drastically changing with sportsperson anticipating technological features in their attire apart from basic requirements of functionality and comfort. It thus becomes mandate for sportswear designers and manufacturers to conform to the expectations of their consumers and come up with technology-laced sportswear that can serve best of both worlds by offering comfort, protection and other functional attributes along with serving as a personal trainer, activity tracker and can monitor physiological parameters of the sportspersons (Figure 13.4). The new generation of sportswear exhibits such smart features that would be considered fantasy a decade ago.

Smart sportswear like fitness pants feature built-in haptic vibrations and signal the wearer to be agile or hold position as per pulse generated at the stress-prone zones like hips, knees and ankles. The smart pants can be synched to wearer's phone via bluetooth and provide additional feedback through the companion app.

Ambiotex's smart shirt intended for athletes is designed with integrated sensors and clip-on box for recording athlete's data pertaining to heart rate variability, threshold, fitness and stress levels.

Sabine Seymour's smart bra is designed with an integrated invisible biometric sensor and can be effectively worn to track heart rate and workout routines during sporting and gyming activities.

Compression sleeve by Komodo technologies is beneficial for sportspersons and heart patients as the sleeve is capable of monitoring heart rate activity via ECG technology. The sensors integrated in the sleeve can monitor wearer's body temperature, sleeping patterns, workout intensity, air quality and UV rays.

A smart shirt laced with bluetooth smart sensor can be paired with fitness apps like Map My Run, Run Keeper and Strava to capture the real-time data and monitor physical activity and physiological parameters of sportspersons such as intensity and recovery, consumed calories fatigue level and sleep patterns.

Athos deals in range of training clothes intended for gym goers. The ensemble is designed with micro-EMG sensors that can detect muscle movement and in turn transfer the workout data such as sportsperson's muscle effort, heart rate and breathing to a smartphone. The app serves as a personal trainer for fitness freaks by providing insights on correct exercise protocol and injury avoidance.

Hexo skin smart sportswear is designed with embedded textile sensors for monitoring cardiac, respiratory phenomenon, physical training, sleep patterns and mundane activities of an individual involved in rigorous sports. The visualization, reporting and analysis of data become very convenient with Hexo skin as the smart clothing is equipped with an accelerometer to quantify body movements and track heart rate to be viewed in real time. Furthermore, it prevents the sportsperson from over training by determining lung capacity for each activities performed and measurement of stress and training fatigue.

Boltt is a sports tech brand pioneer in design and development of consumer-centric health and fitness clothing and smart shoes laced with stride sensors and activity tracker. The real-time audio feedback and customized workout suggestions generated by the brand's advanced AI ecosystem can provide customized health and fitness coaching to wearers.

An on-demand inflatable-deflatable textile tubing network in a jacket developed by Sympatex can provide the extra warmth to wearer when inflated while the jacket can be deflated to release the air held inside the tubes into the environment as it senses an elevation in body temperature of wearer's activity level.

A Cyberia survival suit intended for arctic environment serves as a personal GPS and monitors the wearer's physical conditions. The suit derives its sensing and monitoring capabilities from an array of sensors and connecting electrodes embroidered onto textile substrate.

Qardio Core, an ECG monitor, is a hardware fitness tracker capable of providing continuous medical-grade data by incorporated sensors in the tracker. The smart tracker is designed for people indulging in active lifestyles but with a family history of chronic diseases. The doctors can analyze the obtained data from sensors to monitor the health record and can act instantly in event of any abnormal data.

Exercise routine for elderly people can be managed by smart-knitted cardigan designed by Dutch designer Pauline van Dongen. The ordinary-looking cardigan is equipped with four stretch sensors comprising conductive yarns and can transmit the information to an app for generating feedback. The obtained feedback serves as a guide for physiotherapist to suggest the best exercising options as per age and physical stamina of the wearer (29–32).

## 13.9 CONCLUSION

The convergence of technology and textiles is cornerstone for bright future prospects for smart textiles with ongoing innovations further broadening their application areas. Smart textiles encompass a spectrum of materials that showcase characteristics of smart fabrics, namely the sensing ability in accordance with environmental changes, adapting to varied conditions and providing interactive functionality.

Furthermore, smart textiles play a crucial role in the advancement of wearable technology, fulfilling the ever-increasing demand of consumers for practical and life-saving solutions in the field of health and fitness. The integration of smart textiles in wearable electronics opens up new avenues for seamless blending of form, functionality and interaction and enhancing the user's daily lives and improvising well-being and lifestyle.

Also, inclination of common masses towards fitness and leisure activities prompts them to buy functional yet aesthetic athleisure. The consumers primarily seek performance, health, hygiene, comfort and safety aspects in their sports attire. Furthermore, smart features incorporated in sportsperson's clothing can enhance their performance by monitoring their physiological parameters and keep them updated on their workout patterns. The quest for technology-laced, sensor-integrated smart sportswear is gaining momentum with the advent of technology and smart apps to monitor wearer's health and well-being.

The future of smart textiles is promising as it is envisioned that trailblazing innovations and developments in this domain are seamlessly blending into everyday lives, making everything from the clothing to sportswear and aerospace, home textiles to protective wear smarter and more intuitive.

Nevertheless, the design and development of smart textiles pose some unprecedented challenges as far as incorporation of sustainable principles and large-scale manufacturing is concerned. Various ways and means are being explored to bring down carbon footprints and waste associated with smart textile production processes. The conscientious efforts not only respond to global environmental concerns but also prompt the consumers to transition towards a greener smart and wearable technology.

## REFERENCES


1. Spain's PYKRS Launches Smart Jacket PYKRS X-TREME Parka, 2024, Technical Textiles, https://www.technicaltextile.net/news/spain-s-pykrs-launches-smart-jacket-pykrs-x-treme-parka-293655.html.
2. German Tech Uses Smart Textiles for Intensive Sensory Perception, 2024, Technical Textiles, https://www.technicaltextile.net/news/german-tech-uses-smart-textiles-for-intensive-sensory-perception-294346.html.
3. Bahadir S K, Koncar V and Kalaoglu F, 2016, *Smart Shirt for Obstacle Avoidance for Visually Impaired Persons, Smart Textiles and Their Applications*, Woodhead Publishing Series in Textiles, Woodhead Publishing, pp. 33–70.
4. Classification and Application of Smart Fabric Textiles, 2025, https://www.testextextile.com/classification-and-application-of-smart-fabric-textiles/.
5. Thomas V, Guy B, Frederick B and Jan Va, Integration of stretchable and washable electronic modules for smart textile applications, *The Journal of The Textile Institute*, 103(10), pp. 1127–1138, https://doi.org/10.1080/00405000.2012.664866.
6. Mozhdeh G and Masoud L, 2017, Overview of wearable electronics and smart textiles, *The Journal of the Textile Institute*, 108, p. 63652, https://doi.org/10.1080/00405000.2016.1177870.
7. Ben D, 2024, MIT's 4D-knit dress changes shape in response to heat, https://www.dezeen.com/2024/03/22/solament-down-less-down-jacket-infrared-heat/?li_source=LI&li_medium=bottom_block_1.
8. Rima S, 2024, The down-less down jacket converts infrared light to heat, https://www.dezeen.com/2023/02/21/high-tech-clothing/.
9. Japan's Toray Develops Diaper-Embedded Urination Sensor, 2024, Technical Textiles, https://www.technicaltextile.net/news/japan-s-toray-develops-diaper-embedded-urination-sensor-293693.html.
10. Arzum E, Ece E, Huseyin S, Esma Y and Meltem M, 2024, Recent developments in wearable biosensors for healthcare and biomedical applications, *Trends in Analytical Chemistry*, 171, p. 117.
11. Kunal S, Jayant K and Pintu P, 2019, Recent advancements in wearable & smart textiles: An overview, *Materials Today: Proceedings*, 16, pp. 1518–1523.
12. Thakkar H, 2022, Innovations and developments in smart textiles, *Journal of Textile Science and Engineering*, 12(6), ISSN: 2165-8064.
13. Tyagi G K, Jhanji Y and Bhardwaj A, 2024, Health monitoring with smart textiles, *Asian Technical Textiles*, 7, pp. 36–40.
14. Dhir Y J, 2025, Crusading into the innovative and sustainable concepts of sportswear design & development, in *Sustainable Sportswear & Activewear*, edited by Rajkishore Kayak, Elsevier, 77–129, ISBN: 978-0-443-18872-5.

15. Dhir Y J, 2025, Diverse sustainable aspects in sportwear design and development, in *Sustainable Sportswear & Activewear*, edited by Rajkishore Kayak, Elsevier, 133–150, ISBN: 978-0-443-18872-5.
16. Jaglan A and Dhir Y J, 2021, Smart sportswear – The activity trackers & physiological parameter monitors for sports enthusiasts, *Journal of Textile Association*, 82(3), 150–154.
17. Dhir Y J, 2023, *Fashion Accessories – A Complete Guide to Raw Materials, Construction Methods and Styles*, Woodhead Publishing in India, ISBN: 8195404871.
18. New Textile Revolution: Smart Textiles and Wearable Electronics, 2025, Datatex, https://datatex.com/new-textile-revolution-smart-textiles-and-wearable-electronics/.
19. Zewe A, 2022, Smart textiles sense how their users are moving, https://news.mit.edu/2022/smart-textiles-sense-movement-0707.
20. Woven Futures: Unveiling the Top Smart Textile Trends, 2025, Verified market reports, https://www.verifiedmarketreports.com/blog/top-7-smart-textile-trends/.
21. New Smart Fibre Can Emit Light, Generate Electricity: Journal Science, 2024, Technical Textiles, https://www.technicaltextile.net/news/new-smart-fibre-can-emit-light-generate-electricity-journal-science-294468.html.
22. Achieving Sustainability in Smart Wearable Electronic Textiles, 2024, Market intelligence on advanced materials, https://www.textilemedia.com/smart-textiles-and-wearables/latest-news/achieving-sustainability-in-smart-wearable-electronic-textiles/.
23. US' Researchers Create Fabric Touch Sensor with Machine Learning, 2024, https://www.technicaltextile.net/news/us-researchers-create-fabric-touch-sensor-with-machine-learning-294728.html.
24. Ferri A, Plutino M R and Rosace G, 2019, Recent trends in smart textiles: Wearable sensors and drug release systems, *AIP Conference Proceedings*, https://doi.org/10.1063/1.5123575.
25. Erdumlu N. and Saricam C, 2013, Wicking and drying properties of conventional ring and vortex spun cotton yarns and fabrics, *Journal of The Textile Institute*, 104(12), pp. 1284–1291.
26. Singh M K and Nigam A, 2013, Effect of various ring yarns on fabric comfort, *Journal of Industrial Engineering*, pp. 1–7, https://doi.org/10.1155/2013/206240.
27. Sengupta A K and Murthy H V S, 1985, Wicking in ring spun vis a vis rotor spun yarns, *Indian Journal of Textile Research*, 10(4), pp. 155–157.
28. Chattopadhyay R and Chauhan A, 2004, Wicking behavior of compact and ring spun yarns and fabrics in One day seminar on Comfort in Textiles, Department of Textile Technology, IIT Delhi, New Delhi, 2004, October 16, p. 20.
29. Lord P R, 1974, A comparison of the performance of open end and ring spun yarns in terry toweling, *Textile Research Journal*, 44, pp. 516–522.
30. Jhanji Y, Gupta D and Kothari V K, 2015, Comfort properties of plated knitted fabrics with varying fibre types, *Indian Journal of Fibre & Textile Research*, 40(3), pp. 11–18.
31. YJhanji, D G and Kothari V K, 2014, Moisture management properties of plated knit structures with varying fiber types, *The Journal of The Textile Institute*, https://doi.org/10.1080/00405000.2014.934044.
32. Ghosh S and Kaur A, 2014, Wicking and absorbency behavior of plain knitted fabrics, *Paper Presented at the International Conference on Emerging Trends in Traditional and Technical Textiles*, 11–12 April, 2014, (pp. 26–28), ISBN-979835156700, NIT Jalandhar.

# Index

Note: **Bold** page numbers refer to tables and *italic* page numbers refer to figures.

## D

## E



## F

## G

## X



## Y



## Z

For Product Safety Concerns and Information please contact our EU representative GPSR@taylorandfrancis.com Taylor & Francis Verlag GmbH, Kaufingerstraße 24, 80331 München, Germany